**Microbiology Monographs**
Volume 12

Series Editor: Alexander Steinbüchel
Münster, Germany

# Microbiology Monographs

Volumes published in the series

Masahiro Fujishima
Editor

# Endosymbionts in *Paramecium*

Springer

*Editor*
Professor Dr. Masahiro Fujishima
Department of Environmental Science and Engineering
Graduate School of Science and Engineering
Yamaguchi University
Yoshida 1677-1
Yamaguchi 753-8512
Japan
e-mail: fujishim@yamaguchi-u.ac.jp

*Series Editor*
Professor Dr. Alexander Steinbüchel
Institut für Molekulare Mikrobiologie und Biotechnology
Westfälische Wilhelms-Universität
Corrensstr. 3
48149 Münster
Germany
e-mail: steinbu@uni-muenster.de

ISBN: 978-3-540-92676-4 e-ISBN: 978-3-540-92677-1
DOI: 10.1007/978-3-540-92677-1
Springer Dordrecht Heidelberg London New York

Library of Congress Control Number: 2008943978



*Cover design*: SPi Publisher Services

Printed on acid-free paper

Springer is part of Springer Science+Business Media (www.springer.com)

# Preface

Endosymbiosis is a primary force in eukaryotic cell evolution. Recent studies of algal evolution have shown that this phenomenon has occurred several times and has yielded a wide diversity of eukaryotic cells. Despite the importance of this phenomenon, however, molecular mechanisms for induction of endosymbiosis between different microorganisms are not so well known. How can a symbiont invade the host cytoplasm? How can the symbiont avoid digestion by the host's lysosomal enzymes? How can the symbiont grow synchronously with the host cell? How are symbionts transferred to daughter cells when the host cell divides? Through endosymbiosis, what merit does the host cell obtain to adapt to new environments and to augment its survivable environments? To elucidate these questions, experiments for reestablishment of the endosymbiosis by symbionts isolated from the symbiont-bearing host cells and the symbiont-free host cells are indispensable. In many endosymbiotic communities, however, both the endosymbionts and the aposymbiotic host cells have already lost the ability to survive and grow independently.

Ciliated protista belonging to species of *Paramecium* are extremely valuable cells that allow experiments on the reestablishment of the endosymbiosis, and they frequently bear prokaryotic or eukaryotic or both endosymbionts in the cell. To date, about 60 bacteria (Görtz and Fokin and Fokin and Görtz, this volume), various *Chlorella* species (Hoshina and Imamura, Kodama and Fujishima, Kato and Imamura, Miwa, and Sommaruga and Sonntag, this volume), and yeasts are known as endosymbionts of *Paramecium* species. Probably, the prominent phagocytic activity of the *Paramecium* cell is a cause for acquisition of various endosymbiotic organisms. The importance of *Paramecium* species for studies of such symbioses is that these species retain the ability to grow without the endosymbionts. Although most endosymbiotic bacteria of *Paramecium* species cannot grow outside the host cell as a result of their reduced genome size, even when they are isolated from the host cells they can maintain their infectivity to new host cells for a limited time (Görtz and Fokin, Fokin and Görtz, Fujishima, and Schrallhammer and Schweikert, this volume). Consequently, reestablishment of endosymbiosis between the symbiotic bacteria-free host cells and the symbionts isolated from the host cells is easily done by mixing them. After endosymbiosis, the bacteria alter the host gene expression; the host thereby acquires resistance against various stresses, providing an excellent opportunity for us to elucidate not only the infection processes but also to assess the associations leading to eukaryotic cell evolution (Fujishima, this volume). On the other hand, *P. bursaria* is the only species of *Paramecium* that forms symbiotic relationships with green algae belonging to the genus *Chlorella*. The algae-free host cells and the symbiotic algae retain the ability to grow without a partner. Their mutual endosymbiosis is readily reestablished by mixing them. The finding of four checkpoints for the reestablishment of the symbiosis with the algae reveals that *P. bursaria* is an excellent model for studying the infection process of the algae and the evolution of eukaryotic cells through secondary endosymbiosis between protozoa and algae (Kodama and Fujishima, this volume).

Following earlier published reviews (Fokin 2004; Görtz 1983, 1986, 1988; Preer 1948; Preer et al. 1974; Quackenbush 1988; Sold 1974), great progress has been made in the study of endosymbiosis in *Paramecium*. This book is the first monograph on the endosymbionts of the ciliates, and comprises nine chapters. We know of other topics, some of which were not included in this book, with rapidly developing investigations. We made efforts to cite papers from such research areas in related chapters. It is our hope that the omissions displease neither the researchers involved in those regrettably excluded topics nor readers who might be interested in them.

We express our sincere thanks to all authors who contributed their enlightening chapters to this volume *Endosymbionts in Paramecium*. Their excellent contributions were submitted in a timely fashion. Working with the professionals at Springer on publishing this volume of the series *Microbiology Monographs* was a great pleasure; we especially appreciate the valuable suggestions and support of Dr. Jutta Lindenborn.

Yamaguchi and Münster
November 2008

Masahiro Fujishima
Alexander Steinbüchel

## References

Foin IS (2004) Bacterial endocytobionts of ciliophora and their interactions with the host cell. Int Rev Cytol 236:181–249

Görtz H-D (1983) Endonuclear symbionts in ciliates. Int Rev Cytol 14(Suppl):145–176

Görtz H-D (1986) Endonucleobiosis in ciliates. Int Rev Cytol 102:169–213

Görtz H-D (1988) Endocytobiosis. In: Görtz H-D (ed) *Paramecium*. Springer, Berlin, pp 393–405

Preer JR (1948) A study of some properties of the cytoplasmic factor "kappa" in *P. aurelia*, variety 2. Genetics 33:349–404

Preer JR, Preer LB, Jurand A (1974) Kappa and other endosymbionts in *Paramecium aurelia*. Bacteriol Rev 38:113–163

Quackenbush RL (1988) Endosymbionts of killer paramecia. In: Görtz H-D (ed) *Paramecium*. Springer, Berlin, pp 406–418

Sold AT (1974) Intracellular particles in *Paramecium*. In: van Wagtendonk WJ (ed) *Paramecium*: a current survey. Elsevier, Amsterdam, pp 377–442

# Contents

# Origins of Algal Symbionts of *Paramecium bursaria*

**Ryo Hoshina and Nobutaka Imamura**

**Contents**



**Abstract** The green paramecium, *Paramecium bursaria*, has evolved a mutualistic relationship with green algae that has fascinated microbiologists for over a century. Classical approaches to determining the identities of these algae have indicated that the symbionts are close relatives of the genus *Chlorella*, in which two differentiable algal groups, the "American" (NC64A as the representative strain) and the "European" (Pbi), have been recognized. The identities of these

R. Hoshina (✉)
Department of Biomedical Science, College of Life Sciences, Ritsumeikan University, Noji Higashi 1-1-1 Kusatsu, Shiga, 525-8577, Japan
e-mail: wwhoseena@hotmail.com

N. Imamura
Department of Pharmacy, College of Pharmaceutical Sciences, Ritsumeikan University, Noji Higashi 1-1-1 Kusatsu, Shiga, 525-8577, Japan

M. Fujishima (ed.), *Endosymbionts in Paramecium*, Microbiology Monographs 12,
DOI: 10.1007/978-3-540-92677-1_1, © Springer-Verlag Berlin Heidelberg 2009

algal partners were finally revealed using molecular phylogenetic and genetic structural analyses. Consequently, both the "American" and the "European" algae are thought to belong to *Chlorella sensu stricto*; however, each is equivalent to a species distinct from other known *Chlorella* spp. The majority of *P. bursaria* symbionts are assigned to either the "American" or the "European" group, but in a few exceptional cases there exist paramecium associated with *C. vulgaris*, with *Coccomyxa* sp., and the double-symbiont association of *Choricystis minor* with an unidentified "*Chlorella*-like" alga. All these symbionts are polyphyletic, even among the three *Chlorella* spp., suggesting that they have arisen as paramecian symbionts from different origins. The "American" and "European" algae are readily distinguishable from each other and other algae by clearly distinct internal transcribed spacers, or by the presence of intronic insertions into unique positions in ribosomal DNA. These introns are distinctive and connote an ancient situation in which both the "American" and the "European" algae cohabitated in a single ciliate. Given the hosts' genetic divergence, it allows for the development of an evolutionary scenario for *P. bursaria* with respect to algal acquisition and switching.

## 1 Introduction

The unicellular ciliate *Paramecium* (Peniculia, Oligohymenophorea) is one of the most thoroughly studied protozoans. Considerable research has been conducted regarding the morphology, physiology, genetics, crosses, nuclear reorganization, phylogeny, and the concept of species within *Paramecium*. On the basis of its diversity, researchers have divided *Paramecium* into several subgenera. The latest classification by Fokin et al. (2004), based on morphometric, biological, and molecular characters, lists four subgenera, one of which is "*Chloroparamecium*," which contains only one species, *P. bursaria*. However, although the authors respect the ultrastructural characteristics of the micronucleus and the nuclear reorganization process, the most prominent characteristic of *P. bursaria* should be the retention of green algae historically called "zoochlorella" (the general term for green balls inhabiting various protozoans and invertebrates that do not express taxonomic affiliation). *P. bursaria*, itself a single-celled protozoan, maintains several hundred algal cells within its own cytoplasm, lending it a green color: thus, *P. bursaria* is known as the green *Paramecium*. The alga lives inside the ciliate providing it with photosynthate, while the ciliate provides the alga with protection from other protozoans or viruses, and chauffeurs it to brightly lit areas for optimum photosynthesis; therefore, *P. bursaria* is a ciliate that, like corals or lichens, has established a mutualistic relationship with an alga.

The ciliate packages hundreds of algal cells in individual perialgal vacuoles. Because *P. bursaria* is regarded as a predatory protozoan, it is often thought that the algae are temporary symbionts engulfed as a result of its quotidian phagocytic behavior. In fact, many protozoa are thought to acquire algae, albeit transiently, in this manner (Reisser and Widowski 1992). However, as shown in more detail below, *P. bursaria* symbionts are continuously inherited; that is, daughter cells inherit the same algal symbionts that were retained by the mother cell (Siegel 1960). In contrast, algae-free *P. bursaria* (i.e., white paramecia) can be generated from normal green paramecia by culturing in constant darkness or by exposure to certain herbicides; the cells retain the ability to grow in the absence of symbiotic algae. Symbiotic algae can also be cultured separately, although they are occasionally nutritionally fastidious. Algae-free *P. bursaria* can reabsorb cultured algae isolated from green *P. bursaria*, from another protozoan, or from free-living algae through phagocytosis. Most of the algae are digested, but some manage to escape and take up residence within the host cytoplasm, where new perialgal vacuoles are created from the digestive vacuole (Meier and Wiessner 1989; Kodama et al. 2007; see Kodama and Fujishima, this volume). The symbiotic association between *P. bursaria* and its photobiotic algae is an excellent model system for studying endosymbiosis leading to the condition known as "secondary symbiosis," which has fascinated protozoologists and phycologists for over a century.

In addition to studying this mutualistic relationship, the identification of these naturally occurring symbiotic algae and determination of their origins has been attempted from the perspectives of morphology, physiology, chemical composition, isozyme profiling, virus sensitivity, and so on. However, in the previous two decades, the status of these chlorophytic algal symbionts has been in a state of flux. Recent advances in molecular biology and comparative DNA sequence analyses have revealed a greater diversity among these organisms than previously believed, which has resulted in a revision of hierarchy and greatly altered our ideas regarding the concept of species.

This chapter will review our current knowledge regarding the origins of *P. bursaria* algal symbionts from the perspective of DNA phylogeny. We will describe the molecular characteristics of these symbionts and ways to distinguish them, and we will propose an evolutionary context for *P. bursaria* with regard to their acquisition and switching of algal partners.

## 2 Classical Approaches to the Identification of Symbionts

Although the systematics of the chlorophytic algae remain somewhat disorganized, owing to the natural tendencies of biologists, researchers have determined the progenitor, identified with the known species, and described as the new species for the *Paramecium* symbiont. The similarity of symbiotic algae to *Chlorella* spp. has long been noted and, therefore, taxonomic methods have been developed for use in conjunction with *Chlorella* classification studies.

### 2.1 Species Recognition within the Genus *Chlorella*

*Chlorella vulgaris* is the type species of the genus. Since Beijerinck, a Dutch microbiologist, first described this species, more than 100 species have been described from freshwater, seawater, soil surfaces, and other substrates, such as tree trunks, rocks, and building walls. *Chlorella* is a ubiquitous organism, generally characterized as a spherical green ball lacking flagella, typically 4–10 μm in diameter, with a minimal set of organelles: a single nucleus, chloroplast, and mitochondrion and some peroxisomes and vacuoles. The lack of obvious morphological characters combined with an exclusively asexual reproductive life cycle by means of autosporulation have caused considerable problems in the taxonomic description and identification of *Chlorella* spp. (Kessler and Huss 1992; Krienitz et al. 2004). Combined classification criteria using biochemical and physiological characteristics have provided the most widely acknowledged and practicable system for species delineation in *Chlorella*. These parameters include temperature growth limits, pH, NaCl density, vitamin requirements, nitrate reduction, hydrogenase activity, generation of secondary carotenoids, cell wall sugar composition, and DNA base composition (GC content). These types of studies reduced a plethora of *Chlorella* spp. into 19 characteristic taxa inclusive of the *P. bursaria* symbiont (Huss et al. 1999). While each taxon manifests small differences that separate it from other taxa, this distinction is often only a matter of degree. For example, *C. sorokiniana* can be separated from *C. vulgaris* by the possession of hydrogenase activity and thermophily, whereas *C. lobophora* is only distinguishable from *C. vulgaris* on the basis of its lower saline tolerance. In an age where DNA sequence differences can differentiate between unique species that superficially resemble each other, the species (or subspecies) recognitions that were developed on the basis of nonmolecular characteristics are still deserving of admiration. However, the considerable range in GC content (43–78%) suggests that the genus *Chlorella* represents an assemblage of morphologically similar taxa of polyphyletic origins rather than a natural genus (Kessler 1976). Heterogeneity in GC content has also been observed among the various *C. sorokiniana* strains (up to 15% different; Huss and Jahnke 1994). Some taxa have been suggested to be more similar to non-*Chlorella* spp. For example, on the basis of chemical features, *C. fusca* var. *fusca*, *C. fusca* var. *rubescens*, and *C. fusca* var. *vacuolatus* were thought to be closely related to the morphologically distinct genus *Scenedesmus*, which forms colonies consisting of cylindrical cells aligned in a flat plate, in which the outermost cells each have two long spines. These three taxa were later transferred to the genus *Scenedesmus* as new species; namely, *S. abundans*, *S. rubescens*, and *S. vacuolatus*.

## 2.2 History of *P. bursaria* Symbionts

Reisser and Widowski (1992) reported that the genus *Zoochlorella* was introduced in 1882 by Brandt, who so designated algae isolated from green hydra (*Z. conductrix*) and sponges (*Z. parasitica*). The *P. bursaria* symbiotic algae were later assigned to *Z. conductrix* by Beijerinck in 1890. Increasing knowledge concerning these endosymbiotic algae resulted in the genus *Zoochlorella* losing its taxonomic value. Since then, *P. bursaria* symbiotic algae have been identified on the basis of morphological similarities to taxa such as *C. vulgaris*, *C. sorokiniana*, *C. ellipsoidea*, and because of its poor growth on nitrate *C. protothecoides* (currently *Auxenochlorella protothecoides*). The symbiotic alga was renamed as an independent species, *C. paramecii*, by Loefer in 1936. However, the original paper by Loefer did not provide a valid description, for which reason Shihara and Krauss (1965) later rejected *C. paramecii*, and instead described a new species, *C. variabilis*, on the basis of type strains no. 130 from the Indiana Algal Culture Collection [which was later moved to the Culture Collection of Algae at the University of Texas (UTEX) and this strain is not currently available] and no. 211/6 from the Cambridge Collection [which later moved to the Culture Collection of Algae and Protozoa (CCAP), UK, this strain is not currently available either], which were claimed to be Loefer's *C. paramecii*. However, strain 211/6 showed distinct biochemical and physiological characteristics compared with the other strains of *P. bursaria* symbionts (Douglas and Huss 1986; Kessler and Huss 1990), raising questions as to whether it was a symbiont at all. Furthermore, a similarly named *P. bursaria* symbiont strain (211-6) is listed within the Culture Collection of Algae at the University of Göttingen (SAG), Germany. Consequently, these species names are no longer in use. In stead, the symbiotic algae of *P. bursaria* are regarded as unlabeled organisms, having the taxonomic autonomy of a discrete species.

Modern observations of these symbiotic algae are represented by the studies carried out by German researchers (E. Kessler, W. Reisser, V.A.R. Huss, and their coworkers). These researchers examined algal strains isolated from host *Paramecium* collected in Germany and the USA, containing representative Pbi and NC64A strains [NC64A(M) and NC64A(P) appear in some literature; the former is regarded as the true NC64A and the latter is unlikely to be of symbiotic origin; Douglas and Huss 1986]. Their studies demonstrated that the "American" and "European" algal strains each possess certain distinct characteristics. The general conclusion of taxonomic studies is that there is no reason to distinguish the symbiotic algae from the genus *Chlorella*. Rather, it is reasonable to assume that these symbionts are derived from free-living *Chlorella* spp., but with some evolved characteristics. The morphology of symbiotic algae is not distinguishable from that of other *Chlorella* spp., even under an electron microscope. *P. bursaria* symbionts are smooth-walled spheres (usually 5–8 μm in diameter) that lack flagella and typically release four autospores. The single chloroplast is saucer-, cup-, or girdle-shaped, and the pyrenoid is typically penetrated by a thylakoid membrane. Ruthenium red stains the "American" algal cell wall red and the "European" cell wall pink, these results being incongruent with those reported by Takeda (1995). These characteristics are attributed to *Chlorella*, in particular *C. vulgaris* and *C. sorokiniana* (Reisser et al. 1988b; Reisser and Widowski 1992). The prominent physiological characteristics of the symbionts include external maltose excretion (Brown and Nielsen 1974), the requirement for thiamine and other vitamins, and, in some strains, a reduced nitrogen source (Douglass and Huss 1986; Kessler and Huss 1990; Reisser et al. 1988b; Reisser and Widowski 1992). Physiological assessment is perhaps unreliable. Hydrogenase activity was not detected in either NC64A or Pbi in a study by Reisser et al. (1988b), but was detected in other studies (Douglas and Huss 1986; Kessler and Huss 1990). SAG 211-6 and UTEX-130 are supposed to be derived from a single isolate, yet they differ in their requirement for thiamine, use of nitrate, lower pH limit for growth, and GC content (Douglas and Huss 1986; Kessler and Huss 1990). These incongruent data may reflect different techniques used by researchers or differences in the culture environment. Albers et al. (1982) proposed that the physiological characteristics of the symbiotic algae may also transmute during long-term culture. The Japanese *P. bursaria* symbiotic alga F36-ZK is a comparatively "fresh" strain isolated from the host in 2002. This strain is genetically identical to NC64A (in rDNA level), which was isolated before 1968. However, in NC64A, Reisser et al. (1988b) reported a low photosynthate excretion rate (4.17% of total fixation) at a pH of 4.8, which may more closely resemble the conditions of the inner perialgal vacuole, whereas the freshly isolated F36-ZK released nearly half of its total photosynthate fixed at pH 5 (Kamako and Imamura 2006). Long-term cultured NC64A showed low levels of nitrate reductase activity, whereas F36-ZK showed no nitrate reductase activity (Kamako et al. 2005). Thus, physiological data require careful interpretation when determining whether or not certain characteristics are genuine or constant features of symbiotic algae.

Analyses of symbiotic algae resulted in the discovery of the *Chlorella* virus. A virus specifically infects the symbiotic alga NC64A and other "American" symbionts, but does not infect "European" symbionts or free-living *Chlorella*. Likewise, a distinct virus specifically infects Pbi and other "European" symbionts (Reisser et al. 1988a). Thus, these viruses were designated the NC64A and Pbi viruses. Reisser et al. (1990) confirmed that each virus is able to identify its host species on the basis of a key factor present in the algal cell wall. Kapaun et al. (1992) and Takeda (1995) analyzed the cell wall sugar composition of symbiotic algae and found that these organisms were characterized by a rigid cell wall composed of glucosamine, indicating a close relationship to *C. vulgaris*, *C. sorokiniana*, and *C. kessleri*. Takeda (1995) provided a more extensive description, reporting that Japanese and Chinese symbionts resemble the "American" strain (i.e., NC64A), but differ from the "European" (i.e., Pbi) algae. Their cell wall index of 2.2.2 (glucosamine base – ruthenium red minus – anisotropy negative, see Takeda 1991) indicates a relationship with *C. kessleri*; however, the distinct proportion of compositional sugars indicates that symbiotic algae belong to a new species. Thus, the question of whether or not symbiotic algae represent an independent species has become an increasingly important focus in more recent studies. Isozyme distribution patterns for several enzymes demonstrated the uniformity within the "American" versus "European" algae, and multiformity among different groups (i.e., "American," "European," and free-living *Chlorella* spp.; Linz et al. 1999). Similar results were also obtained through the analysis of universal primer PCR fragmentation patterns (Kvitko et al. 2001).

Two symbiotic algae with different characteristics, PbDu and PbOp, were introduced from Germany (Reisser et al. 1988b). These strains are characterized by larger cells (6–10 μm in diameter), and possess a parietal chloroplast with a pyrenoid located in the thickest part of the chloroplast. The cell walls show the typical trilaminar structure and an inner polysaccharide layer, and are not stained by ruthenium red. Additionally, many of their physiological characteristics were similar to those of the *C. fusca* group (i.e., *Scenedesmus* spp.), but these strains, unfortunately, have never appeared in later studies.

As demonstrated by the studies described above, identifying the progenitor of the *P. bursaria* symbiont is, with respect to *Chlorella* identification and/or systematics, extremely complicated. For this reason, DNA sequencing analyses have long been anticipated (Friedl and Zeltner 1994; Kessler and Huss 1990).

## 3 Molecular Phylogeny of the Symbionts

With the development of DNA sequencing techniques, PCR, and cloning methods, a larger-than-expected diversity has been recognized among small protistan organisms. Molecular phylogenies based on DNA sequence comparisons have exposed many instances of coincidental morphology among organisms of different lineages, and provided insight into their true genetic relationships. Of course, molecular phylogeny has had a considerable impact on biological systematics overall, and the green algae have not escaped this trend.

Huss and Sogin (1990) reported the first molecular phylogenetic analyses of *Chlorella* spp. based on small subunit ribosomal RNA gene (SSU rDNA) sequences, wherein they verified that algae of distinct lineages were grouped within the genus *Chlorella*. On the basis of an SSU rDNA phylogeny, Friedl (1995) proposed the novel class Trebouxiophyceae as a sister class to Chlorophyceae. This class is represented by the taxon formally known as Microthamniales, which includes the symbiotic lichen alga *Trebouxia*, and some autosporic coccoids (i.e., *Chlorella* spp., *Nannochlorum* spp., and the pathogenic colorless alga *Prototheca*). With increasingly available algal sequence data, it has become clear that the "*Chlorella*-like" coccoids are an extremely polyphyletic taxa; an SSU rDNA phylogeny showed that *Chlorella* spp. are dispersed within two classes of Trebouxiophyceae and Chlorophyceae (Huss et al. 1999). *C. fusca* and its varieties were confirmed as close relatives of *Scenedesmus* within Chlorophyceae, whereas *C. sorokiniana*, *C. kessleri*, and *C. lobophora* were clustered with *C. vulgaris* (type species) within Trebouxiophyceae. Consequently, Huss et al. (1999) restricted the true *Chlorella* to four species.

The current family, Chlorellaceae, was recently revised by Krienitz et al. (2004) and consists of the two sister lines of the *Chlorella* clade (including *C. vulgaris*) and the *Parachlorella* clade. The genus *Parachlorella* was established on the basis of the type species *P. beijerinckii*, which is characterized by a gelatinous indusium, and *C. kessleri* was transferred into this genus on the basis of phylogenetic similarity. Although the family Chlorellaceae is phylogenetically defined, it contains algae with variable characteristics. Even within the *Chlorella* clade, many morphologically different taxa, such as *Actinastrum* (slender fusiform cells joined at a common center and radiating in all directions), *Diacanthos* (ovate cells with a single thick bristle on each cell pole), *Didymogenes* (similar to *Diacanthos*, but with a rather flat coenobia), *Micractinium* (colonial with several spines not arranged oppositely), *Dictyosphaerium* (colonized cells on branched threads), and *Meyerella* (lacks a pyrenoid) are phylogenetically related. It is noteworthy that some of these algae can alternate between these irregular forms and spherical forms (Krienitz et al. 2004; Luo et al. 2006) depending upon culture conditions. Additionally, molecular phylogenies within the *Chlorella* clade have demonstrated the paraphyletic relationships among *Chlorella* spp. (Hoshina et al. 2008a; Krienitz et al. 2004), indicating that "*Chlorella*-like" forms have occurred frequently throughout evolutionary history. The true *Chlorella* is currently defined as a spherical unicellular green alga lacking any motile stage (without flagella), an exclusively asexual reproductive system (autosporulation), possessing a single nucleus, a chloroplast with a pyrenoid whose matrix is bisected by the thylakoid membrane, a mitochondrion, glucosamine-based cell wall components, no secondary carotenoids, and a phylogenetic affiliation to the *Chlorella* clade as an essential prerequisite. While some outliers have already been transferred to different genera [e.g., *Auxenochlorella* (*C. protothecoides*), *Watanabea* (*C. reniformis*), and *Mychonastes* (*C. homosphaera*)], others still await taxonomic reassignment (e.g., *C. minutissima*, *C. ellipsoidea*, and *C. saccharophila*). Given this extremely complex background, it is understandably difficult to identify the true *P. bursaria* symbionts without DNA sequence comparison.

## 3.1 Four plus One Symbiont Groups Inferred from rDNA Sequences

The initial rDNA sequencing of the symbiotic algae began with Japanese strains. Kamako et al. (2005) isolated the symbiotic algae from several *P. bursaria* strains and established algal cultures, and whose DNA sequences of SSU rDNA wherein three group I introns, the internal transcribed spacer (ITS) 1, the 5.8S rDNA, and the ITS2 were identical (Hoshina et al. 2004). Phylogenetic analyses based on the SSU rDNA (exon) sequence suggested a relationship to the true *Chlorella* spp. Subsequently, Hoshina et al. (2005) expanded the study materials to include symbionts originating from the USA and European countries, as well as China and Australia. Even though the symbiotic algae came from dispersed geographic locations, they detected only two patterns of rDNA sequences; these are characterized by different length owing to the different number of insertions into the SSU rDNA (three-intron or single-intron). Each algal group had a highly uniform SSU rDNA-ITS sequence, with very little genetic variation detected within the three-intron group (e.g., a single polymorphic nucleotide position in each SSU rDNA exon and ITS1, and two positions in ITS2), whereas no variation was observed within the one-intron group. Between these algal groups, sequences differed by only seven or eight nucleotides in the SSU rDNA (exon), whereas ITS sequences reached an unambiguous level (approximately 20% difference). Characteristic variations of classical approaches have suggested that paramecian symbionts can be separated into "American" and "European" groups (Sect. 2.2). The three-intron group contained NC64A, the authentic "American" strain. Therefore, the three-intron species (i.e., the American, East Asian, and Australian algae) belong to the "American" group. Taking into account the geographical congruencies (German and British), Hoshina et al. (2005) tentatively termed the single-intron group as "European." Shortly afterwards, Gaponova et al. (2007) supported the above sequence-based extrapolation of the "American" and "European" groups. They obtained rDNA PCR products of two different lengths for the paramecian symbionts, in which each length corresponded to either the "American" (three-intron) or the "European" (single-intron) groups. Importantly, the representative "European" strain Pbi was analyzed and was allied with the other "European" strains. Summerer et al. (2008) analyzed additional strains from Austria, in which the ITS1 sequences matched those of the "European" group. Finally, Hoshina and Imamura (2008a) discovered two exceptional algae that did not possess any intron in the SSU rDNA sequence, and thus did not belong to either of these groups. In the SSU rDNA exon region, one alga (*P. bursaria* CCAP 1660/10 symbiont) differed by 15 nucleotides from NC64A ("American") and by nine nucleotides from the "European" group. The second alga (*P. bursaria* CCAP 1660/13 symbiont) differed from all other paramecian symbionts by more than 4% (by 78 nucleotides from strain NC64A, by 76 nucleotides from the "European" group, and by 79 nucleotides from the former alga). Thus, the classification of these organisms resulted in a total of four groups.

Whether all symbionts within a single *P. bursaria* belong to single species is related to the identity of the symbiont. Hoshina et al. (2005) introduced a PCR-based technique that amplifies both algal and ciliate rDNA separately from template DNA extracted from a whole *P. bursaria* extract. Although the authors used a direct sequencing method to decipher these PCR products, they obtained algal sequences with electropherograms showing definite single peaks from each *P. bursaria* strain. Thus, the algal symbionts within a single *P. bursaria* cell are thought to be clones. Summerer et al. (2008) established several algal strains from a single *P. bursaria* in which the DNA sequences were identical, thus supporting the notion of cloned *Paramecium* symbionts. Vertical inheritance over many generations where the symbionts reside inside another organism is generally thought to unify the symbionts (Douglas 1995, 1998; Frank 1997). However, Nakahara et al. (2004) reported an unusual *P. bursaria* strain in which an obviously smaller (1.2–2.00 μm × 1.9–3.0 μm) alga was coexistent with an unidentified "*Chlorella*-like" alga. They observed this strain over a period of 5 years, since collection of the ciliate in 1992, which may indicate that this apparently eccentric situation is actually a stable condition. Small algae were enclosed within perialgal vacuoles; however, the vacuoles varied in size and two to dozens of the small algae were aggregately enclosed. The cells were spherical to ellipsoidal or kidney-shaped, possessed a single saucer- or cup-shaped chloroplast without pyrenoids, and released two to four autospores. SSU rDNA phylogeny identified this small alga as *Choricystis minor*, and is noted as a "plus one" symbiont group rather than a separate fifth group. rDNA sequences of paramecian symbiotic algae are shown in Table 1.

**Table 1** *Paramecium bursaria* symbionts of those ribosomal DNA (rDNA) sequences are available

| Symbiont-type | *P. bursaria* | Syngen | Genotype[a] | Algal strain | Collection site | Available region | Accession number | Remarks |
|---|---|---|---|---|---|---|---|---|
| "American" | | 1 | | ATCC 50258/ (Ac21ckb21)[b] | New York, USA | SSU-ITS1-5.8S-ITS2-LSU | AB206549, AB236862, DQ057340-41, AY876292, AY876294 | NC64A |
| | | 2 | | ATCC 30562 | Ohio, USA | SSU-ITS1-5.8S-ITS2 | AB206550 | Syngen 2-3 |
| | | | | SAG 211-6 | USA | SSU, ITS2 | AB260893, AB301072, AY876290 | "*Chlorella paramecii*"? |
| | | | | N-1-A | USA | SSU (partial) | AY876293 | |
| | OK1 | 1 | D | OK1-ZK | Aichi, Japan | SSU-ITS1-5.8S-ITS2-LSU | AB162912, AB437257 | |
| | So13 | 1 | D | So13-ZK | Nagano, Japan | SSU-ITS1-5.8S-ITS2 | AB162913 | |
| | F36 | 1 | D | F36-ZK | (Crossbreed, Japan×Japan) | SSU-ITS1-5.8S-ITS2 | AB162914 | |
| | KM2 | 1 | D | KM2-ZK/pbKM2 | Shimane, Japan | SSU-ITS1-5.8S-ITS2 | AB162915, EF030567, EF030584 | |
| | Dd1 | 1 | D | Dd1-ZK | Ibaraki, Japan | SSU-ITS1-5.8S-ITS2 | AB162916 | |
| | Bnd1 | 1 | D | Bnd1-ZK | Hiroshima, Japan | SSU-ITS1-5.8S-ITS2 | AB162917 | |
| | HB2-2 | | | HB2-2-1 | Hiroshima, Japan | SSU-ITS1 | AB191205 | Registered as *Chlorella vulgaris* |
| | shiP-7 | | | shiP-7-A4 | Miyazaki, Japan | SSU-ITS1 | AB191206 | Registered as *Chlorella vulgaris* |
| | takaP-3 | | | takaP-3-A2 | Oita, Japan | SSU-ITS1 | AB191207 | Registered as *Chlorella vulgaris* |
| | Cs2 | 1 | D | (Uncultured)[c] | Shanghai, China | SSU-ITS1-5.8S-ITS2 | AB206546 | |
| | MRBG1 | | D | (Uncultured)[c] | Melbourne, Australia | SSU-ITS1-5.8S-ITS2 | AB219527 | |

| | | | | | | | | |
|---|---|---|---|---|---|---|---|---|
| "European" | SW1 | | C | SW1-ZK/ (uncultured)[c] | Black Forest, Germany | SSU-ITS1-5.8S-ITS2-LSU | AB206547, AB437244-56 | |
| | | | | 241-80 | Göttingen, Germany | SSU (partial) | AY876291 | SAG 241.80? |
| | CCAP 1660/11 | 2 | C | (Uncultured)[c] | Cambridge, UK | SSU-ITS1-5.8S-ITS2 | AB206548 | |
| | CCAP 1660/12 | | A | (Uncultured)[c] | Cambridge, UK | SSU-ITS1-5.8S-ITS2 | AB260894 | |
| | | | | PbW | Wildbichl, Austria | SSU (partial), ITS1 | EF030566, EF030583 | |
| | | | | PbPIB | Piburger See, Austria | SSU (partial), ITS1 | EF030565, EF030582 | |
| | | | | Pbu | Russia | SSU (partial), ITS1 | EF030562, EF030579 | |
| | | | | OCH/(OCH cr4/ OCH cr6)[b] | Karelia, Russia | SSU (partial), ITS1 | EF030561, EF030578, AY876295-97 | |
| | | | | OC-1 | Karelia, Russia | SSU (partial) | AY876298 | OS-1? |
| | | | | OC-6 | Karelia, Russia | SSU (partial) | AY876299 | OS-6? |
| *Chlorella vulgaris* | CCAP 1660/10 | 2 | B | (Uncultured)[c] | ? | SSU-ITS1-5.8S-ITS2-LSU | AB260895 | |
| *Coccomyxa* sp. | CCAP 1660/13 | | A | (Uncultured)[c] | Cambridge, UK | SSU-ITS1-5.8S-ITS2-LSU | AB260896 | |
| *Choricystis minor*[d] | | | | OL2-1 | Florida, USA | SSU | AB109544 | |

*SSU* small subunit, *ITS* internal transcribed spacer, *LSU* large subunit

[a]SSU rDNA-base typification of the host ciliate (Hoshina et al. 2006)

[b]Mutant strain obtained from NC64A/OCH (Gaponova et al. 2007)

[c]Algal DNA sequence directly obtained from whole paramecium extract (Hoshina et al. 2005; Hoshina and Imamura 2008a)

[d]Coexistent with "*Chlorella*-like" alga (Nakahara et al. 2004)

## 3.2 Tree Analyses

The SSU rDNA tree shown in Fig. 1 demonstrates that algal symbionts of *P. bursaria* belong to two lineages within Trebouxiophyceae. Three of four symbiotic groups, namely, the "American," "European," and *P. bursaria* CCAP 1660/10 alga, belong to a single clade (i.e., the Chlorellaceae), but one strain, *P. bursaria* CCAP 1660/13 alga, appears in a markedly distinct clade. This second clade includes species of *Coccomyxa* (a lichen symbiont), *Paradoxia multiseta*, and an enigmatic endophytic alga found in *Ginkgo biloba* (a primitive gymnosperm) protoplasts (Trémouillaux-Guiller et al. 2002). The sea anemone *Anthopleura elegantissima* symbiont (Lewis and Muller-Parker 2004) also has a sister relationship to this clade, on the basis of the authors' preliminary analysis. CCAP 1660/13 alga is also clearly distinct from the coexistent small alga, *Choricystis minor*. SSU rDNA analyses have revealed that the CCAP 1660/13 symbiont and the remaining chlorellacean symbionts are polyphyletic. It therefore seems highly likely that symbiont acquisitions have occurred independently in *P. bursaria*. In a study of hydra symbionts, Huss et al. (1993/1994) reported results similar to those described above. Three of four symbionts (HvT, Ssh, and Esh) were phylogenetically close to each other, but the remaining symbiont (Jsh) occupied a markedly distinct position, prompting these authors to conclude that the hydra have experienced at least two independent symbiotic events. Ssh appears within the *Chlorella* clade, whereas the SSU rDNA sequence of Jsh corresponds to that of the hydra symbiont CCAP 211/61 (Fig. 1).

Making a general survey of the SSU rDNA tree, we find that the class Trebouxiophyceae contains many symbiotic and parasitic species that are distributed along seemingly biased lineages. There may exist lineages prone to association with other organisms. One such lineage, the *Chlorella* clade, includes symbiotic algae from several other ciliate protozoans (Summerer et al. 2008), although it is uncertain whether these symbiotic associations are stable.

Within Chlorellaceae, all symbionts belong to the *Chlorella* clade. SSU rDNA is the only sequence for which homologs are available among a wide variety of trebouxiophytes. However, because it is rather highly conserved, the differences in sequence are too few for the dissection of members within this clade. Highly variable ITS sequences have traditionally been used within some of the lower taxonomic categories, for example, closely related lichen symbionts such as *Trebouxia*, *Coccomyxa*, and *Asterochloris* (Cordeiro et al. 2005; Piercey-Normore and DePriest 2001; Zoller and Lutzoni 2003). Meanwhile, Coleman (2003) proposed that the ITS2 sequence within relatively well conserved regions (i.e., helices II and III) can be used for family- or order-level analyses, and of course for subspecies-level analysis. The use of this sequence as a molecular marker for higher-level phylogenies has been strongly encouraged in recent years (Schultz et al. 2006; Selig et al. 2008). Around the same time that Krienitz et al. (2004) conducted phylogenetic analyses of the Chlorellaceae using ITS2 sequences, Hoshina et al. (2004) suggested the use of ITS2 for molecular comparisons of symbiotic and nonsymbiotic *Chlorella* spp., on the basis of the fact that its secondary structures are conserved to a greater extent than ITS1. Since then, ITS2 sequence data for chlorellacean species have been

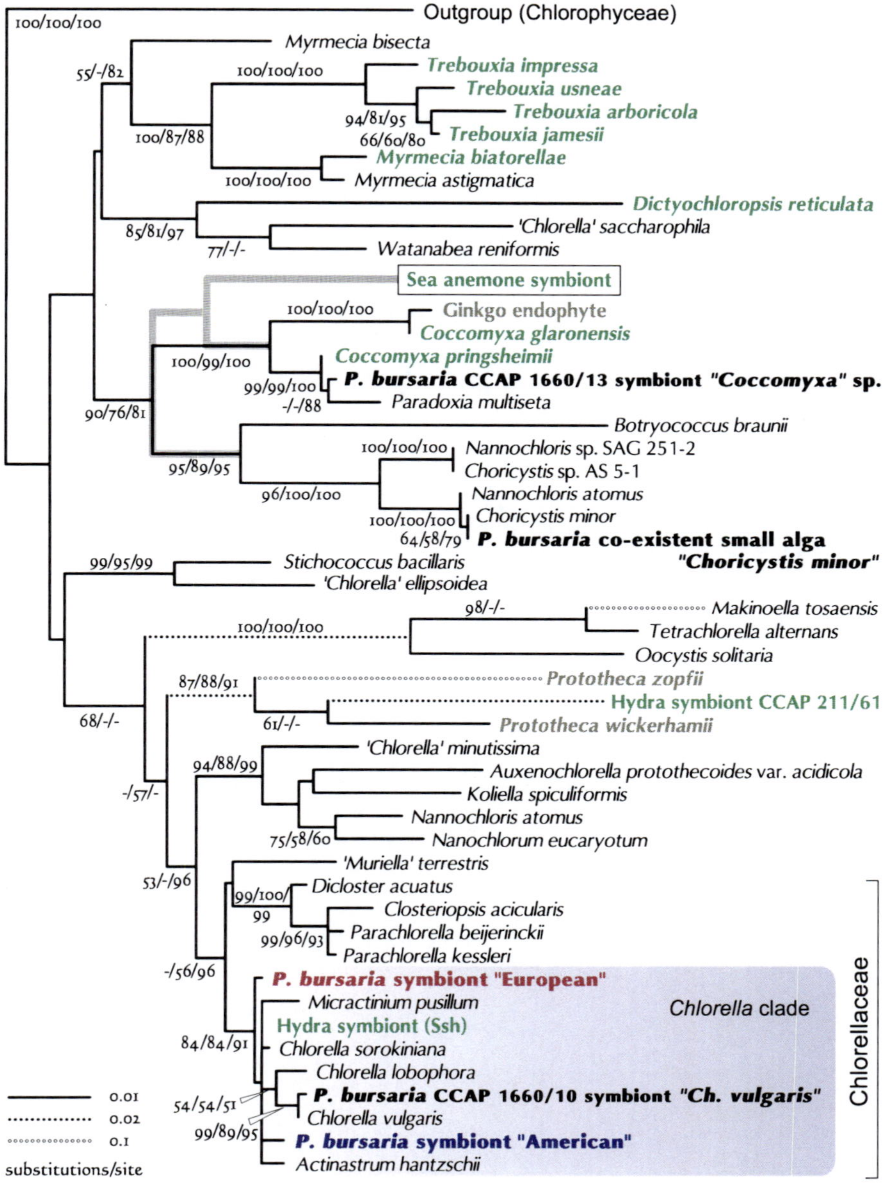


**Fig. 1** Maximum-likelihood (ML) tree (GTR + G + I) constructed from an analysis of small subunit ribosomal RNA (rRNA) gene sequences (length, 1,687 bp). Numbers at each node represent bootstrap probabilities of ML/maximum-parsimony (MP)/neighbor-joining (NJ) analyses; only values above 50% are shown. *Paramecium bursaria* symbionts, including small coexistent algae, are in *boldface*. Other algae existing as photobionts within organisms are denoted in *green*, and parasitic organisms (including achlorophyllous algae with a parasitic phase) are shown in *khaki*. The sea anemone symbiont (AY577786, shorter sequence, shown in the *box*) was added to the original phylogenetic analyses. (The tree was modified from Hoshina and Imamura 2008a)

increasing rapidly (Fawley et al. 2005; Hoshina et al. 2005; Hoshina and Imamura 2008a; Luo et al. 2006; Müller et al. 2005).

Despite still limited taxa, Fig. 2 shows an overall ITS2 tree of trebouxiophytes based on the conserved helices II and III. A general survey of the ITS2 tree supports results gained through SSU rDNA analyses (Fig. 1), while emphasizing genetic dissimilarities among strains or species. Unfortunately, ITS2 data for *Paradoxia* and the ginkgo endophyte related to the CCAP 1660/13 symbiont based on SSU

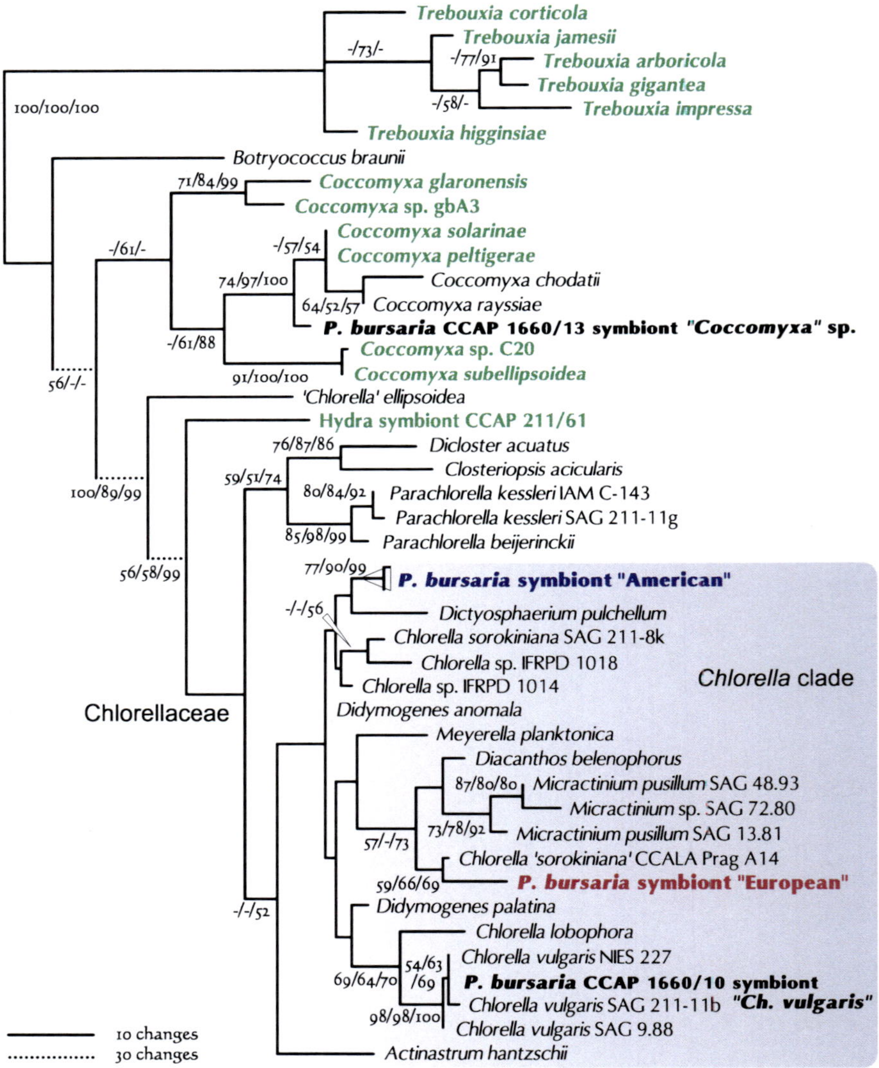


**Fig. 2** One of four most parsimonious trees (462 steps, consistency index 0.486) based on internal transcribed spacer (ITS) 2 helices II and III (length, 140 bp). Numbers at each node represent bootstrap probabilities of ML/MP/NJ analyses; only values above 50% are shown. Other algae existing as photobionts within organisms are denoted in *green*. (The tree was modified from Hoshina and Imamura 2008a)

rDNA (Fig. 1) are lacking. However, the CCAP 1660/13 alga could be regarded as a species of *Coccomyxa* from the viewpoint that this alga falls into a clade embedded in *Coccomyxa* spp. Species of *Coccomyxa* occupy a variety of ecological niches. Taking the members of this tree as examples, *C. chodatii* and *C. rayssiae* are free-living, whereas the other species are associated with ascolichens or basidiolichens (Zoller and Lutzoni 2003). Another is a member of the biofilm communities found on house facades and roof tiles (Karsten et al. 2005). It therefore would not be surprising to find a member of *Coccomyxa* in association with a ciliate.

The divergence of the *Chlorella* clade, including the "American," "European," and CCAP 1660/10 algal symbionts remains unclear; most of the branches are without bootstrap support (Fig. 2). However, it appears very clear that the CCAP 1660/10 symbiont is closely linked to *C. vulgaris*. Müller et al. (2005) investigated the intraspecific diversity of almost 30 strains of *C. vulgaris* available from public culture collections, in which they detected six slightly different sequence variations (variants A through F) for the ITS1-5.8S rDNA-ITS2 region. The sequence from the CCAP 1660/10 alga corresponds to variant D and should, accordingly, be regarded as *C. vulgaris*. In contrast, it was obvious that the three symbiotic groups are not closely related, and that neither the "American" nor the "European" algae have closely related relatives among the free-living algae. Species identification within *Chlorella* is difficult given their highly similar and character-deficient morphology. In fact, genetic differences are more likely to be found than differences in morphology or physiological characteristics. On the basis of sequence analyses, the two *C. sorokiniana* strains (SAG 211-8k is an apparently authentic culture) emerge at different phylogenetic positions [61 per 265 aligned sites (23.0%) were different] within the *Chlorella* clade; this is not surprising given the difference in GC content among *C. sorokiniana* strains (Sect. 2.1). As mentioned in Sect. 2.2, the symbiotic algae forming the "American" and "European" groups have occasionally been regarded as *C. vulgaris* and/or *C. sorokiniana*; however, the symbionts are quite different from either of these species on the basis of ITS2 sequence comparisons. The "American" group has 62/64 nucleotide differences per 260 aligned sites in comparison with the authentic *C. vulgaris* strain, SAG 211-11b, and 63/65 nucleotide differences per 265 sites when compared with the apparently authentic *C. sorokiniana* strain, SAG 211-8k; polymorphic sites among the "American" algae are indicated in Fig. 3. Similarly, the "European" group differs by 74 nucleotides (per 259 sites) from *C. vulgaris* and by 73 nucleotides (per 256 sites) from *C. sorokiniana. C. sorokiniana* strain CCALA Prag A14 could be regarded as the "European" progenitor in a phylogenetic sense, albeit as a weakly supported branch; however, there still exists a large nucleotide difference of 49 per 247 sites. And without doubt, each symbiont group differs from *Parachlorella kessleri*.

## 3.3 Multiple Origins of Symbiosis

In mutualistic symbiotic associations in which the inhabitant lives inside the exhabitant, the exhabitant works to suppress the inhabitant's sexual reproduction (Herre et al. 1999) and genetic change (Doebeli and Knowlton 1998; Law and Lewis 1983), as

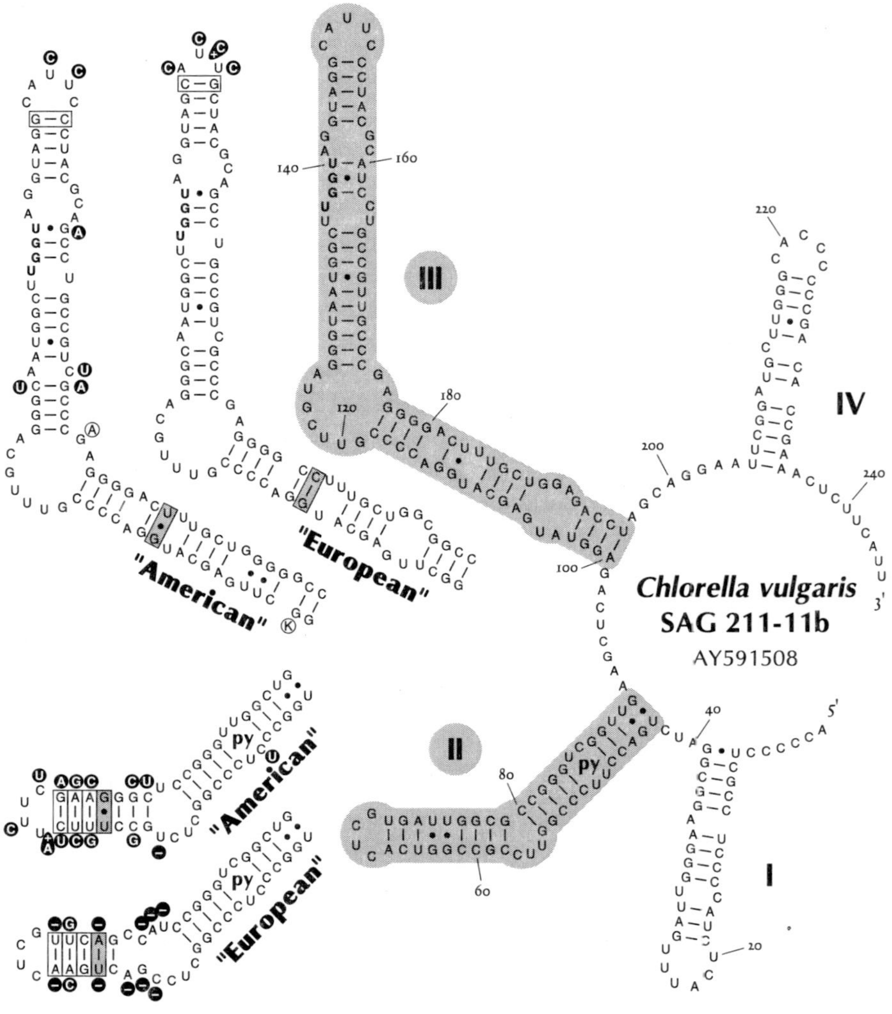


**Fig. 3** ITS2 structural diagram of *Chlorella vulgaris* type strain SAG 211-11b accompanied by helices II and III regions (*highlighted*) of the "American" and "European" groups of *P. bursaria* symbionts. Green algae possess conserved ITS2 motifs (Mai and Coleman 1997), including four-fingered hand motifs (four helices), pyrimidine-pyrimidine bulge (indicated by *py*) on helix II, and the conserved sequence TGGT (*UGGU* in *bold-face*) on the 5′ side of helix III. Known nucleotide polymorphisms among "American" symbionts (a different sequence is observed in strain ATCC 30562 alone; see Table 1) are indicated by *blank circles*. Otherwise, the sequences are identical among "European" symbionts. Nucleotide differences between *Dictyosphaerium pulchellum* (as the phylogenetically closest taxon, see Fig. 2) and "American" algae and between *C. sorokiniana* CCALA Plag A14 and "European" algae are indicated with *filled circles*. Compensatory base changes (CBCs; nucleotide changes on both sides of paired bases) and hemi-CBCs (changes on only one side of nucleotide pair) in helices II and III between the "American" and "European" symbionts are marked with *blank squares* and *gray squares*, respectively. *C. vulgaris* has four CBCs (at nucleotide pair positions 45-89, 62-75, 63-74, and 103-193) and a hemi-CBC (at 129-171) compared with the "American" symbiont, and five CBCs (at 45-89, 62-75, 64-73, 103-193, and 147-153) and two hemi-CBCs (at 112-183 and 129-171) compared with the "European" symbiont. See Hoshina and Imamura (2008a) for the full ITS2 structures of the "American" and "European" symbionts

well as providing it with a nutritionally and abiotically stable environment. Lower rates of genetic change (i.e., nucleotide substitution) of inhabitants have been recognized in various taxa, including cyanobacteria in lichens (Paulsrud and Lindblad 1998), chlorophytes in lichens (Piercey-Normore and DePriest 2001; Zoller and Lutzoni 2003), and dinoflagellates in foraminifera (Gast and Caron 1996). These results can be summarized as follows: (1) the rate of nucleotide substitution is lower in symbiotic inhabitants than in free-living relatives; and (2) the rate of nucleotide substitution is lower in inhabitants than that of their corresponding exhabitants when they have coevolved since their ancient association. Is there a possibility that a single symbiont diverged to create the current "American," "European," and CCAP 1660/10 (i.e., *C. vulgaris*) clades? Coevolution can be demonstrated indirectly by parallel cladogenesis (Page and Hafner 1995); however, these algal relationships are rather polyphyletic and more likely derivative (Fig. 2). The differences in algal symbiont ITS2 sequences (22.6–26.6%) are more than 10 times those of their exhabitants (four nucleotides in 181 aligned sites; Hoshina et al. 2006). These data suggest that multiple symbiont origins are more likely than a single symbiont that diverged to create three clades. The occurrence of closely related but multiple independent origins is also observed in lichen symbionts (Piercey-Normore and DePriest 2001; Zoller and Lutzoni 2003), and it is thought that *P. bursaria* has repeatedly acquired or replaced its photobiont over its evolutionary history, with at least four such events.

## 3.4 ITS2 Structural Analyses

The spacer region ITS2 can be regarded as a gene, assuming that it plays an important role in generating the mature 5.8S ribosomal RNA (rRNA) (3′) and large subunit (LSU) rRNA (5′) termini (Hadjiolova et al. 1994). ITS2 is a comparatively variable molecule in terms of its primary sequence, but its secondary/tertiary structure is much more conserved, even among higher taxonomic groups. For example, green algae and land plants share a common ITS2 secondary structure (Mai and Coleman 1997). In some respects, evolution of the secondary structure of ITS2 appears to occur much more slowly than speciation. Recently, the presence of at least one compensatory base change (CBC) has been correlated with the occurrence of two different species (biological species concept where fertile offspring are produced). That is to say, the presence of a CBC between two organisms indicates that they are distinct at the species level or higher (Coleman 2000). This hypothesis has been supported in various eukaryotic groups (Behnke et al. 2004; Coleman and Vacquier 2002; Young and Coleman 2004), and previous studies have verified the correlation of the presence/absence of a CBC and mutation level between organisms classified within the same genus (Müller et al. 2007). Comparing the secondary structures of the highly conserved ITS2 helices II and III (Coleman 2000, 2003), we always found CBCs between algal symbionts from the "American," "European," and other clades (Fig. 2). For example, there are five CBCs and two hemi-CBCs (compensatory change on only one side of the pair, indicating the next speciation level to the CBC) between strains of "American" symbionts and *Dictyosphaerium pulchellum*; this comparison does not include the variable nucleotide placed at the base of helix III (indicated as "K" in Fig. 3). Similarly, we found one CBC between the "European" symbionts and the *C. sorokiniana* strain CCALA Prag A14. In total, there are (1) four CBCs and two hemi-CBCs between the "American" versus "European" symbionts, (2) four CBCs and one hemi CBC between "American" versus CCAP 1660/10 symbionts, and (3) five CBCs and two hemi-CBCs between "European" versus CCAP 1660/10 symbionts (Fig. 3). Using the CBC differences as a measure for distinguishing species, we see that the *P. bursaria* symbiotic algae are more clearly different species than are the known chlorellacean algae, on the basis of the currently available sequence data.

## 3.5 Taxonomic Status of Symbionts

Of the numerous studies that have been conducted on paramecian symbiotic algae, those that have dealt with taxonomic aspects have traditionally referred to two algal groups, the "American" and "European" algae, both considered to be related to the genus *Chlorella*. However, four different endosymbiotic algal groups, at the species level or higher, can be recognized on the basis of compensatory base differences within the ITS2 secondary structure and phylogenetic analyses. One symbiotic species is definitely *C. vulgaris*, which is the conserved type species of the genus *Chlorella*. The "American" and "European" algae are also phylogenetically included within the *Chlorella* clade; therefore, they meet the criteria for inclusion within *Chlorella*, but are clearly separated from other described species. This raises questions regarding the number of species within the genus *Chlorella*. Two unidentified *Chlorella* spp. (IFRPD 1014 and 1018) subjected to ITS2 analyses also appear to be discrete species on the basis of the CBCs in ITS2; one to three CBCs in helix II are found among *Chlorella* spp. IFRPD 1014, 1018, and *C. sorokiniana* 211-8k, forming a clade (Fig. 2). The cryptic species *C. sorokiniana* CCALA Prag A14 is also included. The exact identity of these species requires further study of the free-living and symbiotic *Chlorella* spp. The algal symbiont from *P. bursaria* CCAP 1660/13 falls within a totally different clade, perhaps belonging to the genus *Coccomyxa*.

Classical identification studies, as described in Sect. 2.2, provided partially correct answers in terms of *Chlorella* affiliation within *P. bursaria* symbionts. Viral infection analyses were probably the most effective technique in the elucidation of relationships among symbiotic algae from American and European *P. bursaria*, symbiotic algae of other protozoa, and free-living algal species. In a viral infection study, Reisser et al. (1990) noted that, "If viruses infecting additional symbiotic and also free-living *Chlorella* spp. are discovered it will be possible to use viruses to classify *Chlorella* sp. (*P. bursaria* symbionts) in the same way that bacteriophages are used in bacterial classification." Unfortunately, no such viruses have been identified to date; expressed another way, each "American" or "European" alga is definitely a discrete species from any other free-living or symbiotic *Chlorella* spp.

# 4 Symbiont rDNA Group I Introns

"American" and "European" symbionts possess characteristic insertions (approximately 300–600 nucleotides each) called the "group I intron" within the nuclear rRNA genes (Gaponova et al. 2007; Hoshina et al. 2004, 2005; Hoshina and Imamura 2008a, b). Thus, length polymorphism of PCR products containing intron(s) may be a useful tool for the identification of symbionts. For example, PCR with the primer pair CHspeRmaeF/chSsotoR will produce different length products of approximately 1,200 bp for the "American" and "European" algae, whereas they produce fragments of only 900 bp for non-intron-containing chlorophytes. The SR-8/INT-5R primer pair will produce a nearly 1,200-bp fragment for the "American" algae, but only a 500-bp fragment for the "European" and other non-intron-containing algae. Nested PCR is easily performed using template DNA extracted from whole *Paramecium* extracts, and the products are sufficient for direct sequencing (for a detailed protocol, see Hoshina et al. 2005).

Group I introns are a distinct group of RNA molecules that function as enzymes, splicing themselves out of precursor RNA transcripts and ligating exons together. Today, group I introns are divided into five major subgroups (IA–IE) according to their structural characteristics and phylogeny. In the nuclear genome, group I introns reside only in rDNA for unknown reasons, and these belong to either subgroup IC or subgroup IE. These introns have a wide and highly sporadic distribution in nature, which is thought to be the result of their characteristic tendency to invade the genomes of different species. Introns may spread via homing or reverse splicing. In homing, the intron encodes a homing endonuclease gene at the terminus of a peripheral helix (when the intron itself takes a higher conformation). Homing endonuclease recognizes 14–40 specific nucleotides of double-stranded DNA, and then cleaves the target site and inserts the intron. Reverse splicing is the inverse process of intron splicing at the RNA level. Direct evidence of this pathway is still lacking; however, recent intronic phylogenetic analyses have provided strong evidence for lateral transfer events via reverse splicing. This pathway also requires four to six specific nucleotides that the intron recognizes as the optimal site for self-insertion. Group I intron movement nearly always occurs at homologous sites owing to these sequence-recognition mechanisms. Consequently, the number of insertion sites is fairly limited, and introns at homologous sites are generally more closely related to each other than are introns at different sites (for a review, see Haugen et al. 2005).

## 4.1 Uniqueness of Intron Insertion Position

"American" symbionts are particularly rich in group I introns. They contain eight introns, the largest number among green algae, with four subgroup IC introns and four subgroup IE introns within the nuclear rDNA region (Fig. 4a). As mentioned already, group I intron insertion sites are generally limited; however, five introns occupy unique genic sites (S1367, L200, L1688, L2184 and L2437; the numbering reflects their homologous positions in the *Escherichia coli* rRNA gene: "S" represents SSU rRNA, "L" represents LSU rRNA). Position S943 has been recognized for subgroup IC introns, but is a novel position for IE introns. Introns of the "European" symbiont (two IE introns) are also characteristic. One intron (L2449) inserts itself at a unique (as IE) site, whereas the intron at S651 has been identified in only one organism, *Chlorella* sp. AN 1-3 (collected from the surface of ice). These introns are excluded from the conventional concept of group I intron spread; how, then, did these introns gain such unique positions?

## 4.2 Phylogenetic Relationships among Introns

Although group I introns often have greater rates of sequence divergence, they still encode sufficient phylogenetic information to estimate their evolutionary relationships (Bhattacharya et al. 2001; Haugen et al. 2005). In phylogenetic analyses,

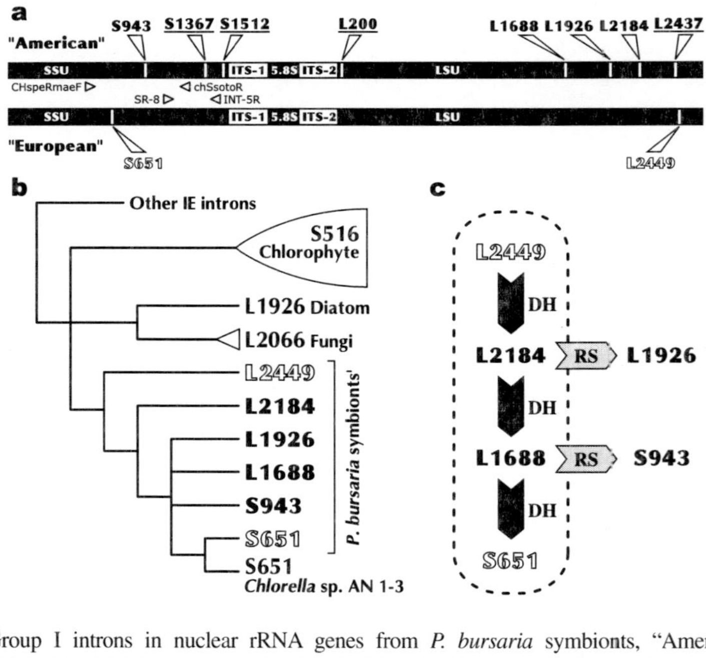


**Fig. 4** Group I introns in nuclear rRNA genes from *P. bursaria* symbionts, "American" and "European" algae. The introns are rendered in *boldface* for "American" algae and are *outlined* for "European" algae. **a** Insertion sites of group I introns. Subgroup IC introns are *underlined*, whereas IE introns are not. The numbering reflects their homologous positions in the *Escherichia coli* rRNA gene. The primer pairs available to detect the presence/absence of introns are indicated (see text). **b** A part of subgroup IE phylogeny concerning the symbionts' introns. **c** A possible evolutionary context for the subgroup IE introns of the symbionts with six heterologous insertion sites. This scheme is based on a hypothetical transposition pathway involving degenerative homing (*DH*) and reverse splicing (*RS*)

"American" IC introns form a clade with chlorophyte introns, with insertions at nine heterologous sites (Hoshina and Imamura 2008b). The evolutionary study of group I introns is most advanced in fungi; however, there are no reports regarding a situation similar to that described above. Therefore, the mechanism of dispersal is still unresolved. The IE introns commonly found in both "American" and "European" symbionts show much greater specificity. Despite the presence of six heterologous insertion sites, they form a monophyletic clade with the *Chlorella* sp. AN 1-3 intron (S651), independent of other intron groups (Fig. 4b). "European" L2449 has a structurally archaic status in this clade (R. Hoshina and N. Imamura, unpublished data) accompanying the phylogenetic relationship (initially divergent within the clade). This indicates that the other introns of the symbionts originated with this L2449 intron. Furthermore, some of the introns share common internal guide sequences, which are necessary for misdirected transfer (i.e., transposition) via reverse splicing. Other introns, however, share similar sequence fragments further upstream and downstream of the insertions. Perhaps the dispersion of these introns can be explained by a combination of the primitive hominglike pathway (we dub it "degenerative homing") and reverse splicing (Fig. 4c). If these putative pathways are correct, infections of the vertical line from L2449 to S651 may have terminated within the *P. bursaria* cell. This idea requires us to postulate that there was a period during which the two different symbionts, namely, the "American" and "European," lived sympatrically and simultaneously within a *P. bursaria* cell. Such cell-cell contact within a small space may accelerate the lateral transfer of group I introns (Bhattacharya et al. 1996; Friedl et al. 2000).

## 5 Host-Symbiont Specificity

As discussed already, symbiotic algae are not completely consistent among all *P. bursaria*; however, their relationships show marked conservation when compared with those of other protists. Summerer et al. (2008) examined host-symbiont relationships in other ciliate protozoans collected from two lakes in Austria. With the exception of *P. bursaria* (whose symbiont was assigned to the "European" group), the symbionts of these ciliates in a lake were closely related *Chlorella* spp., sharing greater than 97% identity in the ITS1 sequence. Ciliates in different lakes, however, possessed different algae, even within the same species of ciliate. These facts imply that these ciliates may value an environmentally habituated symbiont ahead of permanent symbiosis. However, Hoshina et al. (2004) reported six *P. bursaria* symbionts from five different locations in Japan, spanning a distance of 1,000 km, shared identical 4,000-bp sequences, including ITSs and introns; these strains were subsequently assigned to the "American" group. Hence, the symbiont identities of *P. bursaria* are thought to have a higher regional dependence. The following discussion examines the geographic distribution of *P. bursaria* symbionts and the relationship to host genetic variation.

## 5.1 Distribution

Table 1 shows the available nuclear rDNA sequence information for *P. bursaria* symbiotic algae. For the "American" or "European" groups, SSU rDNA with group I intron insertions (i.e., insertion positions or insertion numbers) will indicate the symbionts' identities, whereas highly divergent ITS sequences clearly indicate species identity through sequence similarity searches, such as FASTA or BLAST. Figure 5 shows the distribution of the "American" and "European" algae plotted on a world map. The "European" group is limited to the western European areas of England, Germany, Austria, and northern Europe, Karelia region (Russia). The "American" group is rather widely distributed in the USA (eastern and other unspecified regions), Japan, Shanghai (China), and Melbourne (Australia). The number of samples for them, however, is very meager. Although *P. bursaria* is a pandemic species, much of the temperate zone remains unrepresented. There are a few harvesting records from Africa and South America (Fokin et al. 2004). Further sampling from the underrepresented areas is very much needed.

It may be added that the "American" and "European" algae are also referred to as the "southern" and "northern" ecotypes (*sensu* Kvitko et al. 2001).

## 5.2 Relationships of the Algae and the Hosts' Genetic Variations

The SSU rDNA sequences of *P. bursaria* fall into four genotypes (A–D; Hoshina et al. 2006). *P. bursaria* may be regarded as an "old" species (Stoeck et al. 1998),

**Fig. 5** Distribution of the "American" (*solid symbol*) and "European" (*empty*) *P. bursaria* symbionts as determined by ribosomal DNA sequencing. Two different symbol sizes indicate a record of one to three (smaller) or more than five (larger) samples. Unspecified localities in the USA and Russia, in spite of their vast national territories, are excluded. See also Table 1

possessing large differences in the locus. Genotypes C and D differ in terms of 23 substitutions and four indels observed in genotype D. Genotype C has a highly mutated helix (corresponding to the *Tetrahymena* SSU rRNA E23_5 helix; Strüder-Kypke et al. 2001) with eight nucleotide changes compared with other genotypes. Genotypes A and B are closely related, with only two transitions detected.

Hoshina et al. (2005) suggested the possibility that symbiotic algae depend not only on their localities, but also on their hosts' genotype. In fact, all genotype D hosts have "American" symbiotic algae in common. However, "European" algae inhabit genotype A and genotype C ciliates, a separate genotype A host possesses *Coccomyxa* alga, and a genotype B host possesses *C. vulgaris* as a symbiotic partner (Table 1). Thus, there are consistent correlations between host and symbiont genotypes, although they are only partially understood.

*Paramecium* spp. usually include several sexually separated sibling groups termed "syngens" that are morphologically uniform. Such a species complex is not unique to *Paramecium*. Plenty of examples have been reported in other protists, as well as in plants and animals. Although it is often unclear whether each syngen represents a distinct species, those within the *P. aurelia* complex were raised to species status (Sonneborn 1975). Six syngens (1–6) are found within *P. bursaria*, with four to eight multiple mating types (sexes) in each (Bomford 1966; Wichterman 1986). Although four genotypes of *P. bursaria* exist, their variety is limited to syngens 1, 2, or "unknown." Further analyses of host genetic diversity are required for a better understanding of the relationship between reproductive isolation and speciation in *P. bursaria*, and between the host and the symbiont.

## 6 Concluding Remarks and Evolutionary Considerations

With the exception of primary plants, which evolved from a single endosymbiotic event in which a heterotrophic protozoan engulfed and enslaved a cyanobacterium (i.e., glaucophytes, rhodophytes, and chlorophytes, inclusive of land plants stemming from them), oxygenic phototrophs are thought to have been established by means of secondary or tertiary symbioses, in which heterotrophic protozoa enslaved eukaryotic phototrophs (i.e., euglenophytes, chlorarachniophytes, heterokontophytes, cryptophytes, haptophytes, and dinoflagellates). Inouye and Okamoto (2005) postulated a four-step evolutionary process for plastid acquisition via secondary symbiosis: (1) the kleptochloroplast stage, in which a temporary symbiont is obtained via phagocytotic behavior (*klepto* is an ancient Greek word meaning "stolen" or "cheated"); (2) the persistent symbiont stage featuring synchronized cell division of the host and symbiont; (3) lateral gene transfer to the host nucleus stage, during which the symbiont's organelles disappear but a vestigial nucleus or "nucleomorph" is retained, as in cryptophytes and chlorarachniophytes; and (4) establishment of the plastid (e.g., heterokontophytes, haptophytes, and euglenophytes). The host-symbiont relationship in *P. bursaria* involves a number of complicated and advanced systems. Despite the presence of several hundred symbiotic algae in a single host cell, the host maintains each symbiont in a perialgal vacuole and controls algal cell division (Reisser et al. 1983). *P. bursaria* receives photosynthate from the symbiont and, in return, provides a comfortable habitat. Some symbionts, that is, the "American" and "European" algae, have evolved dramatically. The algae are moderately to highly dependent on their hosts to fulfill their specialized nutrient requirements, and for protection against viral attack. When symbionts are artificially removed from *P. bursaria*, the host will reabsorb the exsymbionts as well as free-living *Chlorella* spp. However, if the exsymbionts and free-living algae are simultaneously presented, *P. bursaria* will selectively absorb the exsymbionts (Summerer et al. 2007). Neither the "American" nor the "European" symbiont is capable of surviving in the natural environment; therefore, the symbiotic association of *P. bursaria* appears to be in the transitional phase between persistent symbiont and lateral gene transfer stage. As the host-symbiont association becomes permanent, genes from the symbiont nucleus are either transferred gradually to the host nucleus or lost, which may occur when the gene products are no longer required (van Dooren et al. 2001). Decreased symbiont autonomy may perhaps be due to the loss of genes necessary for living as an independent organism. Complete genome sequencing for the authentic "American" strain *Chlorella* sp. NC64A is under way (genome project ID 18711) with the type species of the genus, *C. vulgaris* (18715), and is expected to provide exciting evolutionary insight. This is the first genome project focused on a trebouxiophyte.

Lichens and corals, examples of symbiotic association with algal photobionts, spend a phase of their life cycles without symbionts, and must acquire fresh algae to complete their life cycle. In contrast, *P. bursaria* retain their symbionts throughout cell division and sexual reproduction (Siegel 1960). Consequently, the symbiotic relationship in paramecia appears to be permanent. Although symbiont-free *P. bursaria* have been found in the natural environment (Tonooka and Watanabe 2002), such individuals are likely at a disadvantage in terms of survival, growth rate under low-nutrient conditions (perhaps equivalent to the natural habitat), and may also be less tolerant to reactive oxygen species (Sommaruga and Sonntag, this volume). The host may explore appropriate alga as new symbionts through phagocytosis. Algal selection via simultaneous double-symbiont associations may be an alternative opportunity for algal switching. Evolutionarily related group I introns may indicate a period in which the "American" and "European" type algae lived simultaneously in a ciliate (Sect. 4.2), which has been known to occur, as discussed in Sect. 3.1. Although sampling remains patchy and our knowledge regarding the diversity of *P. bursaria* or their symbionts is fragmentary (Sect. 5), the previously discussed findings allow us to outline an evolutionary framework with respect to algal acquisition and subsequent switching. Ancestral *P. bursaria* may have obtained the "American" and "European" species separately, and then retained both for some period of time. One individual "chose" the "American" alga as a symbiotic partner, and then expanded in distribution. Another individual host cell "chose" the "European" symbiont, and subsequently diversified into a lineage with a weakened host-algal partnership and two events of accidental algal switching (Fig. 6). Future studies may discover additional irregular symbioses and, perhaps, additional symbionts specific to *P. bursaria*.

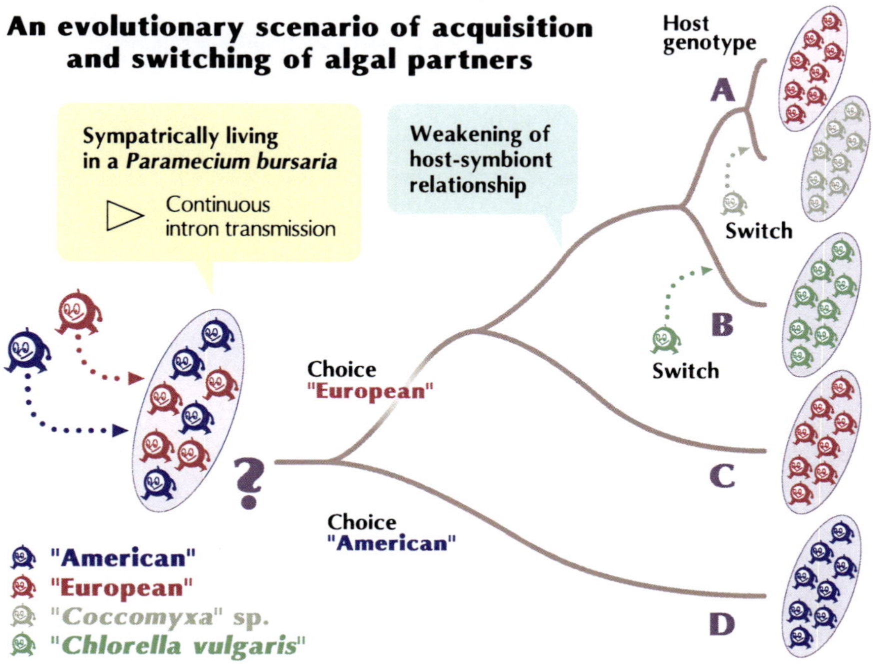


**Fig. 6** An evolutionary scenario for *P. bursaria* with respect to algal acquisition and switching. Ancestral *P. bursaria* may have gained the "American" and "European" algae separately, and then retained both species for a time. Then, one individual "chose" the "American" alga as a symbiotic partner, whereas a second individual "chose" the "European" alga; the latter may have subsequently diverged into a lineage with a weakened host-algal partnership in which accidental algal switching occurred twice

Symbiotic algae are the main constituents of the *Paramecium* protoplasm, and a total of four species have been recognized, including *C. vulgaris* and a *Coccomyxa* sp. However, the complicated taxonomic history of the majority of *P. bursaria* symbionts has resulted in the identification of two algae with different characteristics: the "American" and "European" groups. Molecular phylogeny and gene structure analysis have demonstrated that these groups are distinct from each other and do not belong to any known species. The ITS2 tree and its structural analyses also revealed various *Chlorella* spp., including an inauthentic *C. sorokiniana*. Coccoids with distinct SSU rDNA sequences are readily found in nature, even those belonging to the *Chlorella* clade, and with each different sequence may be regarded as the equivalent to a "species" (Fawley et al. 2004). Therefore, the number of described microalgae is a tiny portion of the diverse species awaiting discovery. Despite the fact that we have entered the era of DNA sequencing, allowing us to determine true affiliations among species, there seems to be no end to the number of irresponsible species identifications that subsequently generate a chain reaction of misidentification and further disorder. Furthermore, in virology, strange terminology, such as "NC64A virus" and "Pbi virus," continues to be used. The underlying problem is clear. Each "American" and "European" symbiont must be assigned a species name as soon as possible to avoid further confusion.

**Acknowledgements** We thank the editor, M. Fujishima, and the series editor, A. Steinbüchel, for providing us with the opportunity to write on this theme. We also thank Prof. I. Miwa (Ibaraki Univ.) for kindly providing copies of some old papers. Some figures were taken from a previous publication (Hoshina and Imamura 2008a) with permission from Elsevier.

## References

Albers D, Reisser W, Wiessner W (1982) Studies on the nitrogen supply of endosymbiotic chlorellae in green *Paramecium bursaria*. Plant Sci Lett 25:85–90

Behnke A, Friedl T, Chepurnov VA, Mann DG (2004) Reproductive compatibility and rDNA sequences analyses in the *Sellaphora pupula* species complex (Bacillariophyta). J Phycol 40:193–208

Bhattacharya D, Cannone JJ, Gutell RR (2001) Group I intron lateral transfer between red and brown algal ribosomal RNA. Curr Genet 40:82–90

Bhattacharya D, Friedl T, Damberger S (1996) Nuclear-encoded rDNA group I introns: origin and phylogenetic relationships of insertion site lineages in the green algae. Mol Biol Evol 13:978–989

Bomford R (1966) The syngens of *Paramecium bursaria*: new mating types and intersyngenic mating reactions. J Protozool 13:497–501

Brown JA, Nielsen PJ (1974) Transfer of photosynthetically produced carbohydrate from endosymbiotic chlorellae to *Paramecium bursaria*. J Protozool 21:569–570

Coleman AW (2000) The significance of a coincidence between evolutionary landmarks found in mating affinity and a DNA sequence. Protist 151:1–9

Coleman AW (2003) ITS2 is a double-edged tool for eukaryote evolutionary comparisons. Trends Genet 19:370–375

Coleman AW, Vacquier VD (2002) Exploring the phylogenetic utility of ITS sequences for animals: a test case for abalone (*Haliotis*). J Mol Evol 54:246–257

Cordeiro LMC, Reis RA, Cruz LM, Stocker-Wörgötter E, Grube M, Iacomini M (2005) Molecular studies of photobionts of selected lichens from the coastal vegetation of Brazil. FEMS Microbiol Ecol 54:381–390

Doebeli M, Knowlton N (1998) The evolution of interspecific mutualisms. Proc Natl Acad Sci U S A 95:8676–8680

Douglas AE (1995) The ecology of symbiotic micro-organisms. Adv Ecol Res 26:69–103

Douglas AE (1998) Host benefit and the evolution of specialization in symbiosis. Heredity 81:599–603

Douglas AE, Huss VAR (1986) On the characteristics and taxonomic position of symbiotic *Chlorella*. Arch Microbiol 145:80–84

Fawley MW, Fawley KP, Buchheim MA (2004) Molecular diversity among communities of freshwater microchlorophytes. Microb Ecol 48:489–499

Fawley MW, Fawley KP, Owen HA (2005) Diversity and ecology of small coccoid green algae from Lake Itasca, Minnesota, USA, including *Meyerella planktonica*, gen. et sp. nov. Phycologia 44:35–48

Fokin SI, Przybos E, Chivilev SM, Beier CL, Horn M, Skotarczak B, Wodecka B, Fujishima M (2004) Morphological and molecular investigations of *Paramecium schewiakoffi* sp. nov.

(Ciliophora, Oligohymenophorea) and current status of distribution and taxonomy of *Paramecium* spp. Eur J Protistol 40:225–243

Frank SA (1997) Models of symbiosis. Am Nat 150(Suppl):80–99

Friedl T (1995) Inferring taxonomic positions and testing genus level assignments in coccoid green lichen algae: a phylogenetic analysis of 18S ribosomal RNA sequences from *Dictyochloropsis reticulata* and from members of the genus *Myrmecia* (Chlorophyta, Trebouxiophyceae cl. nov.). J Phycol 31:632–639

Friedl T, Zeltner C (1994) Assessing the relationships of some coccoid green lichen algae and the Microthamniales (Chlorophyta) with 18S ribosomal RNA gene sequence comparisons. J Phycol 30:500–506

Friedl T, Besendahl A, Pfeiffer P, Bhattacharya D (2000) The distribution of group I introns in lichen algae suggests that lichenization facilitates intron lateral transfer. Mol Phylogenet Evol 14:342–352

Gaponova IN, Andronov EE, Migunova AV, Vorobyev KP, Chizhevskaja EP, Kvitko KV (2007) Genomic dactyloscopy of *Chlorella* sp., symbionts of *Paramecium bursaria*. Protistology 4:311–317

Gast RJ, Caron DA (1996) Molecular phylogeny of symbiotic dinoflagellates from planktionic foraminifera and radiolaria. Mol Biol Evol 13:1192–1197

Hadjiolova KV, Normann A, Cavaillé J, Soupéne E, Mazan S, Hadjiolov AA, Bachellerie J-P (1994) Processing of truncated mouse or human rRNA transcribed from ribosomal minigenes transfected into mouse cells. Mol Cell Biol 14:4044–4056

Haugen P, Simon DM, Bhattacharya D (2005) The natural history of group I introns. Trends Genet 21:111–119

Herre EA, Knowlton N, Müller UA, Rehner SA (1999) The evolution of mutualism: exploring the paths between conflict and cooperation. Trends Ecol Evol 14:49–53

Hoshina R, Imamura N (2008a) Multiple origins of the symbioses in *Paramecium bursaria*. Protist 159:53–63

Hoshina R, Imamura N (2008b) Eu-*Chlorella* large subunit rDNA sequences and group I introns in ribosomal DNA of the paramecian symbiotic alga NC64A. Phycol Res 56:21–32

Hoshina R, Kamako S-i, Imamura N (2004) Phylogenetic position of endosymbiotic green algae in *Paramecium bursaria* Ehrenberg from Japan. Plant Biol 6:447–453

Hoshina R, Kato Y, Kamako S-i, Imamura N (2005) Genetic evidence of "American" and "European" type symbiotic algae of *Paramecium bursaria* Ehrenberg. Plant Biol 7:526–532

Hoshina R, Hayashi S, Imamura N (2006) Intraspecific genetic divergence of *Paramecium bursaria* and re-construction of paramecian phylogenetic tree. Acta Protozool 45:377–386

Huss VAR, Jahnke K-D (1994) DNA base modification in *Chlorella*. Eur J Phycol 29:87–92

Huss VAR, Sogin ML (1990) Phylogenetic position of some *Chlorella* species within the Chlorococcales based upon complete small-subunit ribosomal RNA sequences. J Mol Evol 31:432–442

Huss VAR, Frank C, Hartmann EC, Hirmer M, Kloboucek A, Seidel BM, Wenzeler P, Kessler E (1999) Biochemical taxonomy and molecular phylogeny of the genus *Chlorella* sensu lato (Chlorophyta) J Phycol 35:587–598

Huss VAR, Holweg C, Seidel B, Reich V, Rahat M, Kessler E (1993/1994) There is an ecological basis for host/symbiont specificity in *Chlorella/Hydra* symbioses. Endocyt Cell Res 10:35–46

Inouye I, Okamoto N (2005) Changing concepts of a plant: current knowledge on plant diversity and evolution. Plant Biotechnol 22:505–514

Kamako S-i, Imamura N (2006) Effect of Japanese *Paramecium bursa*ria extract on photosynthetic carbon fixation of symbiotic algae. J Eukaryot Microbiol 53:136–141

Kamako S-i, Hoshina R, Ueno S, Imamura N (2005) Establishment of axenic endosymbiotic strains of Japanese *Paramecium bursaria* and the utilization of carbohydrate and nitrogen compounds by the isolated algae. Eur J Protistol 41:193–202

Kapaun E, Loos E, Reisser W (1992) Cell wall composition of virus-sensitive symbiotic *Chlorella* species. Phytochemistry 31:3103–3104

Karsten U, Friedl T, Schumann R, Hoyer K, Lembcke S (2005) Mycosporine-like amino acids and phylogenies in green algae: *Prasiola* and its relatives from the Trebouxiophyceae (Chlorophyta). J Phycol 41:557–566

Kessler E (1976) Comparative physiology, biochemistry, and the taxonomy of *Chlorella* (Chlorophyceae). Plant Syst Evol 125:129–138

Kessler E, Huss VAR (1990) Biochemical taxonomy of symbiotic *Chlorella* strains from *Paramecium* and *Acanthocystis*. Bot Acta 103:140–142

Kessler E, Huss VAR (1992) Comparative physiology and biochemistry and taxonomic assignment of the *Chlorella* (Chlorophyceae) strains of the Culture Collection of The University of Texas at Austin. J Phycol 28:550–553

Kodama Y, Nakahara M, Fujishima M (2007) Symbiotic alga *Chlorella vulgaris* of the ciliate *Paramecium bursaria* shows temporary resistance to host lysosomal enzymes during the early infection process. Protoplasma 230:61–67

Krienitz L, Hegewald E, Hepperle D, Huss VAR, Rohr T, Wolf M (2004) Phylogenetic relationship of *Chlorella* and *Parachlorella* gen. nov. (Chlorophyta, Trebouxiophyceae). Phycologia 43:529–542

Kvitko KV, Migunova AV, Karelov DV, Prokosheva MJu (2001) Molecular taxonomy of virus-sensitive *Chlorella* sp. – symbionts of *Paramecium bursaria*. Protistology 2:96–104

Law R, Lewis DH (1983) Biotic environments and the maintenance of sex – some evidence from mutualistic symbioses. Biol J Linn Soc 20:249–276

Lewis LA and Muller-Parker G (2004) Phylogenetic placement of "zoochlorellae" (Chlorophyta), algal symbiont of the temperate sea anemone *Anthopleura elegantissima*. Biol Bull 207:87–92

Luo W, Pflugmacher S, Pröschold T, Walz N, Krienitz L (2006) Genotype versus phenotype variability in *Chlorella* and *Micractinium* (Chlorophyta, Trebouxiophyceae). Protist 157:315–333

Linz B, Linz A, Migunova AV, Kvitko KV (1999) Correlation between virus-sensitivity and isoenzyme spectrum in symbiotic *Chlorella*-like algae. Protistology 1:76–81

Mai JC, Coleman AW (1997) The internal transcribed spacer 2 exhibits a common secondary structure in green algae and flowering plants. J Mol Evol 44:258–271

Meier R, Wiessner W (1989) Infection of algae-free *Paramecium bursaria* with symbiotic *Chlorella* sp. isolated from green paramecia. J Cell Sci 93:571–579

Müller J, Friedl T, Hepperle D, Lorenz, M (2005) Distinction between multiple isolates of *Chlorella vulgaris* (Chlorophyta, Trebouxiophyceae) and testing for conspecificity using amplified fragment length polymorphism and ITS rDNA sequences. J Phycol 41:1236–1247

Müller T, Philippi N, Dandekar T, Schultz J, Wolf M (2007) Distinguishing species. RNA 13:1469–1472

Nakahara M, Handa S, Watanabe S, Deguchi H (2004) *Choricystis minor* as a new symbiont of simultaneous two-species association with *Paramecium bursaria* and implications for its phylogeny. Symbiosis 36:127–151

Page RDM, Hafner MS (1995) Molecular phylogenies and host-parasite cospeciation: gophers and lice as a model system. Philos Trans R Soc Lond B 349:77–83

Paulsrud P, Lindblad P (1998) Sequence variation of the tRNA (Leu) intron as a marker for genetic diversity and specificity of symbiotic cyanobacteria in some lichens. Appl Environ Microbiol 64:310–315

Piercey-Normore MD, DePriest PT (2001) Algal switching among lichen symbioses. Am J Bot 88:1490–1498

Reisser W, Widowski M (1992) Taxonomy of eukaryotic algae endosymbiotic in freshwater associations. In: Reisser W (ed) Algae and symbioses. Biopress, Bristol, pp 21–40

Reisser W, Meier R, Kurmeier B (1983) The regulation of the endosymbiotic algal population size in ciliate-algae associations. An ecological model. In: Schenk HEA, Schwemmler W (eds) Endocytobiology II. de Gruyter, Berlin, pp 533–543

Reisser W, Burbank DE, Meints SM, Meints RH, Becker B, Van Etten JL (1988a) A comparison of viruses infecting two different *Chlorella*-like green algae. Virology 167:143–149

Reisser W, Vietze S, Widowski M (1988b) Taxonomic studies on endocytobiotic chlorophycean algae isolated from different American and European strains of *Paramecium bursaria*. Symbiosis 6:253–270

Reisser W, Burbank DE, Meints RH, Becker B, Van Etten JL (1990) Viruses distinguish symbiotic *Chlorella* spp. of *Paramecium bursaria*. Endocytobiosis Cell Res 7:245–251

Schult J, Müller T, Achtziger M, Seibel PN, Dandekar T, Wolf M (2006) The internal transcribed spacer 2 database – a web server for (not only) low level phylogenetic analyses. Nucleic Acids Res 34:W704–W707

Selig C, Wolf M, Müller T, Dandekar T, Schultz j (2008) The ITS2 Database II: homology modelling RNA structure for molecular systematics. Nucleic Acids Res 36:D377–D380

Shihara I, Krauss RW (1965) *Chlorella*. Physiology and taxonomy of forty-one isolates. University of Maryland, College Park

Siegel RW (1960) Hereditary endosymbiosis in *Paramecium bursaria*. Exp Cell Res 19:239–252

Sonneborn TM (1975) The *Paramecium aurelia* complex of fourteen sibling species. Trans Am Microsc Soc 94:155–178

Stoeck T, Przybos E, Schmidt HJ (1998) A combination of genetics with inter- and intra-strain crosses and RAPD-fingerprints reveals different population structures within the *Paramecium aurelia* complex. Eur J Protistol 34:410–420

Strüder-Kypke MC, Wright AD, Jerome CA, Lynn DH (2001) Parallel evolution of histophagy in ciliates of the genus *Tetrahymena*. BMC Evol Biol 1:5

Summerer M, Sonntag B, Sommaruga R (2007) An experimental test of the symbiosis specificity between the ciliate *Paramecium bursaria* and strains of the unicellular green alga *Chlorella*. Environ Microbiol 9:2117–2122

Summerer M, Sonntag B, Sommaruga R (2008) Ciliate-symbiont specificity of freshwater endosymbiotic *Chlorella* (Trebouxiophyceae, Chlorophyta). J Phycol 44:77–84

Takeda H (1991) Sugar composition of the cell wall and the taxonomy of *Chlorella*. J Phycol 27:224–232

Takeda H (1995) Cell wall composition and taxonomy of symbiotic *Chlorella* from *Paramecium* and *Acanthocystis*. Phytochemistry 40:457–459

Tonooka Y, Watanabe T (2002) A natural strain of *Paramecium bursari*a lacking symbiotic algae. Eur J Protistol 38:55–58

Trémouillaux-Guiller J, Rohr T, Rohr R, Huss VAR (2002) Discovery of an endophytic alga in *Ginkgo biloba*. Am J Bot 89:727–733

van Dooren GG, Schwartzbach SD, Osafune T, McFadden GI (2001) Translocation of proteins across the multiple membranes of complex plastids. Biochim Biophys (Acta Mol Cell Res) 1541:34–53

Wichterman R (1986) The Biology of *Paramecium*, 2nd edn. Plenum, New York

Young I, Coleman AW (2004) The advantages of the ITS2 region of the nuclear rDNA cistron for analysis of phylogenetic relationships of insects: a *Drosophila* example. Mol Phylogenet Evol 30:236–242

Zoller S, Lutzoni F (2003) Slow algae, fast fungi: exceptionally high nucleotide substitution rate differences between lichenized fungi *Omphalina* and their symbiotic green algae *Coccomyxa*. Mol Phylogenet Evol 29:629–640

# Infection of *Paramecium bursaria* by Symbiotic *Chlorella* Species

**Yuuki Kodama and Masahiro Fujishima**

**Contents**



**Abstract** *Paramecium bursaria* and endosymbiotic *Chlorella* species retain their ability to grow independently, but can reestablish endosymbiosis by mixing. Infection is induced through the host's digestive vacuoles (DVs). Acidosomal and lysosomal fusions to the DVs begin at 0.5 and 2–3 min after mixing, respectively.

---

Y. Kodama
Graduate School of Life and Environmental Sciences, University of Tsukuba, Tennoudai 1-1-1, 305-8572, Tsukuba, Japan

M. Fujishima (✉)
Department of Environmental Science and Engineering, Graduate School of Science and Engineering, Yamaguchi University, Yoshida 1677-1, 753-8512, Yamaguchi, Japan
e-mail: fujishim@yamaguchi-u.ac.jp

M. Fujishima (ed.), *Endosymbionts in Paramecium*, Microbiology Monographs 12, DOI: 10.1007/978-3-540-92677-1_2, 

Pulse-labeling of algae-free paramecia with isolated symbiotic algae for 1.5 min and chasing for various times shows that some algae acquire temporal resistance to lysosomal enzymes in the DVs. They begin to escape from the DVs by budding of the vacuole membrane at 30 min after mixing. Then each small vacuole enclosing a green alga differentiates to a perialgal vacuole (PV), which gives protection from the host lysosomal fusion, and which translocates beneath the host cell surface. Algal cell division in the PV begins at about 24 h after mixing. Infection experiments with infection-capable and infection-incapable algae indicate that the infectivity is based on their ability to localize beneath the host surface after escaping from DVs. Algal proteins synthesized during photosynthesis serve some important functions to prevent expansion of the PV and to attach under the host surface, and to protect the PV from host lysosomal fusion. Although molecular mechanisms for these phenomena remain to be elucidated, accumulated evidence suggests that the symbiotic *Chlorella* sp. have a very long evolutionary history. This chapter mainly addresses studies of DV differentiation, the infection process of the algae, characteristics of the PV membrane, and related phenomena.

## 1 Introduction

Among ciliate *Paramecium* species, only *Paramecium bursaria* can maintain endosymbiotic algae in the cytoplasm. In fact, algae-free *P. bursaria* cells are rare in natural environments. Typically, *P. bursaria* cells harbor several hundred symbiotic algae in their cytoplasm (Fig. 1a); the association of *P. bursaria* with the symbiotic *Chlorella* sp. is a mutual symbiosis. For instance, the host can supply algal cells with nitrogen components and $CO_2$ (Albers and Wiessner 1985; Albers et al. 1982; Reisser 1976, 1980). Furthermore, when within the host, the host protects algae from infection of the *Chlorella* virus (Kawakami and Kawakami 1978; Reisser et al. 1988; Van Etten et al. 1983; Yamada et al. 2006). Also, algal carbon fixation is enhanced in the host (Kamako and Imamura 2006; see Kato and Imamura, this volume). The algae can supply the host with a photosynthetic product, maltose (Brown and Nielsen 1974; Reisser 1976, 1986). The algae in the host show a higher rate of photosynthetic oxygen production than in the isolated state, thereby guaranteeing an oxygen supply for the host (Reisser 1980). Algae-bearing *P. bursaria* can grow better than non-algae-bearing cells (Görtz 1982; Karakashian 1963, 1975); the alga has a photoprotective role for the host (Hörtnagl and Sommaruga 2007; see Sommaruga and Sonntag, this volume). Photosynthetic products of symbiotic *Chlorella* are related closely to the expression of circadian rhythms in host *P. bursaria* (Miwa et al. 1996; Tanaka and Miwa 1996, 2000; Miwa et al. 1996; see Miwa, this volume). Moreover, timing of cell divisions of both the algae and the host cells is well coordinated (Kadono et al. 2004; Takahashi et al. 2007).

Algae-free *P. bursaria* can be produced easily from algae-bearing cells using one of several methods: rapid fission (Jennings 1938); cultivation in darkness

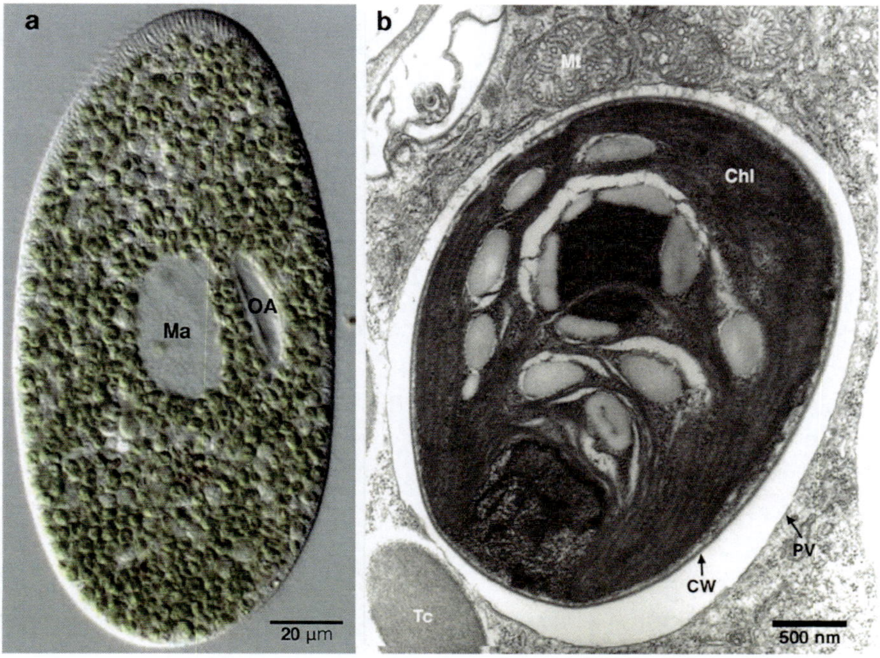


**Fig. 1** Light and transmission electron micrographs of *Paramecium bursaria*: **a** Differential interference contrast (DIC); **b** Transmission electron microscopy (TEM); *Ma,* macronucleus; *OA*, oral apparatus; *Chl, Chlorella*; *PV,* perialgal vacuole membrane; *CW,* cell wall; *Mt,* mitochondrion; *Tc*, trichocyst. (**b** provided by Y. Kodama and I. Inouye, University of Tsukuba)

(Karakashian 1963; Pado 1965; Weis 1969); X-ray irradiation (Wichterman 1948); treatment with 3-(3,4-dichlorophenyl)-1,1-dimethylurea (DCMU), a blocker of electron flow in photosystem II (Reisser 1976); treatment with the herbicide paraquat (Hosoya et al. 1995; Tanaka et al. 2002); or treatment with cycloheximide (Kodama and Fujishima 2008; Kodama et al. 2007; Weis 1984). Irrespective of mutual relationships between *P. bursaria* and symbiotic algae, the algae-free cells and the symbiotic algae retain the ability to grow without a partner. Furthermore, endosymbiosis between the algae-free *P. bursaria* cells and the symbiotic algae isolated from the algae-bearing *P. bursaria* cells is easily reestablished by mixing them (Karakashian 1975; Siegel and Karakashian 1959). Therefore, the symbiotic associations between these eukaryotic cells are excellent models for studying cell-to-cell interaction and the evolution of eukaryotic cells through secondary endosymbiosis between different protists. However, the mechanisms and timings used by the algae to escape from the host DV and to protect themselves from host lysosomal fusion were not revealed for a long time. In *P. bursaria*, each symbiotic alga is enclosed in a PV derived from the host DV, which is assumed to have the ability to protect the alga from lysosomal fusion (Gu et al. 2002; Karakashian and

Rudzinska 1981) (Fig. 1b). Recently, important cytological events needed for establishing endosymbiosis and their timings in the infection process were clarified by pulse-labeling with symbiotic *Chlorella* cells isolated from the algae-bearing *P. bursaria* for 1.5 min; then chasing for various times (Kodama and Fujishima 2005, 2007, 2008, 2009a, 2009b; Kodama et al. 2007). This review specifically examines these four checkpoints for establishing stable endosymbiosis between *P. bursaria* and the symbiotic *Chlorella* species.

## 2 Differentiation of DVs of *P. bursaria*

Infection of the symbiotic algae to the host *P. bursaria* cells is performed through the host's phagocytosis. To investigate the infection process, we must first know the differentiation process of the host's DVs in phagocytosis.

### 2.1 Morphological Classification of DVs and Timing of Appearance of Each DV

To classify stages of DVs that appear during infection by *Chlorella* sp. and to determine the timing of the appearance of each stage, symbiotic algae isolated from algae-bearing *P. bursaria* cells were mixed with algae-free paramecia at densities of $5 \times 10^7$ algae/ml and $5 \times 10^3$ paramecia/ml under fluorescent light (1,500 lx) at 25 ± 1°C and fixed with 4% (w/v) paraformaldehyde as shown in Fig. 2 (Kodama and Fujishima 2005). The DVs during the infection process are classified into eight different stages on the basis of their morphological characteristics and algal color changes in the DVs (Fig. 3). The DV-I vacuole has a rounded vacuole membrane containing green algae. Its membrane is clearly visible under a DIC microscope. In contrast, DV-II has a contracted vacuole, which renders the vacuole membrane barely visible; the algae are green. In DV-III, the vacuole has increased size, making the vacuole membrane visible; the algae are discolored faint yellow or green, or both. DV-III, is further classified into three substages: DV-IIIa contains green algae only; DV-IIIb contains both discolored faint yellow and green algae; and DV-IIIc contains discolored faint yellow algae only. In the final stage, DV-IV, the membrane is again contracted, as in DV-II, rendering the vacuole membrane barely visible under a microscope; the algae are green or brown, or both. Unlike DV-II, this vacuole was not observed in cells before 2 min after mixing at 25 ± 1°C but after 20–30 min. DV-IV was also further classified into three substages: DV-IVa contains green algae only; DV-IVb contains both green and brown algae; and DV-IVc contains brown algae only. DVs containing single green *Chlorella* (SGC) were observed in cells fixed 30 min after mixing, but all SGCs present in cells before 30 min after mixing were digested for 30 min. The algae-free cells were mixed with the isolated symbiotic

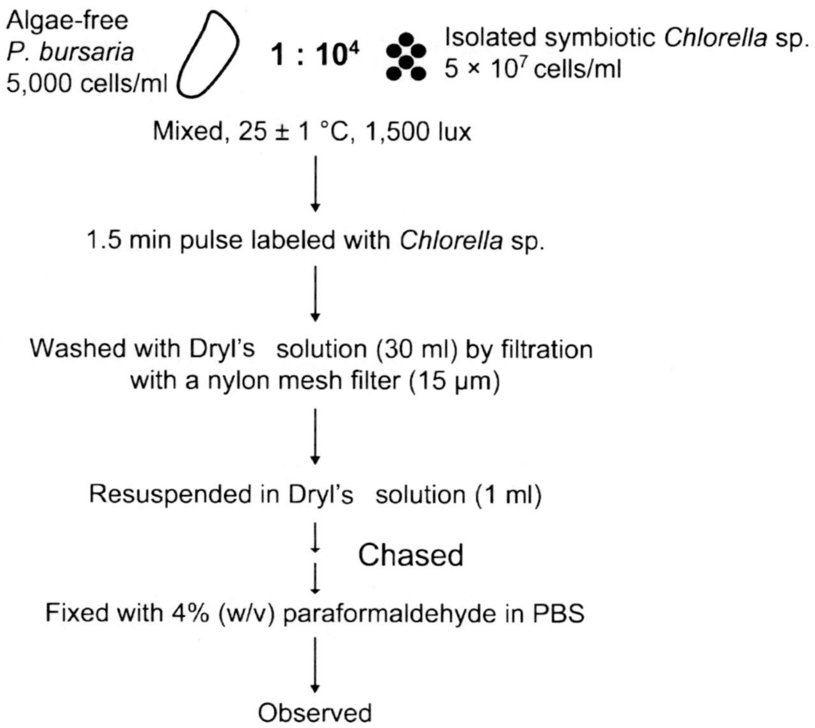


**Fig. 2** Pulse-labeling and chasing with isolated *Chlorella* sp. Algae-free *P. bursaria* cells were mixed at a density of $5 \times 10^3$ per milliliter with isolated *Chlorella* sp. at $5 \times 10^7$ algae/ml under a fluorescent light (1,500 lx) for 1.5 min at 25 ± 1°C. The ciliate–algae mixture was transferred to a centrifuge tube equipped with a 15-μm pore size nylon mesh and filtered. By pouring fresh modified Dryl's solution (MDS) (Dryl 1959) ($KH_2PO_4$ was used instead of $NaH_2PO_4 \cdot 2H_2O$) into this tube, the paramecia were washed and algal cells outside the paramecia were simultaneously removed through the mesh. The paramecia retained in the mesh were transferred to a centrifuge tube and resuspended in MDS, and then chased for various times under a fluorescent light (1,500 lx) at 25 ± 1°C. The cell suspension was fixed by mixing it with 4% (w/v) paraformaldehyde in phosphate-buffered saline (PBS) (137 mM NaCl, 2.68 mM KCl, 8.1 mM $Na_2HPO_4 \cdot 12H_2O$, 1.47 mM $KH_2PO_4$, pH 7.2) at various time points, and the cells were observed under a DIC microscope

algae and fixed at 10-s intervals for 60 s to determine the timing of the DV-II appearance. Actually, DV-II appeared in cells fixed at 30 s after mixing. To determine the timings of DV-III and DV-IV appearance, the algae-free cells were pulsed with isolated algae for 1.5 min, washed, chased, and fixed at every 1-min interval after mixing. Both DV-III and DV-IV appeared in cells at 2–3 and at 20–30 min after mixing, respectively. Because of changes in the number of each stage of DV, the majority of stage DV-I vacuoles become DV-II, DV-IIIa, DV-IIIb, and then DV-IVb, in that order. Other changes, for example, from DV-IIIa to DV-IVa, from DV-IVa to DV-IVb, and from DV-IIIb to DV-IIIc, were few if they existed at all (Kodama and Fujishima 2005, 2007, 2009a).

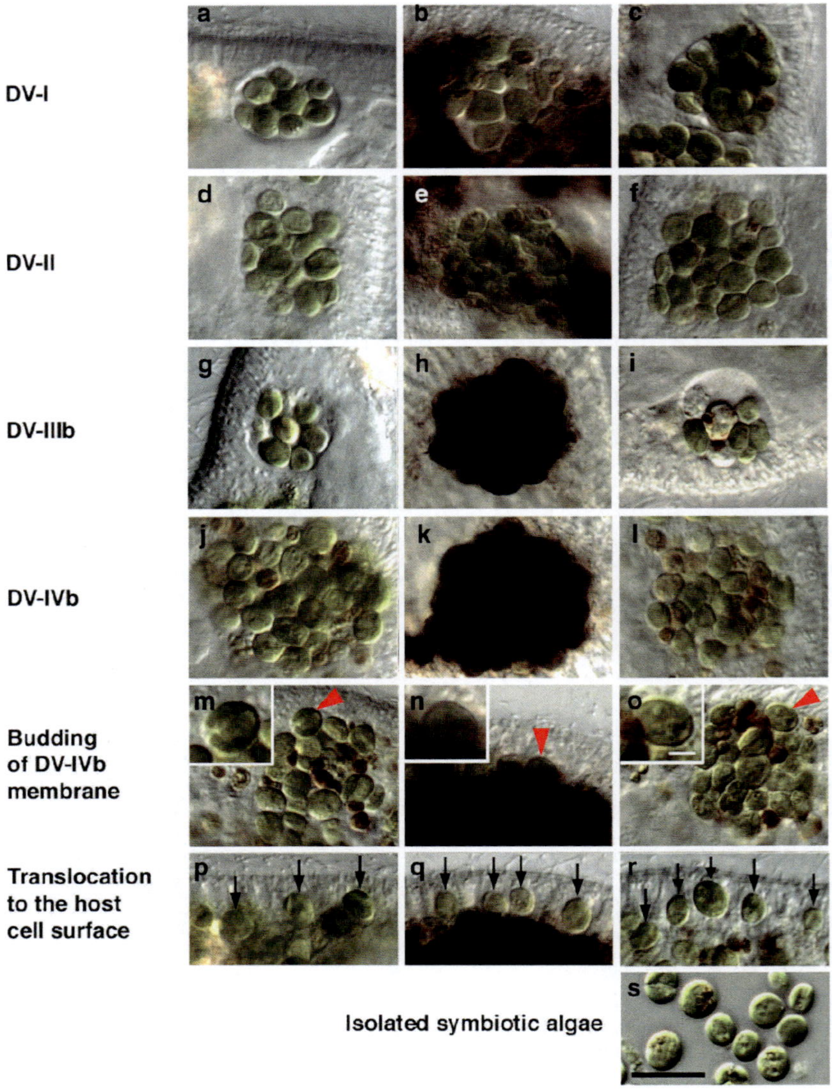


**Fig. 3** DIC micrographs of the infection process of symbiotic *C. vulgaris* cells to the algae-free *P. bursaria* cells. *Chlorella*-free paramecia were mixed with isolated symbiotic algae and fixed at 0.5 min (**a–c**), 1 min (**d–f**), 10 min (**g–i**), 30 min (**j–l**), and 3 h (**m–r**) after mixing. **a**, **d**, **g**, **j**, **m**, **p** Cells non-treated with Gomori's solution. **b**, **e**, **h**, **k**, **n**, **q** Cells treated with Gomori's solution. **c**, **f**, **i**, **l**, **o**, **r** Cells treated with Gomori's solution lacking sodium β-glycerophosphate, a substrate for the acid phosphatase (AcPase) (control experiment). **s** Isolated symbiotic algae treated with Gomori's solution lacking the substrate. Experiments were repeated more than ten times; the results were reproducible. **a–c** DV-I; **d–f** DV-II; **g–i** DV-IIIb; **j–l** DV-IVb; **m–o** An alga is just escaping by budding of the DV-IVb membrane (*red arrowhead*). *Insets* in **m–o** show enlarged photomicrographs of the escaping alga. **p–r** show algae attached just beneath the host cell surface (*black arrows*). DV-I (**b**) and DV-II (**e**) are AcPase activity-negative; DV-IIIb (**h**) and DV-IVb (**k**, **n**) are AcPase activity-positive. The single green *Chlorella* (SGC) that escaped from the host digestive vacuoles (DVs) and translocated just beneath the host cell surface are AcPase activity-negative (**q**, *black arrows*). *Bars* 10 μm (**s**) and 2 μm (*inset* in **o**). (Updated from Kodama and Fujishima 2009a)

## 2.2 Timing of Acidosomal Fusion to DVs

To determine the time taken for the intravacuolar pH to change, yeast cells labeled with the pH indicator dyes bromcresol green (BCG), bromophenol blue, or Congo red were given to the algae-free *P. bursaria*. The color changes of the ingested yeasts were observed (Fig. 4). The Intravacuolar pH rapidly decreases from 6.4–7.0 to 2.4–3.0 at 0.5–1.0 min by acidosomal fusion to DVs; this occurs simultaneously with morphological differentiation into a DV-II vacuole. However, the intravacuolar pH begins to increase again before differentiation into a DV-III vacuole (Kodama and Fujishima 2005).

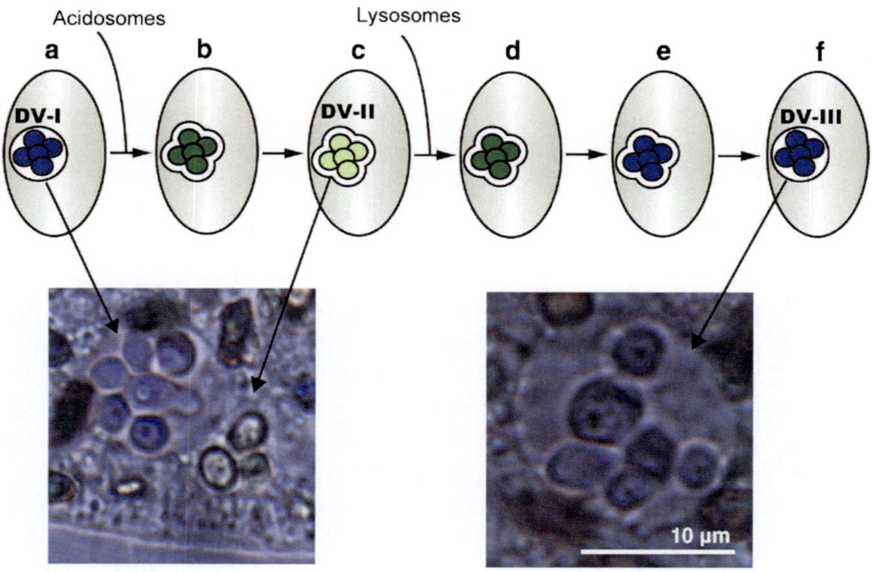


**Fig. 4** Color changes of bromcresol green (BCG)-labeled yeast cells in DVs of *P. bursaria*. Algae-free paramecia and BCG-labeled yeast cells were mixed and the color of the yeast in the host DVs was observed in living cells. **a** DV-I – soon after mixing, the yeast cells are blue, indicating that the intravacuolar pH is 6.4–7.0. **b–e** DV-II: **b** Blue-green yeast cells at 0.5–1 min after mixing, indicating that the intravacuolar pH is 4.4–5.0; **c** Yellow-green yeast cells at 1–2 min after mixing, indicating that the intravacuolar pH is 2.4–3.0; **d** Blue-green yeast cells at 2–3 min after mixing, indicating that the intravacuolar pH has risen to 4.4–5.0; **e** DV-II bearing blue yeast cells at 2–3 min after mixing, indicating that the intravacuolar pH rises to 6.4–7.0 before morphological differentiation to DV-III. **f** DV-III bearing blue yeast cells at 3 min after mixing, indicating that the intravacuolar pH had risen to 6.4–7.0. The changes in yeast color indicate that acidosomes fuse between the stages depicted in *panels* **a** and **b**, and lysosomes fuse between the stages depicted in *panels* **c** and **d**. Photomicrographs are of DVs ingesting BCG-labeled yeasts. *Bar* 10 µm. (Updated from Kodama and Fujishima 2005)

### 2.3 Timing of Lysosomal Fusion to DVs

Partially digested algae appear first in DV-IIIb at 2–3 min after mixing; the increase in pH in acidified DV-II begins in this late DV-II. These results suggest that lysosomal fusion might start before 2–3 min after mixing. Gomori's staining shows that DV-I and DV-II vacuoles are AcPase-activity-negative, but all substages of DV-III and DV-IV are positive (Fig. 3) (Kodama and Fujishima 2008, 2009a). These results suggest that the timing of lysosomal fusion occurs at or immediately before DV-III at 2–3 min after mixing.

## 3 Fates of *Chlorella* Cells in Infection

Each alga that is maintained in the host cell is surrounded by a PV membrane in the host cytoplasm (Karakashian et al. 1968; Meier and Wiessner 1989; Meier et al. 1984; Reisser 1981). This indicates that DVs containing SGC appear from DVs containing many green algae in the early infection process. To determine the timing of the appearance of the SGCs, algae-free paramecia were pulsed with symbiotic algae for 1.5 min, chased, then fixed at 0.05, 0.5, 1, 1.5, 2, 3, 6, 9, 24, 48, and 72 h after mixing. The percentages of cells with SGC, single digested *Chlorella* (SDC), DV-IIIa, DV-IIIb, DV-IVa, and DV-IVb are portrayed in Fig. 5a (Kodama and Fujishima 2005). All SGCs that were present in the host cytoplasm before 30 min after mixing are digested. One hour after mixing, however, SGCs reappear in the host cytoplasm. Figure 5a depicts that the SGCs appearing after 0.5 h are derived from DV-IVa or DV-IVb because no green algae are present in other DVs. At 24 h, the SGCs begin to multiply by cell division, indicating that these algae had established endosymbiosis. In contrast to results of an earlier study (Meier and Wiessner 1989), Fig. 5a shows that the algal escape from the DV-IVa or the DV-IVb vacuole occurs after acidosomal and lysosomal fusion to the DV. They are all digested in DV-III when boiled algae are added to algae-free paramecia (Fig. 5b).

As mentioned above, the SGCs that appeared at 30 min after mixing with the algae-free paramecia and succeeded in establishing endosymbiosis in the host cell appeared to originate from either DV-IVa or DV-IVb. After 30 min after mixing with *Chlorella* sp., five kinds of cells appeared: those with no algae, those with digested algae only, those with DV-IVa, those with DV-IVb, and those with SGCs (Fig. 6a). The frequency of appearance of these cells was examined by fixing cells at various time points after a 1.5-min pulse label with isolated *Chlorella* sp. (Fig. 6b). When DVs of multiple types were present in the same cell, the cell was classified according to the DV that was oldest, following the scheme shown in Fig. 6a. As shown in Fig. 6b, the proportion of cells with SGCs was 0% at 0.5 h and increased as time elapsed. In contrast, the percentage of cells with DV-IVb decreased. At 72 h after mixing, all cells either contained SGCs or were without any

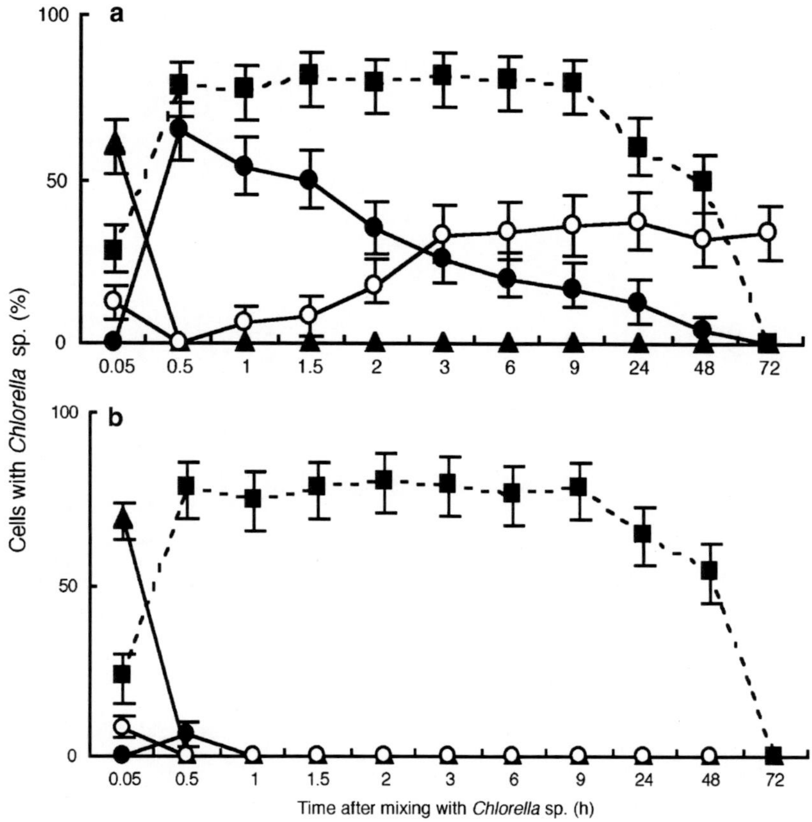


**Fig. 5** Fates of living and boiled *Chlorella* sp. during the infection process. Isolated living (**a**) or boiled (**b**) *Chlorella* sp. and algae-free paramecia were mixed, washed, chased, and fixed at 0.05, 0.5, 1, 1.5, 2, 3, 6, 9, 24, 48, and 72 h after mixing. The percentages of cells with SGC, single digested *Chlorella* (SDC), DV-IIIa, DV-IIIb, DV-IVa, and DV-IVb were determined. All SGCs that appeared before 0.5 h after mixing were digested by 0.5 h. *Triangles* cells with DV-IIIa and DV-IIIb, *filled circles* DV-IVa or DV-IVb, *open circles* SGC, *squares* SDC. For each fixing time interval, 100–300 cells were observed. *Bar* 90% confidence limit. (From Kodama and Fujishima 2005)

algae. Dividing SGCs were observed at and after 24 h of mixing. At 72 h after mixing, about 35% of the cells contained SGCs. On the other hand, less than 5% of cells contained DV-IVa, if any. This indicates that the majority of the SGCs maintained in the host cell originated from DV-IVb. The reproducibility of these results was confirmed twice.

These observations suggest a new mechanism by which *Chlorella* sp. can avoid digestion. First, some of the algae show temporary resistance to the host lysosome enzymes in DV-III and DV-IV vacuoles. Second, the algae bud off from DV-IVb into the host cytoplasm. Finally, the algae lose their temporary resistance to the host lysosomal enzymes but are protected from lysosomal fusion by the PV membrane. The infection process of the symbiotic algae is depicted in Fig. 7.

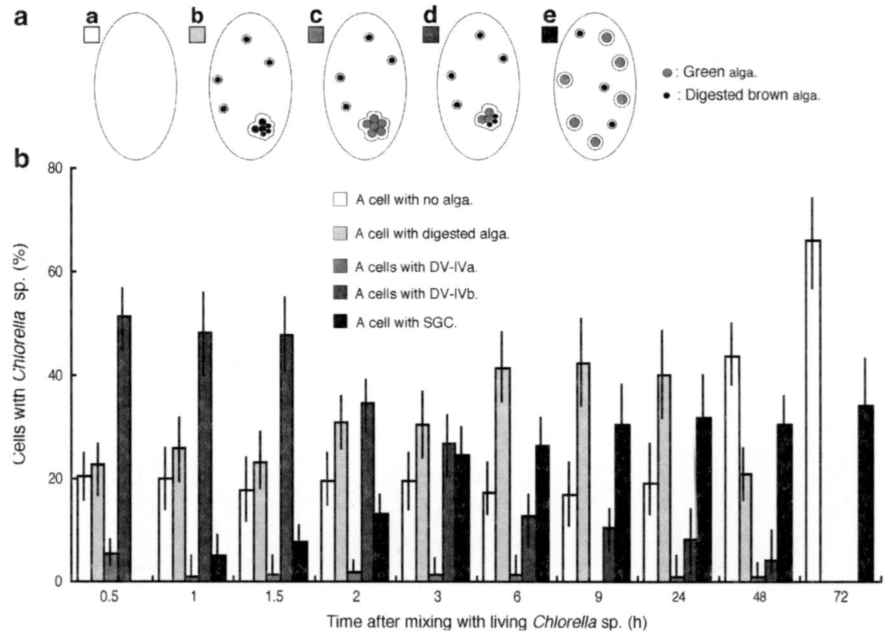


**Fig. 6** Source of SGCs that can establish endosymbiosis. *Chlorella* sp.-free cells were mixed with isolated symbiotic algae, washed, chased, and fixed at 0.5, 1, 1.5, 2, 3, 6, 9, 24, 48, and 72 h after mixing. **a** Cells fixed at 0.5 h after mixing were classified into five types according to the stages of their DVs: **a** a cell with no algae, **b** a cell with digested algae, **c** a cell with DV-IVa, **d** a cell with DV-IVb, **e** a cell with SGC. When a cell had several types of DVs, i.e., types b-e are seen together, the cell was classified in the order b < c < d < e. For example, when a cell has DVs with digested algae and SGC, the cell was classified as type e. **b** Summary of DV types found during the infection process. At each fixing time interval, 100–230 cells were observed. *Bar* 95% confidence limit. (From Kodama and Fujishima 2005)

## 3.1 Acquisition of Temporal Resistance to Lysosomal Enzymes in the Host DVs

It is known that different types of algae infecting *P. bursaria* clones exhibit different infection ratios and host dependencies (Nakahara et al. 2003; Nishihara et al. 1999). This suggests a possibility that different *Chlorella* species or strains used together for infection might result in different algal fates in DV-IVb. To remove this possibility, a OS1g1N clone of *P. bursaria* that was produced by infection of algae-free *P. bursaria* OS1w cells with symbiotic algal clone 1 N cells was used in experiments (Kodama et al. 2007). This algal clone 1 N was obtained from isolated symbiotic algae from an OS1g cell and identified as *C. vulgaris* from observation of its morphological characteristics using light and electron microscopy and by performing a similarity search with the 18S ribosomal DNA sequence from the cells (M. Nakahara, unpublished data). All SGCs that can establish endosymbiosis arose from DV-IVb vacuoles, as

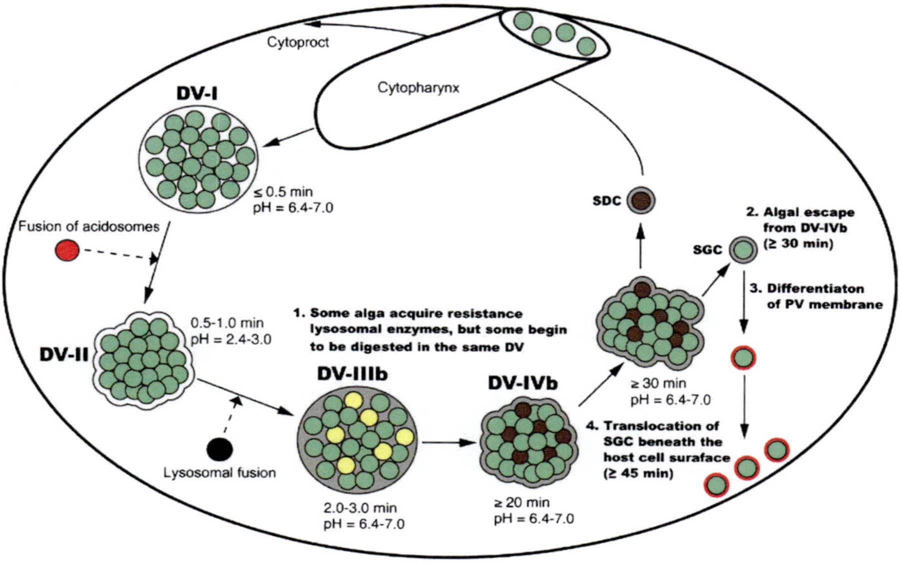


**Fig. 7** The process of infection of algae-free *P. bursaria* cells by symbiotic *Chlorella* sp.. The spherical DV-I vacuole containing green algae differentiates to a condensed DV-II vacuole by fusion of acidosomes and acidification. Then the vacuole differentiates to a swollen DV-III vacuole by fusion of primary lysosomes. The AcPase activity-positive area by Gomori's staining is shown as a *gray* area in the DV. Some algae exhibit resistance to the lysosomal enzymes and can retain a green color and their original morphology. The remaining algae are partially digested and show a *yellow* color in the same DV. The internal pH of the DV-III vacuole increased. DV-III is further classified into three substages: DV-IIIa contains green algae only; DV-IIIb contains both discolored by partial digestion and green algae; and DV-IIIc contains discolored algae only. Then, the condensed DV-IV vacuole differentiates. DV-IV is classified into three substages: DV-IVa contains green algae only; DV-IVb contains both green and brown algae; DV-IVc contains brown algae only. SGCs and SDCs appear through DV-IVb vacuole membrane budding. This phenomenon occurs despite the fact that the alga is intact or partially digested. Algae in the buds remained covered by a *gray* thin layer by Gomori's staining. The SGCs translocate just beneath the host cell surface, anchor there, and initiate algal cell division to establish endosymbiosis. Algae attached beneath the cell surface have no AcPase activity, suggesting that the vacuole membrane wrapping the algae differentiates to the PV membrane (*red circle*) immediately after budding from the DV membrane. (Updated from Kodama and Fujishima 2005, 2007, 2009a)

portrayed in Fig. 6b. Approximately 50% of cells 30 min after mixing have DV-IVb vacuoles (Fig. 6b, Kodama and Fujishima 2005); most of the DVs at this time are AcPase activity-positive ones (Kodama and Fujishima 2009a), which shows that the DV-IVb vacuoles are AcPase activity-positive, and that some of the algae in DV-III and DV-IV vacuoles acquired lysosomal enzyme resistance in the DVs, although the remaining algae are digested in the same DV. Boiled algae are all digested in DV-III when they are added to algae-free paramecia (Fig. 5b). Furthermore, algae fixed with

2.5% (v/v) glutaraldehyde or with 5.0% (v/v) formaldehyde are also digested in DV-III (Kodama and Fujishima 2005, Kodama et al. 2007). Consequently, only living algae can develop temporary resistance to the lysosomal enzymes in the host DV. To date, what factor determines their fates in the host DV remains unknown. However, the following have been revealed (Kodama et al. 2007). Although genetically identical symbiotic algal cells, strain 1 N, are used for the infection, DV-IVb vacuoles appear. Light microscopy shows that the algal fate does not depend on the cell cycle stage or the location in the DV. Electron microscopy shows that the nondigested algae are not protected using a preexisting PV membrane in the DV-IVb vacuoles (Fig. 8). Moreover, this phenomenon is observed in the presence of cycloheximide and puromycin, which are known to inhibit algal and host protein synthesis, respectively.

## 3.2 Escape from the Host DVs by Budding of the Membrane

Some algae are freed from the DV-IVb vacuoles by budding of the DV membrane 30 min after mixing with the host cells. This phenomenon occurred even in cases where the alga was intact or partially digested. Furthermore, the budding is induced not only by living *Chlorella* sp., but also by boiled algae or fixed algae (Kodama and Fujishima 2005). Yeasts also escaped from the *P. bursaria* DVs (Suzaki et al. 2003; Y. Kodama and M. Fujishima, unpublished results). However, India ink, polystyrene latex spheres of 0.81-μm diameter, and the food bacterium *Klebsiella pneumoniae* do not induce this phenomenon (Kodama and Fujishima 2005). The bacterium *Pseudomonas* sp. and the yeasts *Rhodotorula rubra* and *Yarrowia lipolytica* ingested by *P. bursaria* can be maintained in host PVs, indicating that these organisms also induce budding of the host DVs (Görtz 1982; Suzaki et al. 2003). Molecular mechanisms for the budding and escape from the DV are not known.

## 3.3 Differentiation of PV Membrane

To understand the timing of differentiation of PV from the host DV, algae-free *P. bursaria* cells were fed symbiotic *C. vulgaris* cells for 1.5 min, washed, chased, and fixed at various times after mixing. Then, AcPase activity in the vacuoles enclosing the algae was detected using Gomori's staining (Fig. 3). This activity appears in 3-min-old vacuoles; all DVs containing algae demonstrate the activity at 30 min. Algal escape from the DVs begins at 30 min by budding of the DV membrane (Fig. 3m). In the budded membrane, each alga is surrounded by a layer of Gomori's thin positive staining (Fig. 3n, red arrowhead). The vacuoles involving a SGC move quickly and attach immediately beneath the host cell surface (Fig. 3p, arrow). Such vacuoles are Gomori's staining-negative (Fig. 3q, arrow; Kodama and

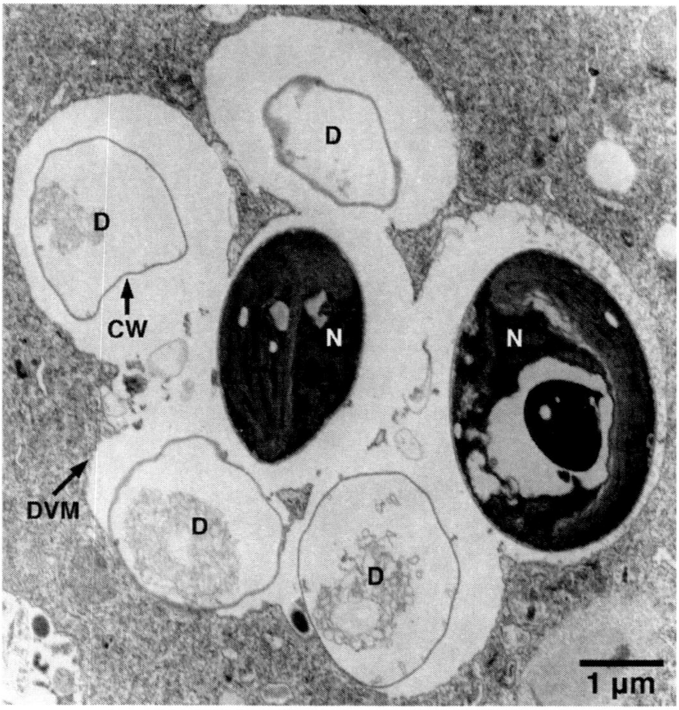


**Fig. 8** Transmission electron micrograph of a DV-IVb vacuole. Three hours after mixing with *C. vulgaris*, algae-free *P. bursaria* were fixed for TEM observation. Note that the DV-IVb vacuole is condensed. Partially digested (*D*) and nondigested (*N*) algae are present together in the same DV. The nondigested algae are not separated from the digested algae by a membrane representing a PV membrane. *DVM*, DV membrane; *CW*, cell wall. (Provided by Y. Kodama and I. Inouye, University of Tsukuba)

Fujishima 2009a). These observations indicate that the PV membrane differentiates soon after the algal escape from the host DV, as presented in Fig. 7.

## 3.4 Translocation and Attachment Beneath the Host Cell Surface

The SGCs that are able to escape from DV-IVb vacuoles by budding of the DV membrane translocate beneath the host surface and are embedded among the trichocysts. This phenomenon is observed only with living and not with boiled algae (Kodama and Fujishima 2005). A similar phenomenon has also been reported for the yeasts *Rhodotorula rubra* and *Yarrowia lipolytica* ingested by *P. bursaria* (Suzaki et al. 2003) and also in symbiotic algal cells of various protists (Reisser 1986). These observations indicate that the PV membrane functions not only for protection of the symbiont from lysosomal fusion, but also for stable attachment of the symbiotic algae to the host cell.

Tonooka and Watanabe (2002, 2007) collected a natural aposymbiotic strain of *P. bursaria*. Infection experiments revealed that this strain showed unstable symbiosis with *Chlorella* sp. The algae aggregated at the posterior region of the host, resulting in aposymbiotic cell production after cell division. Crossbreeding analyses between this strain and a normal strain showed that all F1 progenies showed stable symbiosis with the algae, but some F2 progenies, through sibling crosses between symbiotic F1 progenies, showed unstable symbiosis resembling that of the original aposymbiotic strain. These results indicate that the attachment of the PVs beneath the host cell surface involves a genetically controlled unknown host factor.

### 3.5 Cell Division of Algae After Establishment of Endosymbiosis

The timing of algal cell division in the infection process remained unknown for a long time. Algae-free paramecia at the stationary phase of growth were pulse-labeled by isolated symbiotic algae for 1.5 min, chased, and fixed at 1, 3, 6, 9, 24, and 48 h after mixing; the mean number of green algae per *Paramecium* cell was counted (Kodama and Fujishima 2005). The mean number of green algae was 9.4 algae/cell at 1 h, decreasing to 4.0 algae/cell at 3 h, and remaining constant until 9 h. However, it began to increase to 5.3 algae/cell at 24 h. In addition, dividing algae were frequently observed beneath the host cell surface at 24 h. Thus, algae begin cell division at about 24 h after mixing with the host cells. On the other hand, *Paramecium* cells apparently did not multiply by binary fission until 72 h after mixing the algae.

## 4 Different Behaviors in Infection Between Infection-Capable and Infection-Incapable *Chlorella* Species

In *P. bursaria*, only three free-living *Chlorella* and *Parachlorella* species can establish endosymbiosis: *C. vulgaris*, *P. kessleri*, and *C. sorokiniana*. Actually, *C. ellipsoidea*, *C. saccharophila*, *C. luteoviridis*, *C. zofingiensis*, and *C. mirabilis* are infection-incapable species (Takeda et al. 1998). Symbiotic *Chlorella* species derived from *Stentor polymorphus* or *Spongilla fluviatilis* were digested (Bomford 1965). What is the difference between the infection behaviors of these *Chlorella* species?

### 4.1 Infectivity of Various Symbiotic and Free-Living *Chlorella* Species

Takeda et al. (1998) reported that "infection-capable" *Chlorella* species, including symbiotic ones, are distinguishable by the presence of glucosamine as a chemical component in their rigid walls (alkali-insoluble part of the cell wall), whereas the rigid walls of "infection-incapable" species contained glucose and mannose. They suggested that the presence of glucosamine in the rigid wall of the alga seems to be a prerequisite for determination of the symbiotic association between *P. bursaria* and *Chlorella* species. Algae-free *P. bursaria* cells were mixed with 15 strains of cultivated *Chlorella* species and observed for the establishment of endosymbiosis at 1 h and 3 weeks after mixing to determine the relationship between the infectivity of various *Chlorella* species and the nature of their cell wall components. Only two free-living strains – *C. sorokiniana* strain C-212 and *P. kessleri* strain C-531 – were maintained in the host cells. In contrast, free-living *C. sorokiniana* strain C-43, *P. kessleri* strain C-208, *C. vulgaris* strain C-27, *C. ellipsoidea* strains C-87 and C-542, *C. saccharophila* strains C-183 and C-169, *C. fusca* var. *vacuolata* strains C-104 and C-28, *C. zofingiensis* strain C-111, *C. protothecoides* strains C-150 and C-206, and a cultivated symbiotic *Chlorella* sp. strain C-201 derived from *S. fluviatilis* could not be maintained. Therefore, it appears that the establishment of endosymbiosis is not only algal-species-specific but is also algal-strain-specific. It is noteworthy that these infection-incapable strains were able to escape from the host DV by budding of the host DV membrane, but they failed to localize beneath the host cell surface and were eventually digested (Fig. 9) (Kodama and Fujishima 2007). Consequently, algal attachment beneath the host cell surface is an indispensable phenomenon for establishment of endosymbiosis.

## 4.2 Lectin Binding Ability of Symbiotic and Free-Living *Chlorella* Species

To confirm the relationship between the infectivity of various *Chlorella* species and the nature of their rigid walls, as suggested by Takeda et al. (1998), various *Chlorella* species were mixed with Alexa Fluor 488-conjugated wheat germ agglutinin (WGA), a lectin derived from *Griffonia simplicifolia* (GS-II) or concanavalin A (Con A), with or without pretreatment with 0.4 N NaOH. As presented in Table 1, however, no relationship was found between their infectivity and stainability with these lectins (Kodama and Fujishima 2007). Our results suggest that the infectivity of *Chlorella* species for *P. bursaria* is not based on sugar residues on their cell wall and on the alkali-insoluble part of the cell wall components, but on their ability to localize immediately beneath the host cell surface after escaping from the host DV, as described in Sect. 4.1. Reisser et al. (1982) and Weis (1980) reported that the infection ratio of algae-free cells decreased if the algae were pretreated with Con A. However, we were unable to observe a statistically significant difference in the infection ratio between the lectin-treated and the untreated *Chlorella* sp. cells (Kodama and Fujishima 2007).

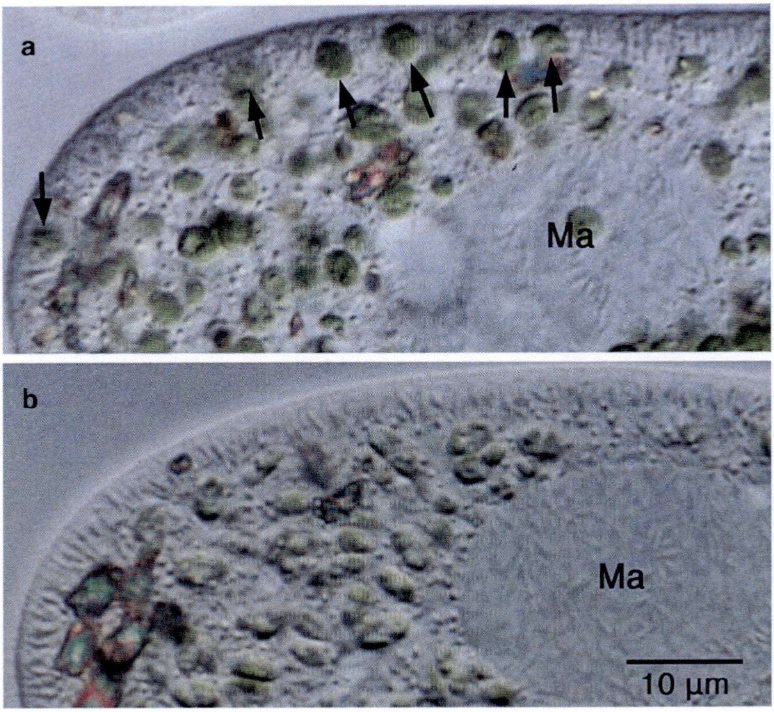


**Fig. 9** Photomicrographs of *P. bursaria* OS1w cells pulse-labeled with symbiotic *Chlorella* sp. cells isolated from *P. bursaria* OS1g cells (**a**) and with *C. saccharophila* strain C-169 cells (**b**). Paramecia were observed 3 h after mixing. Note that the symbiotic *Chlorella* sp. cells are localized beneath the host cell surface (*arrows*), whereas *C. saccharophila* C-169 cells are not. *Ma,* macronucleus. (From Kodama and Fujishima 2007)

## 5 Characteristics of PV Membrane

To date, the following qualitative differences are known to exist between the DV and the PV membrane:

1. The PV encloses a single algal cell (Gu et al. 2002; Karakashian and Rudzinska 1981).
2. The gap separating the algal cell wall and the PV membrane is about 0.05 μm, so the PV membrane is hardly observable under a light microscope (Reisser 1986).
3. The PV diameter does not vary much (2.5–4.5 μm), except during the division of the enclosed alga (Reisser 1992).
4. The PV does not participate in cyclosis, but localizes beneath the host cell surface (Kodama and Fujishima 2005; Reisser 1986).
5. Particle density and its distribution of the PV show few signs hinting at any endocytotic or exocytotic activity (Meier et al. 1984).

**Table 1** Lectin-binding activity of the NaOH-treated and untreated cell walls of *Chlorella* species and their infectivity for algae-free *Paramecium bursaria*

| | | | | Lectin labeling of[a] | | | | | |
|---|---|---|---|---|---|---|---|---|---|
| | Strain (alternative name) | Infectivity | | Nontreated cells | | | NaOH-treated cells | | |
| Species | | T316w | OS1w | WGA | GS-II | Con A | WGA | GS-II | Con A |
| *C. vulgaris* | C-27[b] | | – | – | – | – | + | + | + |
| *C. sorokiniana* | C-212 (211-8k)[c] | + | + | + | – | – | ± | + | ± |
| | C-43[b] | | – | + | – | – | + | + | + |
| *Parachlorella kessleri* (formerly called *C. kessleri*) | C-208 (211-11g)[c] | + | – | – | – | – | – | ± | – |
| | C-531 (211-11h)[c] | + | + | – | – | – | – | ± | – |
| *C. ellipsoidea* | C-87 (211-1a)[c] | – | – | – | – | + | – | + | + |
| | C-542 (211-1a)[c] | – | – | – | – | + | – | + | + |
| *C. saccharophila* | C-183[b] | | – | – | + | + | – | + | + |
| | C-169[b] | | – | – | – | + | – | – | + |
| *C. fusca* var. *vacuolata* | C-104 (211-8b)[b] | | – | – | – | + | – | – | + |
| | C-28[b] | | – | – | – | + | – | – | – |
| *C. zofingiensis* | C-111 (211-14)[b] | | – | – | – | + | – | – | + |
| *C. protothecoides* | C-150 (211-11a)[b] | | – | – | – | + | – | – | + |
| | C-206[b] | | – | – | – | + | – | – | + |
| Symbiotic *Chlorella* sp. | C-201[b] | | – | – | – | + | – | + | + |
| | OS1g[b] | | + | – | – | – | + | + | + |
| | Dd1g[b] | | + | – | – | – | + | + | + |
| | KM2g[b] | | + | – | – | – | + | + | + |
| | Bwk-16 (C[+])[b] | | + | ± | – | ± | + | + | + |
| | 1N[b] | | + | – | – | – | + | + | + |

Updated from Kodama and Fujishima (2007)

*WGA* wheat germ agglutinin, *GS-II Griffonia simplicifolia*, *Con A* concanavalin A

[a]Algal cells were labeled with Alexa Fluor 488-conjugated Con A, WGA, or GS-II. +, 100% of cells with fluorescence; ±, less than 100%; –, 0%. For each experiment, more than 100 algal cells were observed.

[b]Strain used only in Kodama and Fujishima (2007)

[c]Strain used in Kodama and Fujishima (2007) and by Takeda et al. (1998)

6. The PV has no AcPase activity (Karakashian and Rudzinska 1981; Kodama and Fujishima 2009a).
7. The PVs show synchronous swelling by treatment with cycloheximide in the presence of algal photosynthesis (Kodama and Fujishima 2008).

## 5.1 Protection from Lysosomal Fusion

The timing of differentiation of the PV membrane from the host DV membrane is presumed to occur soon after the algal escape by budding of the host DV membrane and before the PV's attachment beneath the host cell surface, as explained in Sect. 3.3. However, no direct evidence demonstrates that the PV membrane does not allow lysosomal fusion to the membrane. The PVs containing SGC of about 3–4-μm diameter are embedded among trichocysts beneath the host cell surface; the surface area of 5–10-μm depth is AcPase activity-negative by Gomori's staining (Kodama and Fujishima 2008, 2009a, 2009b). These observations raise the possibility that the PV membrane has no capability for protection from lysosomal fusion, but can avoid lysosomal fusion by localization at the primary lysosome-less area of the cell. This AcPase activity-negative area can be reduced to less than 3-μm depth if the trichocysts are removed through treatment with 1mg/ml lysozyme. In such cells, some of the PVs were exposed to the AcPase activity-positive area, although such PVs were not stained with Gomori's staining, and the algae in each PV were not digested (Y. Kodama and M. Fujishima 2009b). These results demonstrate that the PV membrane requires no trichocysts for localization beneath the host cell surface, and that the PV membrane, unlike the DV membrane, provides protection from host lysosomal fusion, although molecular mechanisms that would enable the PV membrane to give protection from lysosomal fusion are unknown.

## 5.2 Synchronous Swelling of PVs

Cycloheximide is known to inhibit protein synthesis of symbiotic *Chlorella* of *P. bursaria* preferentially, but it only slightly inhibits host protein synthesis (Kodama and Fujishima 2008; Kodama et al. 2007; Weis 1984). Treatment of algae-bearing *Paramecium* cells with cycloheximide induces synchronous swelling of PVs in the host cell at about 1 day after treatment (Fig. 10b; Kodama and Fujishima 2008). This phenomenon is tentatively designated as "synchronous PV swelling (SPVS)". The space between the symbiotic algal cell wall and the PV membrane expands to about 25 times its usual width 24 h after the treatment. Then, the vacuoles left from beneath the host cell surface become stained with Gomori's staining, and the algae in the vacuoles are digested. Although SPVS is induced only under a constant-light (LL) condition, not under a constant-dark (DD) condition, even under a LL condition, this phenomenon is not induced in paramecia treated with cycloheximide in the presence of the photosynthesis inhibitor DCMU. These results indicate that the algal proteins synthesized in the presence of the algal photosynthesis serve some important function to prevent expansion of the PV and to maintain the ability of the PV membrane for localization beneath the host cell surface and for protection from host lysosomal fusion. The possible mechanisms of induction of the SPVS and digestion of the symbiotic algae are presented in Fig. 11. To date, little is known about qualitative characteristics of the PV membrane.

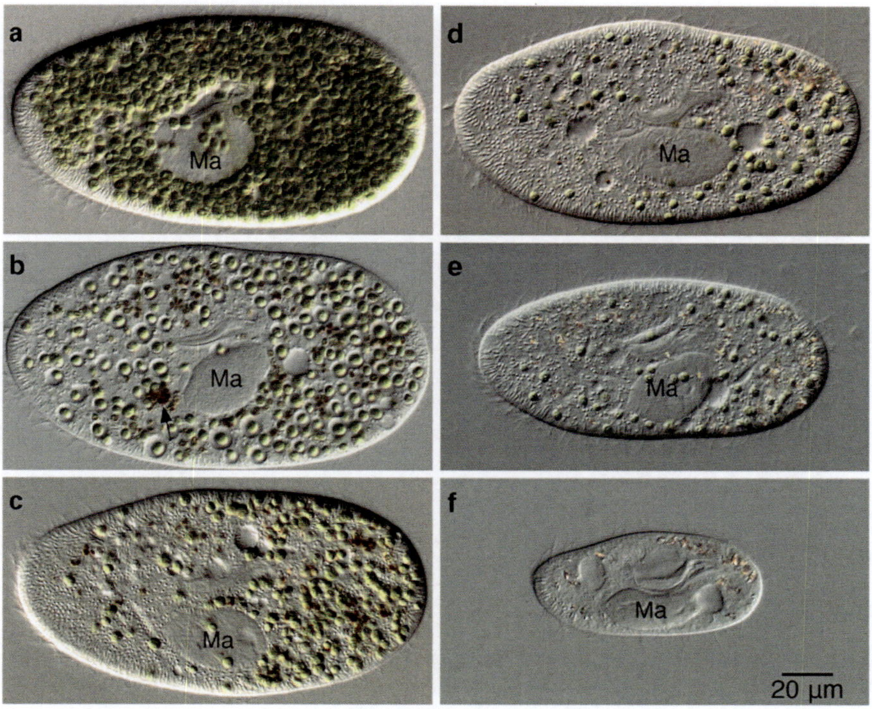


**Fig. 10** Photomicrographs of algae-bearing OS1g1N cells suspended in fresh culture medium containing 10 µg ml$^{-1}$ cycloheximide at $5 \times 10^3$ paramecia/ml at 25 ± 1°C under a constant-light (LL) condition: **a** Before treatment; **b** 1 day after mixing with cycloheximide. All PVs containing green algae swelled synchronously. Furthermore, digested algae appeared in the cytoplasm (*arrow*). **c** 2 days after mixing with cycloheximide, the green algae were numerically reduced; **d** 3 days after mixing; **e** 5 days after mixing; and **f** 7 days after mixing. All algal cells disappeared from the host cytoplasm. The *Paramecium* cells became small. *Ma*, macronucleus. (From Kodama and Fujishima 2008)

The intramembrane particles of the PV membrane are fewer than those of the DV membrane (Meier et al. 1984); lysosomes do not fuse with the PV membrane (Gu et al. 2002; Karakashian and Rudzinska 1981). Identification of SPVS will contribute to studies of PV membrane properties.

## 6 Protection of Symbiotic Algae from *Chlorella* Virus by Endosymbiosis

*Chlorella* virus has been identified as a lytic virus in symbiotic *Chlorella* cells of *P. bursaria* (Kawakami and Kawakami 1978). Later, Van Etten et al. (1982, 1983) found similar viruses in *P. bursaria* and established a laboratory infection system for these viruses using an exsymbiotic algal strain as the host that was originally isolated

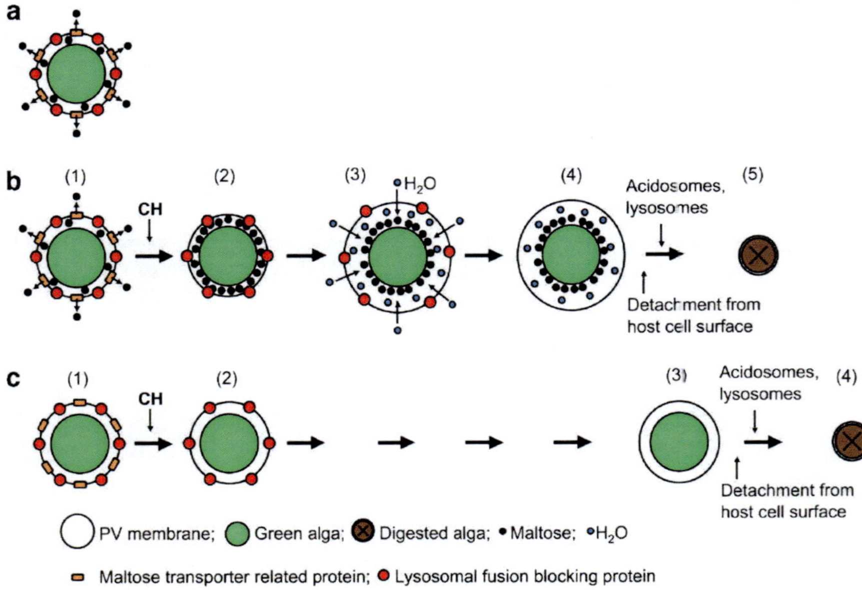


**Fig. 11** Some hypotheses related to induction of synchronous PV swelling (SPVS) and digestion of symbiotic alga after treatment of algae-bearing *P. bursaria* cells with 10 μg ml$^{-1}$ cycloheximide under LL or constant-dark (DD) conditions. As molecules responsible for functions of the PV membrane, two proteins that are synthesized by the algae, excreted outside the algae and localized on the PV membrane, are postulated. The first is a hypothetical maltose-transporter-related protein (*orange squares* on the PV membrane) (Willenbrink 1987), which is synthesized by the alga during photosynthesis and transports maltose from inside the PV membrane to the outside. Loss of this protein in the LL condition induces accumulation of photosynthesized carbohydrates, mainly maltose. The second is a hypothetical lysosomal fusion blocking protein (*red circles* on the PV membrane) that is synthesized by the algae and which has abilities to block lysosomal fusion to the PV membrane and also to attach to unknown structures immediately beneath the host surface, so loss of this protein induces detachment of the PV from the host cell surface and induces fusion with the host lysosomes. The maltose-transporter-related protein disappears rapidly from the PV membrane when algal protein synthesis is inhibited by cycloheximide. On the other hand, the lysosomal fusion blocking protein has a longer turnover time than that of the former; for this reason, this protein remains for some time on the PV membrane when the algal protein synthesis is inhibited by cycloheximide. Under the LL condition (**a**), the symbiotic alga synthesizes mainly maltose by photosynthesis in the host cell (Muscatine et al. 1967) and excretes it into a lumen between the algal cell wall and the PV membrane. The maltose is then transferred outside the PV membrane through the maltose-transporter-related protein on the PV membrane. The maltose-transporter-related protein disappears from the PV membrane (**b**-*2*) when the algal protein synthesis is inhibited by treatment with cycloheximide under the LL condition (**b**). Ayala and Weis (1987) reported that, by treatment with 100 μg ml$^{-1}$ cycloheximide, the rate of carbohydrate secretion by symbiotic algae under the LL condition showed no significant difference between the treated and untreated groups. Consequently, the concentration of the carbohydrates including maltose increases inside the PV membrane, and outside water flows into the PV and induces the SPVS (**b**-*3*). Later, the lysosomal fusion blocking protein disappears from the swollen PV membrane (**b**-*4*). Therefore, the vacuole containing an alga detaches from the host surface; then the host acidosomes fuse to the swollen vacuole and the vacuole contracts by membrane replacement between the acidosomal membrane and the swollen vacuole (Fok et al. 1982; Kodama and Fujishima 2005). Thereafter, lysosomal fusion occurs to the contracted vacuole and the alga is digested (**b**-*5*). As presented in Fig. 10c, the PVs that can avoid the lysosomal

from a *P. bursaria* cell. The virus infects the algal cell by attaching to the algal cell wall. Subsequently, it degrades the cell wall at the attachment point, and the algae are lysed within about 6 h (Meints et al. 1984, Reisser et al. 1988). Reisser et al. (1988) also showed that algae-bearing paramecia grown in the presence of viruses show virus particles only in DVs but never in symbiotic *Chlorella* nor in PVs or in the cytoplasm of the paramecia, indicating that the viruses cannot leave the DVs. Although freshly isolated algae from the host cells do not lyse spontaneously within 14 days, the algae are lysed rapidly by mixing with the viruses. This fact demonstrates that the algae are not kept in a lysogenic status in the host cell, but are instead protected from viral infection. The reason that virus particles cannot enter into PVs with algae when the algae escape from the DV by budding of the DV membrane has not yet been clarified.

It is noteworthy that all the experimental host strains used so far for viral infection were originally endosymbionts of paramecia or hydras, and the virus cannot infect any of the free-living *Chlorella* strains tested to date (Van Etten et al. 1991). The *Chlorella* viruses can be collected in high densities in various countries (Van Etten et al. 1985; Yamada et al. 1993; Zhang et al. 1988). However, the symbiotic algae are protected from the viral infection in the host PVs, and no cells have so far been found which can be regarded as natural hosts (Yamada et al. 1993). Yamada et al. (1993) suggested that unknown free-living *Chlorella* species, which are natural hosts of the viruses, are ubiquitously distributed, and such natural host cells must share some common properties with the infection-capable symbiotic algal strains in the cell wall composition. If this is the case, we can expect that free-living *Chlorella* species become expressed receptors needed for the viral infection on their cell wall when the algae establish endosymbiosis with the *P. bursaria* cell, and that monoclonal antibodies specific for the symbiotic algal cell wall should be obtained. We recently succeeded in developing such a monoclonal antibody, which can react with the symbiotic algal cell wall of all strains of *P. bursaria* examined but not with the cell walls of free-living algal strains (A. Nishijima, Y. Kodama, and M. Fujishima, unpublished results).

## 7 Concluding Remarks and Further Perspectives

About 50 years have passed since infection experiments involving *P. bursaria* cells and the symbiotic *Chlorella* sp. were described by Siegel and Karakashian (1959). However, since that time, *P. bursaria* and its symbiotic algae have come to be used

**Fig. 11** (continued) fusion in the presence of cycloheximide are recontracted. Such vacuoles might be produced by evasion of lysosomal fusion after acidosomal fusion. Under the DD condition (**c**-*1*), cycloheximide treatment induces loss of the maltose-transporter-related protein from the PV membrane (**c**-*2*), but no morphological change is induced. Later, the lysosomal fusion blocking protein disappears from the PV membrane (**c**-*3*). The vacuole detaches from the host cell surface and fuses with acidosomes and lysosomes; then the algae are digested (**c**-*4*). Under the DD condition, the fate of the PV is the same as that in **c**, irrespective of the presence or the absence of cycloheximide. *CH,* cycloheximide. (From Kodama and Fujishima 2008)

by many researchers for studies of endosymbiosis because both can grow separately and establish endosymbiosis soon after their mixing. Nevertheless, the infection process has not been revealed over those many years. Recently, through pulse-labeling with isolated symbiotic algae from *P. bursaria* for 1.5 min and chasing for various times by Kodama and Fujishima (2005), differentiation processes of the DVs, the route of infection, and four important cytological phenomena induced for establishing endosymbiosis were found, as summarized in Fig. 7. Although molecular analyses of these four phenomena have just begun, the results of such analyses will dramatically promote the study of endosymbiosis control mechanisms in the near future. How can the symbiotic algae acquire temporal resistance to lysosomal enzymes in the host DV-III and DV-IV vacuoles? How can algae escape from the host DV by budding of the DV membrane? What are substantial differences between the DV and the PV? What is the transportation system of the PV beneath the host cell surface? What is the signal for induction of algal cell division after localization beneath the host cell surface? Can endosymbiosis be established when the symbiotic algae are not taken up through DVs, but are instead inserted by microinjection? Why, among all of the various *Paramecium* species, can only *P. bursaria* establish endosymbiosis with *Chlorella* species? How can the PV membrane segregate an old cell wall, which is discarded after the algal cell division, and daughter algal cells? Can a single *Paramecium* cell maintain plural *Chlorella* species or plural strains of the same *Chlorella* species in a single *P. bursaria* cell? These problems remain to be elucidated.

**Acknowledgements** The authors are indebted to all coworkers at Yamaguchi University. This work was supported by a Japan Society for the Promotion of Science (JSPS) Research Fellowship for Young Scientists granted to Y.K., and by a grant-in-aid for Scientific Research (B) (no. 17405020) to M.F. from JSPS.

## References

Albers D, Wiessner W (1985) Nitrogen nutrition of endosymbiotic *Chlorella* spec. Endocytobiosis Cell Res 1:55–64

Albers D, Reisser W, Wiessner W (1982) Studies of the nitrogen supply of endosymbiotic chlorellae in green *Paramecium bursaria*. Plant Sci Lett 25:85–90

Ayala A, Weis DS (1987) The effect of puromycin and cycloheximide on the infection of algae-free *Paramecium bursaria* by symbiotic *Chlorella*. J Protozool 34:377–381

Bomford R (1965) Infection of alga-free *Paramecium bursaria* with strains of *Chlorella*, Scenedesmus, and yeast. J Protozool 12:221–224

Brown JA, Nielsen PJ (1974) Transfer of photosynthetically produced carbohydrate from endosymbiotic *Chlorellae* to *Paramecium bursaria*. J Protozool 21:569–570

Dryl S (1959) Antigenic transformation in *Paramecium aurelia* after homologous antiserum treatment during autogamy and conjugation. J Protozool 6(Suppl):25

Fok A, Lee Y, Allen RD (1982) The correlation in DV pH and size with the digestive cycle in *Paramecium caudatum*. J Protozool 29:409–414

Görtz H-D (1982) Infection of *Paramecium bursaria* with bacteria and yeasts. J Cell Sci 58:445–453

Gu FK, Ni L, Chen B, Zhang X (2002) A comparative study on the electron microscopic enzymo-cytochemistry of *Paramecium bursaria* from light and dark cultures. Eur J Protistol 38:267–278

Hörtnagl PH, Sommaruga R (2007) Oxidative and UV-induced photooxidative stress in symbiotic and aposymbiotic strains of the ciliate *Paramecium bursaria*. Photochem Photobiol Sci 6:842–847

Hosoya H, Kimura K, Matsuda S, Kitaura M, Takahashi T (1995) Symbiotic algae-free strains of the green *Paramecium bursaria* produced by herbicide paraquat. Zool Sci 12:807–810

Jennings HS (1938) Sex reaction types and their interrelations in *Paramecium bursaria* I and II. Clones collected from natural habitats. Proc Natl Acad Sci U S A 24:112–120

Kamako S-I, Imamura N (2006) Effect of Japanese *Paramecium bursaria* extract on photosynthetic carbon fixation of symbiotic algae. J Euk Microbiol 53:136–141

Karakashian SJ (1963) Growth of *Paramecium bursaria* as influenced by the presence of algal symbionts. Physiol Zool 36:52–68

Karakashian SJ (1975) Symbiosis in *Paramecium bursaria*. Symp Soc Exp Biol 29:145–173

Karakashian SJ, Rudzinska MA (1981) Inhibition of lysosomal fusion with symbiont-containing vacuoles in *Paramecium bursaria*. Exp Cell Res 131:387–393

Karakashian SJ, Karakashian MW, Rudzinska MA (1968) Electron microscopic observations on the symbiosis of *Paramecium bursaria* and its intracellular algae. J Protozool 15:113–128

Kadono T, Kawano T, Hosoya H, and Kosaka T (2004) Flow cytometric studies of the host-regulated cell cycle in algae symbiotic with green paramecium. Protoplasma 223:133–141

Kawakami H, Kawakami N (1978) Behavior of a virus in a symbiotic system, *Paramecium bursaria*-zoochlorella. J Protozool 25:217–225

Kodama Y, Fujishima M (2005) Symbiotic *Chlorella* sp. of the ciliate *Paramecium bursaria* do not prevent acidification and lysosomal fusion of host digestive vacuoles during infection. Protoplasma 225:191–203

Kodama Y, Fujishima M (2007) Infectivity of *Chlorella* species for the ciliate *Paramecium bursaria* is not based on sugar residues of their cell wall components, but based on their ability to localize beneath the host cell membrane after escaping from the host digestive vacuole in the early infection process. Protoplasma 231:55–63

Kodama Y, Fujishima M (2008) Cycloheximide induces synchronous swelling of perialgal vacuoles enclosing symbiotic *Chlorella vulgaris* and digestion of the algae in the ciliate *Paramecium bursaria*. Protist 159:483–494

Kodama Y, Fujishima M (2009a) Timing of perialgal vacuole membrane differentiation from digestive vacuole membrane in infection of symbiotic algae *Chlorella vulgaris* of the ciliate *Paramecium bursaria*. Protist 160:65–74

Kodama Y, Fujishima M (2009b) Localization of perialgal vacuoles beneath the host cell surface is not a prerequisite phenomenon for protection from the host's lysosomal fusion in the ciliate *Parameium bursaria*. Protist (in press) doi:10.1016/j.protis.2008.11.003

Kodama Y, Nakahara M, Fujishima M (2007) Symbiotic alga *Chlorella vulgaris* of the ciliate *Paramecium bursaria* shows temporary resistance to host lysosomal enzymes during the early infection process. Protoplasma 230:61–67

Meier R, Lefort-Tran M, Pouphile M, Reisser W, Wiessner W (1984) Comparative freeze-fracture study of perialgal and digestive vacuoles in *Paramecium bursaria*. J Cell Sci 71:121–140

Meier R, Wiessner W (1989) Infection of algae-free *Paramecium bursaria* with symbiotic *Chlorella* sp. isolated from green paramecia II: a timed study. J Cell Sci 93:571–579

Meints RH, Lee K, Burbank DE, Van Etten JL (1984) Infection of a chlorella-like alga with the virus, PBCV-1: ultrastructural studies. Virology 138:341–346

Miwa I, Fujimori N, Tanaka M (1996) Effects of symbiotic *Chlorella* on the period length and the phase shift of circadian rhythms in *Paramecium bursaria*. Eur J Protistol 32(Suppl 1): 102–107

Muscatine L, Karakashian SJ, Karakashian MW (1967) Soluble extracellular products of algae symbiotic with a ciliate, a sponge and a mutant *Hydra*. Comp Biochem Physiol 20:1–12

Nakahara M, Handa S, Nakano T, Deguchi H (2003) Culture and pyrenoid structure of a symbiotic *Chlorella* species isolated from *Paramecium bursaria*. Symbiosis 34:203–214

Nishihara N, Murata-Hori M, Yamada T, Kosaka T, Hosoya H (1999) Characteristic variation of cloned endosymbiotic algae in *Paramecium bursaria*. Jpn J Protozool 32:31

Pado R (1965) Mutual relation of protozoans and symbiotic algae in *Paramecium bursaria* I. The influence of light on the growth of symbionts. Folia Biol 13:173–182

Reisser W (1976) Die stoffwechselphysiologischen Beziehungen zwischen *Paramecium bursaria* Ehrbg. und *Chlorella* spec. in der *Paramecium bursaria*-Symbiose I: der Stickstoff- und der Kohlenstoff-Stoffwechsel. Arch Microbiol 107:357–360

Reisser W (1980) The metabolic interactions between *Paramecium bursaria* Ehrbg. and *Chlorella* spec. in the *Paramecium bursaria*-symbiosis III. The influence of different $CO_2$-concentrations and of glucose on the photosynthetic and respiratory capacity of the symbiotic unit. Arch Microbiol 125:291–293

Reisser W (1981) Host-symbiont interaction in *Paramecium bursaria*: physiological and morphological features and their evolutionary significance. Ber Dtsch Bot Ges 94:557–563

Reisser W (1986) Endosymbiotic associations of freshwater protozoa and algae. Prog Protistol 1:195–214

Reisser W (1992) Endosymbiotic associations of algae with freshwater protozoa and invertebrates. In: Reisser W (Ed) Algae and symbioses: plants, animals, fungi, viruses, interactions explored, vol.1.1. Biopress, Bristol, pp 1–19

Reisser W, Radunz A, Wiessner W (1982) Participation of algal surface structures in the cell recognition process during infection of aposymbiotic *Paramecium bursaria* with symbiotic chlorellae. Cytobios 33:39–50

Reisser W, Klein T, Becker B (1988) Studies of phycoviruses I. On the ecology of viruses attacking *Chlorellae* exsymbiotic from an European strain of *Paramecium bursaria*. Arch Hydrobiol 111:575–583

Siegel R, Karakashian S (1959) Dissociation and restoration of endocellular symbiosis in *Paramecium bursaria*. Anat Rec 134:639

Suzaki T, Omura G, Görtz H-D (2003) Infection of symbiont-free *Paramecium bursaria* with yeasts. Jpn J Protozool 36:17–18

Takahashi T, Shirai Y, Kosaka T, Hosoya H (2007) Arrest of cytoplasmic streaming induces algal proliferation in green paramecia. PLoS ONE 2(12):e1352. doi:10.1371/journal.pone.0001352

Takeda H, Sekiguchi T, Nunokawa S, Usuki I (1998) Species-specificity of *Chlorella* for establishment of symbiotic association with *Paramecium bursaria*. Does infectivity depend upon sugar components of the cell wall? Eur J Protistol 34:133–137

Tanaka M, Miwa I (1996) Significance of photosynthetic products of symbiotic *Chlorella* to establish the endosymbiosis and to express the mating reactivity rhythm in *Paramecium bursaria*. Zool Sci 13:685–692

Tanaka M, Miwa I (2000) Correlation of photosynthetic products of symbiotic *Chlorella* with the mating reactivity rhythms in a mutant strain of *Paramecium bursaria*. Zool Sci 17:735–742

Tanaka M, Murata-Hori M, Kadono T, Yamada T, Kawano T, Kosaka T, Hosoya H (2002) Complete elimination of endosymbiotic algae from *Paramecium bursaria* and its confirmation by diagnostic PCR. Acta Protozool 41:255–261

Tonooka Y, Watanabe T (2002) A natural strain of *Paramecium bursaria* lacking symbiotic algae. Eur J Protistol 38:55–58

Tonooka Y, Watanabe T (2007) Genetics of the relationship between the ciliate *Paramecium bursaria* and its symbiotic algae. Inv Biol 126:287–294

Van Etten JL, Meints RH, Kuczmarski D, Burbank DE, Lee K (1982) Viruses of symbiotic chlorella-like algae isolated from *Paramecium bursaria* and *Hydra viridis*. Proc Natl Acad Sci U S A 79:3867–3871

Van Etten, JL, Burbank DE, Kuczmarski D, Meints RH (1983) Virus infection for culturable *Chlorella*-like algae and development of a plaque assay. Science 219:994–996

Van Etten JL, Van Etten CH, Johnson JK, Burbank DE (1985) A survey for viruses from fresh water that infect a eucaryotic-chlorella-like green alga. Appl Environ Microbiol 49:1326–1328

Van Etten JL, Lane LC, Meints RH (1991) Viruses and virus like particles of eukaryotic algae. Microbiol Rev 55: 586–620

Weis DS (1969) Regulation of host and symbiont population size in *Paramecium bursaria*. Experientia 15:664–666

Weis DS (1980) Hypothesis: Free maltose and algal cell surface sugars are signals in the infection of *Paramecium bursaria* by algae. In: Schwemmler W, Schenk EA (eds) Endosymbiosis and cell biology, vol. I. de Gruyter, Berlin, pp 105–112

Weis DS (1984) The effect of accumulation time of separate cultivation on the frequency of infection of aposymbiotic ciliates by symbiotic algae in *Paramecium bursaria*. J Protozool 31:14A

Wichterman R (1948) The biological effects of X-rays on mating types and conjugation of *Paramecium bursaria*. Biol Bull 93:201

Willenbrink J (1987) Die pflanzliche Vakuole als Speicher. Naturwissenschaften 74:22–29

Yamada T, Onimatsu H, Van Etten J (2006) *Chlorella* viruses. Adv Virus Res 66:293–336

Yamada T, Shimomae A, Furukawa S, Takehara J (1993) Widespread distribution of *Chlorella* viruses in Japan. Biosci Biotechnol Biochem 57:733–739

Zhang Y, Burbank DE, Van Etten JL (1988) *Chlorella* viruses isolated in China. Appl Environ Microbiol 54:2170–2173

# Metabolic Control Between the Symbiotic *Chlorella* and the Host *Paramecium*

**Yutaka Kato and Nobutaka Imamura**

**Contents**



**Abstract** Metabolic control, including the transfer of materials between a host and a symbiont, is important for understanding symbiotic relationships. However, sugars, mainly maltose, are the only confirmed class of material transferred from symbionts to *Paramecium bursaria*. An axenic Japanese *Chlorella* symbiont, which had been thought hard to isolate and maintain, was found to irreversibly adapt to its symbiotic milieu. Analysis of its features, such as the unique availability of nitrogenous compounds (e.g., amino acids) and its uncommon stimulation of carbon fixation by the host extract, revealed that three constitutional amino acid transport systems that can be controlled by $Ca^{2+}$ and sugar are present, and that the carbon fixation

Y. Kato
Research Organization of Science and Engineering, Ritsumeikan University,
Noji-Higashi 1-1-1, Kusatsu Shiga, 525-8577, Japan

N. Imamura (✉)
Department of Pharmacy, College of Pharmaceutical Sciences, Ritsumeikan University,
Noji-Higashi 1-1-1, Kusatsu Shiga, 525-8577, Japan
e-mail: imamura@ph.ritsumei.ac.jp

M. Fujishima (ed.), *Endosymbionts in Paramecium*, Microbiology Monographs 12,
DOI: 10.1007/978-3-540-92677-1_3, © Springer-Verlag Berlin Heidelberg 2009

ability of the symbiont depends on the extracellular cation concentration. These novel features of the Japanese symbiont imply metabolic control between the host and the symbiont.

## 1 Introduction

Cells of the green paramecium, *Paramecium bursaria*, contain hundreds of endosymbiotic *Chlorella* cells within perialgal vacuoles in the cytoplasm. *P. bursaria* can live under autotrophic conditions, e.g., in a medium for microalgae containing inorganic minerals and some vitamins but lacking carbohydrates and amino acids (Loefer 1936), clearly indicating that the protozoa and the green algae are mutually beneficial to each other; the protozoa can only obtain the organic compounds they need as an energy source from the symbiotic algae. In exchange for the algal photosynthetic products, the host provides the symbiont with several materials including minerals as nutrients. This exchange of substances probably requires the sharing of information to maintain homeostasis in *P. bursaria*.

Although the transfer of materials has been thoroughly investigated, the details of the transfer in host–symbiont relationships have not been elucidated. Different types of photosynthate, mainly maltose, have been reported as being released from cultured symbiotic algae (Brown and Nielsen 1974; Reisser 1976, 1986). Maltose produced by the symbiont is an essential energy supply for the host during symbiosis under autotrophic conditions. However, given the autotrophic growth habit of algae, determining which compounds are essential for their growth is more difficult.

If all classes of materials around the *Paramecium* cell could pass through freely to algal cells, it may seem that there is no requirement for the supply of materials by the host, but that is not the case. After carbon, nitrogen is the material transferred from the host to the symbiont in the largest amounts and thus has received major research interest. Studies focusing on the nitrogen sources of symbionts have revealed the differential availability of nitrate between the algal cells before and after isolation from the host, and the possibility of amino acid transfer, especially of L-glutamine, from the host to the symbiont (Albers et al. 1982). Plausible hypothetical control methods between the host and the symbiont during symbiosis have been proposed (Reisser 1988). However, the diversity of symbiont features found (Reisser and Widowski 1992) indicated that the relationship between the host and the symbiont may vary on a case-by-case basis, making it difficult to discuss in general terms.

Reisser et al. (1988) stated that “a typical symbiotic *Chlorella* strain common to all *P. bursaria* strains does not exist.” Symbiotic *Chlorella* strains vary and each possesses characteristic features, such as differing availabilities of amino acids, nitrate, and ammonium. Previous studies on symbiotic *Chlorella* were carried out on strains that could be cultured in standard inorganic media. Even these cultivatable strains showed great diversity. Symbiotic *Chlorella* did not show any specific adaptation to the symbiotic milieu except for the pH-dependent release of photosynthate

(Reisser et al. 1988). Recently, phylogenetic analyses determined the different origins of European and American *P. bursaria* (see Hoshina and Imamura, this volume). Consequently, the differing host-symbiont relationships for the species resulted in a diverse range of symbiont features.

The Japanese symbionts of *P. bursaria* (F36-ZK), by way of their inability to utilize nitrate, were thought to be difficult to isolate and cultivate. The many novel features of F36-ZK suggest that it is an extreme example with adaptations to the symbiotic milieu, implying that its features should reflect the symbiotic surroundings. Thus, analyzing the features of F36-ZK could provide important information about the transfer of material and metabolic control between the host and the symbiont. In this chapter, we will first describe the novel features of the Japanese symbiotic *Chlorella*, which appears greatly dependent on the host, and we will then consider the metabolic control between the Japanese symbiont and its host.

## 2 Features of Symbiotic *Chlorella* Species

Symbionts are thought to be special *Chlorella* species or strains. For example, symbiont-free *P. bursaria* can distinguish a symbiotic *Chlorella* from a free-living one and can generally make a stable association (Reisser et al. 1982). This ability to be recognized is a novel feature of symbiotic *Chlorella*. In 1983, a virus that came to play an important role in *P. bursaria* studies was reported (Van Etten et al. 1983). This virus, known as chlorovirus, could infect cultured symbiotic *Chlorella* strains isolated from green hydra or *P. bursaria*, but not free-living *Chlorella* strains. In 1991, Reisser et al. reported that the virus could distinguish between *P. bursaria* collected in Europe and that collected in America. Since then, the geographical origin of *P. bursaria* has been important when considering the features of symbiotic *Chlorella* (Reisser and Widowski 1992). Recently, Hoshina et al. (2005) classified the European and American symbiotic *Chlorella* strains in different groups using phylogenetic studies based on ribosomal DNA. Thus, symbiotic *Chlorella* strains in *P. bursaria* around the world should not be considered identical.

### 2.1 European and American Symbiotic *Chlorella*

Before European and American strains of *Chlorella* were distinguished by chlorovirus, Reisser et al. (1988) compared the morphological and physiological features of symbiotic *Chlorella* strains isolated from European and American *P. bursaria* with those of free-living *Chlorella* strains from an ecological standpoint. They found that carbohydrate release of symbiotic *Chlorella* was higher at acidic pH than at neutral pH, whereas that of free-living *Chlorella* was not significantly dependent on pH. Furthermore, many physiological and cytochemical features were different in individual strains, regardless of whether they were symbiotic or free-living. Because these symbiotic *Chlorella* strains did not show common properties, no underlying features could be found to account for the differences between symbiotic and free-living *Chlorella*. Most free-living *Chlorella*, including the European and American symbionts tested, can utilize nitrate (Reisser and Widowski 1992). This nitrate-utilization ability was a characteristic common to symbiotic and free-living *Chlorella* because the symbiont was isolated on agar plates containing nitrate as the sole nitrogen source. However, these observations were only applicable to cultivable symbiotic *Chlorella* strains.

## 2.2 Japanese Symbiotic *Chlorella*

In contrast to many studies using European and American symbionts, few studies have been conducted on the symbiotic algae in Japanese *P. bursaria*. Takeda (1995) studied six symbiotic strains of Japanese *P. bursaria*. On the basis of cell wall composition comparisons, all the Japanese symbionts appeared identical to each other, but were different from German symbionts. The Japanese symbionts were isolated as a homogeneous colony on an agar plate using nitrate as the sole nitrogen source. Nishihara et al. (1998) reported that the Japanese symbionts always coexisted with bacteria; therefore, no bacteria-free Japanese symbionts could be isolated, regardless of the type of culture medium used. However, Nakahara et al. (2003) succeeded in culturing Japanese symbionts under axenic conditions using a medium containing ammonium nitrate ($NH_4NO_3$) as the sole nitrogen source, although the strains could not be kept for more than a few months.

Symbiotic algae previously associated with *P. bursaria* were isolated and maintained on an inorganic medium using nitrate as the sole nitrogen source (Reisser 1984). The Japanese symbionts were thought to possess characteristic properties that were distinguishable by nitrogen utilization from the cultivatable strains reported. To investigate the Japanese symbionts, axenic strains were established (Kamako et al. 2005). Isolation efforts were first carried out on an agar plate using inorganic nitrogen sources for algal autotrophic growth (C medium: Ichimura 1971). After incubation for several weeks, clones on the initial agar plate were transferred to a flesh agar plate; all colonies were contaminated with bacteria, which varied by colony. This indicated that there was no species-specific relationship between the symbiont and the contaminating bacterial strain; colonies on the agar plate, visible to the naked eye, were not axenic. Agar plates were observed under a microscope after a short period of incubation and small colonies that appeared to be free of bacteria were marked. After further incubation for several weeks, xenic colonies were observed to grow well and became visible to the naked eye, but the marked colonies remained almost the same size (80–100-μm diameter) as when initially measured. After cultures on several media had been attempted, the axenic Japanese symbiotic strain was found to grow on agar containing modified Bold's basal medium (MBBM) (Nichols and Bold 1965). Once the axenicity of the algal strains isolated from three Japanese *P. bursaria* strains had been confirmed, phylogenetic studies based on 18S ribosomal DNA analyses revealed that all the Japanese symbiotic *Chlorella* strains were identical. The most closely related free-living species was considered to be *C. vulgaris*. The Japanese symbionts were classified into an American symbiotic *Chlorella* group; virus sensitivity also supported this classification.

The characteristic constituents of MBBM, Bacto peptone and sucrose, were examined separately to determine which components were required for algal growth; Bacto peptone was found to be essential, while sucrose was not (Kamako et al. 2005). Axenic symbionts showed some growth on C medium owing to organic nitrogen impurities in the agar, which were detected using the ninhydrin reagent. However, xenic clones growing on the medium were visible with the naked eye, perhaps supported by organic nitrogen compounds produced from nitrate by coexisting bacteria. Therefore, for the Japanese symbiotic strains, the availability of nitrate seemed to depend on whether the strain was axenic or not.

The utilization of saccharides by Japanese symbionts was also examined, because knowledge of saccharide utilization was poor. Results of algal growth in media containing every respective sugar showed that the monosaccharides glucose and fructose seemed to be available, while the disaccharides sucrose and maltose were not (Kamako et al. 2005). However, recent experiments using radioactive tracers revealed that these saccharides could not be imported in symbiotic algal cells (Kato and Imamura 2008b). Thus, the growth-stimulating effect of glucose and fructose was not caused by nutrition, and the effect of glucose will be discussed later.

## 2.3 Nitrogen Utilization of Symbiotic *Chlorella*

One of the characteristic features of Japanese symbiotic algae is nitrogen utilization. Preliminary nitrogen-utilization experiments were performed using Bacto peptone, casamino acids, ammonium ($NH_4^+$), nitrate ($NO_3^-$), and nitrite ($NO_2^-$) as the respective sole nitrogen sources. The symbiotic algal strains, American NC64A obtained from the American Type Culture Collection (ATCC) and Japanese F36-ZK, showed similar trends in nitrogen utilization. Both grew most rapidly in the medium containing casamino acids, more slowly with ammonium, and poorly with Bacto peptone, but neither grew with nitrate or nitrite. The nitrogenous compounds in Bacto peptone are mainly oligopeptides with small amounts of free amino acids, while those in casamino acids are free amino acids. Because the nitrogen content was adjusted to equivalent amounts in these experiments, these results indicate that amino acids were more effective for F36-ZK growth than ammonium (Kamako et al. 2005). Kessler and Huss (1990) reported that all four European *Chlorella* symbiotic strains tested were able to use nitrate, while only three could utilize nitrite. Reisser and Widowski (1992) also stated that most *Chlorella* symbionts in European and American *P. bursaria* use nitrate as a nitrogen source. The American symbiotic strain NC64A was reported to grow in a medium containing nitrate as a sole nitrogen source (Reisser et al. 1988), although the NC64A strain from the ATCC could not use nitrate in a recent experiment (Kamako et al. 2005). The growth of the free-living strains *C. kessleri* and *C. vulgaris* was also examined using Bacto peptone, casamino acids, ammonium, nitrate, and nitrite as the sole nitrogen sources. *C. vulgaris* grew well in media containing inorganic nitrogen compounds and casamino acids, but not with Bacto peptone, whereas *C. kessleri* grew well in all media. The utilization of nitrogenous compounds by *Chlorella* strains should depend on the assimilation capacity of inorganic nitrogen and on the transport and metabolism of organic nitrogen. The latter were investigated to better understand the features of nitrogen utilization by F36-ZK.

### 2.3.1 Nitrate and Nitrite Assimilation

Nitrate reductase (NR) and nitrite reductase (NiR) catalyze the first steps of nitrogen assimilation by reducing nitrate to ammonium. Both enzymes can be activated by inducers such as nitrate and glucose. The activities of these enzymes in NC64A and F36-ZK cells were measured under conditions of enzymatic induction. NC64A exhibited immediate NR and NiR activity with nitrate induction, although the activities of the symbiont remained low over 48 h. The NR activity of NC64A was not sufficient to allow survival. On the other hand, F36-ZK did not produce NR under any experimental conditions, but NiR activity was observed with nitrate induction as in free-living strain cells. The difference in response to nitrate induction between F36-ZK and NC64A indicated important differences related to NR expression. The distinct properties of NR and NiR activities of the two symbionts resembled those of two spinach mutants, an NR regulatory gene mutant and an NR structural gene mutant (Ogawa et al. 1994). The regulatory gene mutant showed low NR and NiR activities with nitrate induction, whereas the structural mutant exhibited undetectable NR and potent NiR activities, because NR messenger RNA was considered to play a role as a signal for NiR gene expression although the messenger RNA could not be correctly translated to NR. These similarities suggest that F36-ZK may have a mutation in the NR gene and that NC64A may be a regulatory gene mutant. Symbiotic algae do not need NR in the host cell because *P. bursaria* cannot import nitrate (Alberts et al. 1982); hence, the lack of NR in F36-ZK seems to be an irreversible adaptation to the symbiotic milieu.

### 2.3.2 Amino Acid Utilization

The preliminary results of the experiment described in the previous sections clearly indicated that F36-ZK could use amino acids. To confirm which amino acids can be utilized, the growth of the Japanese symbiont F36-ZK was measured in C medium supplemented with 20 individual amino acids (Kato et al. 2006). The six amino acids L-arginine, L-asparagine, L-glutamine, L-serine, L-alanine, and glycine allowed healthy and remarkable growth of F36-ZK, as summarized in Table 1. The total concentration of these amino acids in Bacto™ casamino

**Table 1** Amino acid utilization and uptake of *Chlorella* spp. (Modified from Kato et al. 2006)

| | | Doubling time (days)[a] | Rate of uptake ($\times 10^{-8}$nmol min$^{-1}$cell$^{-1}$)[b] | |
|---|---|---|---|---|
| Amino acid | | F36-ZK | F36-ZK | C. vulgaris NIES-227 |
| Basic | Arg | 1.9 | 3.30 | 1.34 |
| | Lys | NG | 0.64 | ND |
| | His | NG | 2.46 | ND |
| Acidic | Asp | NG | 0.14 | ND |
| | Glu | NG | 2.08 | ND |
| Polar | Asn | 3.0 | 0.42 | ND |
| | Gln | 4.7 | 6.92 | ND |
| | Ser | 2.9 | 2.28 | ND |
| | Thr | NG | 3.16 | ND |
| | Tyr | NG | 1.08 | ND |
| Nonpolar | Ala | 2.9 | 2.86 | ND |
| | Val | NG | 2.52 | ND |
| | Leu | NG | 3.26 | ND |
| | Ile | NG | 2.34 | ND |
| | Pro | NG | 3.46 | ND |
| | Phe | NG | 3.22 | ND |
| | Met | NG | 2.32 | ND |
| | Trp | NG | 3.16 | ND |
| | Gly | 3.4 | 3.20 | ND |
| | Cys | NG | 5.62 | ND |

NG no growth, ND not detectable (i.e., counts per minute values at 1 and 2 min almost did not change)

[a]Endosymbiotic algae were cultured in C medium added with amino acid (200 μg ml$^{-1}$). The culture conditions were 25°C, 30 μmol photons$^{-1}$m$^{-2}$ s$^{-1}$, light-dark 16 h-8 h. Doubling times were calculated from a period of 3–6 days. Data variations less than 2.0%, $n = 2$.

[b]$1 \times 10^8$ cells ml$^{-1}$ algae were incubated under the condition of 25°C, 90 μmol photons m$^{-2}$ s$^{-1}$. Uptake rates were determined by filtering 100-μl samples at 1 and 2 min after the addition of 1 mM [$^{14}$C]amino acid (specific activity 2–4 (μCi /(μmol(1 ). Data variations less than < 9.5%, n = 2. n.d.: not detectable, i.e., cpm values at 1 and 2 min didn't almost change.

acids is less than 20% (BD Technical Center 2003). No L-glutamine or L-asparagine was present in the media because they were converted to L-glutamic acid and L-aspartic acid, respectively, by acidic hydrolysis. All six amino acids are candidate nitrogen sources for symbionts from the host. Ammonium is an additional candidate, although its efficiency for the growth of F36-ZK should be quite low considering the available amino acid content of casamino acids. The use of three dipeptides containing the utilizable amino acids glycine and L-arginine, namely, glycylglycine, glycyltyrosine, and arginylarginine, was also examined. As expected from the results of F36-ZK growth with Bacto peptone, no algal growth was observed.

Albers et al. (1982) speculated that L-glutamine and ammonium were the key nitrogen compounds transported from the paramecium host to its symbionts. L-Glutamine was then thought to be the most important candidate (Reisser and Widowski 1992). In many reports, only a few amino acids such as L-glutamine and L-glutamic acid were used for symbiotic algal growth experiments. McAuley (1986) studied the relationship between amino acid utilization and uptake of the American symbionts 3N813A and NC64A.

He used eight amino acids, L-arginine, L-glutamine, L-proline, L-serine, L-alanine, glycine, L-lysine, and L-glutamic acid, and reported that NC64A used L-arginine and L-glutamine and imported L-arginine and L-lysine. L-Glutamine stimulated the growth of NC64A (although the cells were described as pale and shrunken), but it was not imported by the organism. In contrast to NC64A, 3N813A imported all amino acids except L-glutamic acid and used every amino acid except L-glutamic acid and L-lysine. From these results, McAuley (1986) concluded that these algal amino acid utilizations could be correlated with their amino acid uptakeabilities.

## 3 Amino Acid Transport of Symbiotic *Chlorella* and Its Regulation Factors

Most free-living *Chlorella* are known to show little amino acid uptake under usual growth conditions, but amino acid uptake can be induced in some cases. For example, glucose induced the amino acid uptake systems of *C. kessleri* for basic and neutral amino acids (Cho et al. 1981). Studies on amino acid transport systems of symbiotic *Chlorella* have not been performed. The major characteristic features of the Japanese symbiont are its developed amino acid transport systems and novel regulation with some materials, as described next.

### 3.1 Amino Acid Uptake by Symbiotic *Chlorella*

Amino acid uptake was examined with radioactive tracers (Kato et al. 2006). The results of amino acid uptake by F36-ZK and free-living *C. vulgaris* NIES-227 are displayed in Table 1. Surprisingly, F36-ZK imported all amino acids at adequate rates without any induction, but *C. vulgaris* imported only L-arginine. Contrary to McAuley's aforementioned conclusion regarding American symbiotic strains, only six amino acids supported the growth of F36-ZK among the 20 amino acids imported. Some amino acids with no growth-stimulating effects, such as L-leucine, L-proline, and L-cysteine, are imported more rapidly than growth-stimulating ones such as L-alanine, for instance. These results clearly indicate that the six growth-stimulating amino acids are used as nitrogen sources via the intracellular metabolic pathway. These six amino acids are thought to generate ammonium ions; L-arginine, L-glutamine, and L-asparagine have a nitrogen-containing group in their side chain, which easily releases nitrogen during metabolism. L-Alanine and L-serine can be converted to glycine, two molecules of which generate each molecule of ammonium, carbon dioxide, and L-serine in photorespiration processes. Compared with amino acids, extracellular ammonium is not a good source of nitrogen, but intracellular ammonium could play an important role as a nitrogen source.

The activities of enzymes responsible for ammonium assimilation in F36-ZK were also measured, but clear results were not obtained. When cells were grown under weak light intensity (50 μmol photons $s^{-1}$ $m^{-2}$) in media containing ammonium as the sole nitrogen source, the total activity of glutamine synthetase (GS), a key enzyme in ammonium assimilation, in F36-ZK cells was either not detectable (Kato et al. 2006) or low compared with that of *C. vulgaris* and NC64A. However, GS activity in F36-ZK cells grown under strong light conditions (150 μmol photons $s^{-1}$ $m^{-2}$) was similar to that observed in *C. vulgaris* and NC64A. Among these algae cultured in media containing ammonium, F36-ZK was the most sensitive organism to the GS inhibitor, methionine sulfoximine (MSX). MSX is a chemical analogue of L-glutamic acid and L-glutamine, and could therefore be imported by F36-ZK but not by other algal strains. Because MSX showed lethal effects against F36-ZK in media containing L-serine or L-arginine as a nitrogen source, it is very likely that GS plays a role in ammonium assimilation from L-serine or L-arginine.

## 3.2 Amino Acid Transport Systems

F36-ZK transported more amino acids than the American *Paramecium* symbionts (McAuley 1986, 1989), whereas free-living *C. vulgaris* transported only one amino acid, L-arginine (Kato et al. 2006). These differences suggest that the amino acid transport systems in symbiotic *Chlorella* F36-ZK evolved over time.

### 3.2.1 Amino Acid Transport Systems in F36-ZK

To analyze the amino acid transport systems, the uptake of all amino acids was performed using radioactive tracer methods by kinetic analyses (Kato and Imamura 2009). From the $K_m$ values, L-arginine and L-tyrosine showed higher affinity, whereas L-glutamine, L-histidine, L-asparagine, and acidic amino acids showed lower affinity to their transporters. For all amino acids, except L-glutamine, $V_{max}$ ranged from 1.02 to 31.4 × $10^{-8}$ nmol $min^{-1}$ $cell^{-1}$. L-Glutamine showed the highest $V_{max}$, 250 × $10^{-8}$ nmol $min^{-1}$ $cell^{-1}$, although the affinity of L-glutamine for its transporter was low. L-Lysine and L-alanine were transported by at least two systems because the kinetic data for their uptake on a Lineweaver–Burk plot were biphasic in nature. The other 18 amino acids had single $K_m$ values, indicating transport by one system.

To sort amino acids into their respective transport systems, competitive experiments were carried out. In other words, the uptake of a radioactive amino acid was measured in the presence of another unlabeled amino acid. The uptake of L-lysine was strongly inhibited by the addition of L-arginine, indicating that these amino acids were transported via the same basic amino acid transport system. The uptake of most amino acids, including L-lysine, L-histidine, L-aspartate, and L-glutamic acid, was inhibited by the addition of neutral amino acids. Therefore, the existence of a common broad transport system that transports 19 amino acids but not L-arginine was revealed and is referred to as a general amino acid transport system. In addition, a Lineweaver–Burk plot suggested the existence of an alanine transport system. Thus, kinetic analyses and competitive experiments disclosed three amino acid transport systems in F36-ZK, a basic amino acid transport system for L-arginine and L-lysine, a general amino acid transport system for 19 amino acids, and a separate L-alanine transport system.

### 3.2.2 Features of Amino Acid Transport Systems

Studies focused on amino acid transport systems of microalgae have been scarce; the uptake of several amino acids by F36-ZK was measured in the pH range of 4.0–8.0 to obtain general information on these amino acid transport systems. Optimal uptake of all neutral amino acids tested, such as L-serine, L-alanine, and L-glutamine, was observed near pH 5.0; L-arginine uptake showed a broad optimal pH range between pH 5.0 and 6.5. A slight positive charge seemed to be required for amino acid transport, because the optimal pH was slightly more acidic than the isoelectric points. These results support an amino acid–proton symport mechanism which has been reported in many organisms (Bush 1993; Cho and Komor 1983; Young et al. 2003). In the case of L-aspartic acid and L-glutamic acid, their uptake was very sensitive to external pH, i.e., uptake increased when the pH became acidic and little uptake was measured above pH 6. Previous studies (Albers et al. 1982; Kato et al. 2006; McAuley 1986) have reported that symbiotic *Chlorella* could not utilize L-glutamic acid; however, since those experiments were carried out in a medium with a pH above 6.3, and L-glutamic acid uptake showed strict pH dependence as mentioned above, the utilization of L-glutamic acid by symbiotic *Chlorella* should be reviewed (Kato and Imamura 2009).

To assess whether the transport system in F36-ZK is active or passive, intracellular and extracellular concentrations of L-arginine (for the basic amino acid transport system), L-serine (for the general amino acid transport system), and L-alanine (for the alanine transport system) were measured. Intracellular concentrations of all amino acids were higher than extracellular ones even at 1 min after the addition of tracers. The differences between the intracellular and extracellular amino acid concentrations increased after 30 min of incubation, indicating that the transport systems were active. All amino acid transport systems in F36-ZK were considered to be active transporters carrying out amino acid–proton symport, on the basis of the following results: (1) a decrease in external proton concentration was observed after the addition of L-serine and L-alanine; (2) a protonophore (carbonyl cyanide *m*-chlorophenyl hydrazone) and an inhibitor of ATP synthesis (sodium azide) strongly inhibited amino acid uptake, while an inhibitor of phosphorylation (vanadate) also inhibited amino acid uptake. Amino acid–proton symport systems were also supported by the above-mentioned pH-dependent amino acid uptake.

In the case of microbes, the uptake of amino acids was reportedly inhibited by ammonium ions (Grenson et al. 1970; Willis et al. 1975). Thus, the uptake of several amino acids by F36-ZK was measured in the presence of nitrate and ammonium. Nitrogen ions from inorganic sources did not affect the uptake of L-arginine, L-serine, or L-glutamine. Furthermore, to examine the effect of cultivation with ammonium, the amino acid uptake of F36-ZK was measured using cells grown with ammonium as the sole nitrogen source. The cells transported all 20 amino acids and, thus, all of the transport systems were expressed under culturing conditions with ammonium. Overall, the expression of the amino acid transport systems was independent of the nitrogen source, i.e., all transport systems are constitutive systems and represent one of the specific features of F36-ZK.

Many microalgae grown photoautotrophically import a few amino acids but possess interesting inducible amino acid transport systems (Cho and Komor 1985; Kato et al. 2006; Kirk and Kirk 1978). For example, amino acid transport systems in free-living *C. kessleri* can be induced by glucose (Cho et al. 1981). The induced general system of *C. kessleri* exhibited broad specificity for ten amino acids (Sauer 1984). In the case of *C. vulgaris*, sugars or glycine induce an amino acid transport system for four neutral amino acids (Plakunov et al. 1995; Seifullina et al. 1995). Therefore, free-living *Chlorella* can build various amino acid transport systems that are usually repressed under photoautotrophic conditions.

## 3.3 Factors Affecting Amino Acid Transport

Although environmental conditions under which the symbiont is living are a very important consideration for *P. bursaria* symbiosis, the underlying mechanistic details are unknown. Examples of extracellular conditions affecting the amino acid uptake of fungal and plant cells have been reported (Cameron and Lejohn 1972; Harrington et al. 1981; Rickauer and Tanner 1986; Smith 1978). For instance, the divalent cations $Ca^{2+}$ and $Mg^{2+}$ have been found to activate amino acid uptake. The mechanism of activation by such cations is still unclear, although several hypotheses have been proposed. The first suggests that divalent cations create a proton motive force for amino acid transport in tobacco cells (Smith 1978); another implicates $Ca^{2+}$ as the activator of the p-type ATPase of *Neurospora crassa* (Lew 1989), and lastly $Ca^{2+}$ was shown to affect the permeability of the cell membrane in *Phaseolus vulgaris* (Rickauer and Tanner 1986). Whatever the mode of activation, divalent cations stimulate amino acid uptake. Although no reports deal with cation transport in the symbiosis of *Paramecium*, the host must supply cations to its symbionts because *Chlorella* requires divalent cations for survival. Studies on amino acid uptake have found that the $Ca^{2+}$ concentration in test cell suspensions affects amino acid uptake (Kato and Imamura 2008a). The novel effects of cations and glucose on amino acid uptake of F36-ZK will be described further.

### 3.3.1 Effects of Divalent Cations on Amino Acid TransportSystems in F36-ZK

Divalent cations such as $Ca^{2+}$ and $Mg^{2+}$ generally activate amino acid uptake as previously mentioned, but there are no specific reports of these effects in *Chlorella*. The effects of divalent cations on amino acid uptake in free-living *C. vulgaris* and in the Japanese symbiont F36-ZK are displayed in Fig. 1. The uptake of L-arginine by *C. vulgaris* was enhanced by $Ca^{2+}$ and $Mg^{2+}$ (Fig. 1a). However, the uptake of L-serine by the symbiont F36-ZK was decreased by the cations, especially by $Ca^{2+}$

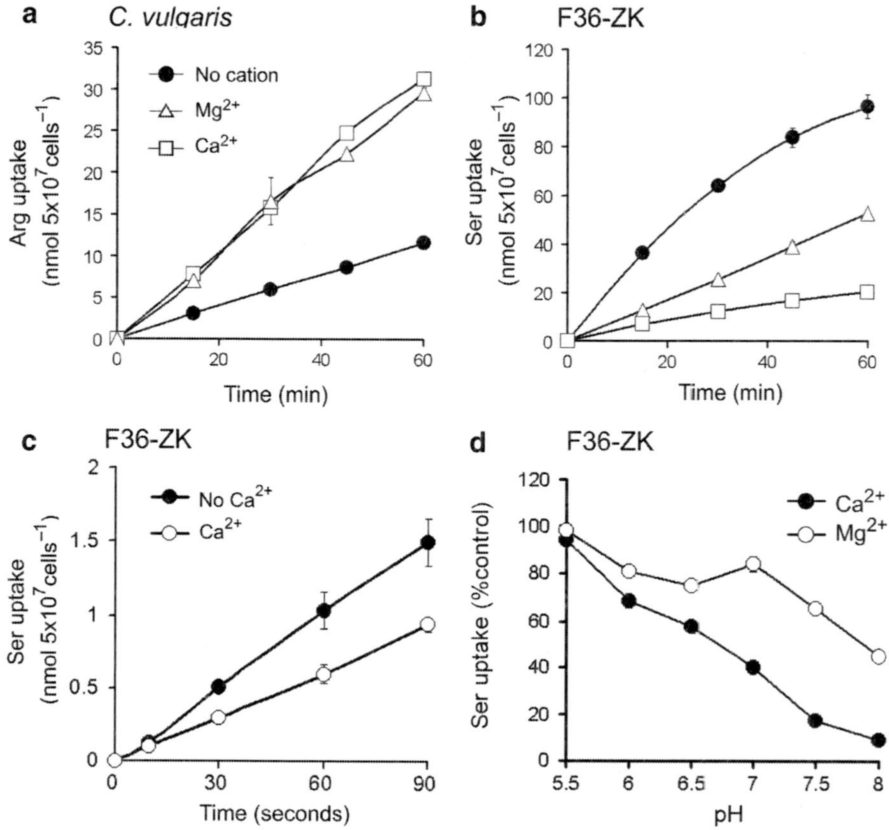


**Fig. 1** Uptake of amino acids by *Chlorella* in the presence of divalent cations (Kato and Imamura 2008a). **a** Arginine uptake by free-living *Chlorella vulgaris* and **b** serine uptake by the *Paramecium* symbiont F36-ZK in the presence of 0.64 mM $Ca^{2+}$ (*squares*), 0.64 mM $Mg^{2+}$ (*triangles*), or no cations (*circles*) at amino acid concentrations of 1 mM. Algae were treated with these cations for 15 min prior to the experiments. *Bars* represent the mean ± the standard deviation (SD) of three (F36-ZK) or four (*C. vulgaris*) replicates. **c** Uptake of serine by the *Paramecium* symbiont F36-ZK in the presence (*open circles*) or absence (*filled circles*) of 1 mM $Ca^{2+}$. Serine was supplied at a concentration of 0.1 mM. *Bars* represent the mean ± SD of three replicates. **d** Uptake of serine by the *Paramecium* symbiont in relation to pH. Cells were pretreated with 1 mM $Ca^{2+}$ (*filled circles*) or 4 mM $Mg^{2+}$ (*open circles*). Serine was supplied at a concentration of 1 mM. Each point is the mean of three experiments

(Fig. 1b). When L-serine and $Ca^{2+}$ were added at the same time, a rapid decrease in L-serine uptake was observed within 30 s (Fig. 1c); the rate of L-serine uptake decreased with a preincubation time lasting up to 30 min. The uptake of L-serine with $Ca^{2+}$ or $Mg^{2+}$ treatment was decreased from pH 5.5 to 8.0 (Fig. 1d).

The phenomenon of divalent cations decreasing amino acid uptake is novel and is one of the features of F36-ZK. To characterize this phenomenon in detail, a dose–response relationship between divalent cation concentrations and decreased

L-serine uptake was measured. The half-maximal effective concentration ($EC_{50}$) of $Ca^{2+}$ was 0.21 mM and the minimum L-serine uptake, which was less than 10% in nontreated F36-ZK, was observed at a $Ca^{2+}$ concentration greater than 1 mM. In the case of $Mg^{2+}$ the minimum L-serine uptake, which was approximately 40% in nontreated F36-ZK, occurred at a $Mg^{2+}$ concentration greater than 5 mM. The effects of monovalent and polyvalent cations on serine uptake by F36-ZK were also examined. Monovalent cations, such as $Li^{+}$, $Na^{+}$, and $K^{+}$, did not significantly affect serine uptake; divalent cations, such as $Ni^{2+}$, $Mn^{2+}$, and $La^{3+}$ (a calcium ion analogue), also decreased serine uptake. $Ca^{2+}$ was the strongest inhibitor among the cations tested.

F36-ZK has three amino acid transport systems. To evaluate the effect of $Ca^{2+}$ on the transporters, the uptake rates of L-arginine, L-serine, L-alanine, and L-glutamine were measured after $Ca^{2+}$ treatment. Little inhibition was observed for L-arginine uptake, but the uptake of L-serine, L-alanine, and L-glutamine was decreased to be less than 20% of that for nontreated F36-ZK. $Ca^{2+}$ inhibited the uptake via the general amino acid transport system, but not via the basic amino acid transport system (Kato and Imamura 2009). The alanine transport system also appeared inhibited by $Ca^{2+}$ because the transport system should function at the concentration of L-alanine used in the experiment. The effects of $Ca^{2+}$ on the uptake of L-serine were also monitored by kinetic analysis. No effect on the $K_m$ value was observed, but the $V_{max}$ value increased about twofold. These results indicate that the noncompetitive inhibition of serine uptake via the general amino acid transport system was $Ca^{2+}$-dependent.

Calcium ions are essential for survival in living organisms. To analyze the mechanism of the above-mentioned process, $Ca^{2+}$ uptake of F36-ZK was measured using radioactive $^{45}Ca^{2+}$. The symbiont F36-ZK imported $Ca^{2+}$ with no observed change in external pH during uptake, indicating that $Ca^{2+}$ uptake occurred without a proton motive force. Free-living *C. vulgaris* also showed similar results. To identify the major membrane transport protein for $Ca^{2+}$ uptake in F36-ZK, experiments were performed using calcium channel blockers and a $Ca^{2+}$-ATPase inhibitor. Calcium channel blockers, especially the hydrophobic compounds verapamil and diltiazem, inhibited F36-ZK's $Ca^{2+}$ uptake, but the $Ca^{2+}$-ATPase inhibitor erythrosine B did not. Although F36-ZK imported $Ca^{2+}$, it was not clear whether the intracellular $Ca^{2+}$ contributed to the inhibition of L-serine uptake. To determine this, extracellular $Ca^{2+}$ was removed using ethylene glycol bis(2-aminoethyl ether)tetraacetic acid (EGTA), a $Ca^{2+}$ chelator, and the inhibition of amino acid uptake by $Ca^{2+}$ was recorded. The uptake of L-serine that was inhibited by $Ca^{2+}$ was restored by the addition of EGTA. In contrast, the uptake of L-arginine in *C. vulgaris* was accelerated by $Ca^{2+}$ in the presence of EGTA. Therefore, extracellular $Ca^{2+}$ inhibited the uptake of L-serine in F36-ZK and intracellular $Ca^{2+}$ accelerated the uptake of L-arginine in *C. vulgaris*.

The detailed mechanism behind these observations remains unclear. The activity of an $H^{+}$-ATPase, an energy source provider for amino acid transport in general (Bush 1993; Cho and Komor 1983; Young et al. 2003), and the membrane permeability of amino acids were changed by $Ca^{2+}$ in other organisms. However, there was no effect of $Ca^{2+}$ on the uptake of L-arginine by F36-ZK, indicating that the inhibition of L-serine uptake was not due to a decrease in $H^{+}$-ATPase activity (Kato and Imamura 2009). The possibility that $Ca^{2+}$affects membrane permeability should be neglected

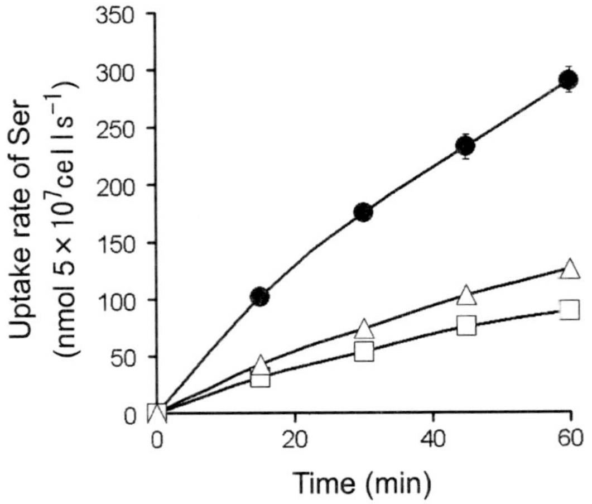


**Fig. 2** Uptake of serine by an general amino acid transporter induced with glucose in *C. kessleri* in the presence of $Ca^{2+}$. Serine uptake by *C. kessleri* in the presence of 0.64 mM $Ca^{2+}$ (*squares*), 0.64 mM $Mg^{2+}$ (*triangles*), and no added cation (*circles*) at a concentration of 1 mM serine. Algae were pretreated with 13 mM glucose for 3 h

because $Ca^{2+}$exhibited opposing effects on membrane permeability, and indeed, efflux of intracellular L serine by $Ca^{2+}$was not observed.

As stated above, the amino acid transporters of free-living *C. kessleri* could be induced by glucose and the induced amino acid transport system could import L-serine in the manner of general amino acid transport of F36-ZK (Sauer 1984). The general amino acid transport system of F36-ZK is inhibited by $Ca^{2+}$, and it would be interesting to determine whether the transport system induced in *C. kessleri* can be inhibited by the cation. It should be noted that $Ca^{2+}$ and $Mg^{2+}$inhibited the induced amino acid transport system in *C. kessleri*, as shown in Fig. 2 (Kato and Imamura 2008a). Thus, free-living *C. kessleri* have an amino acid transport system similar to that of F36-ZK, but it is not expressed under usual conditions.

### 3.3.2 Effects of Glucose on Amino Acid Transport Systems in F36-ZK

In a symbiotic *P. bursaria* cell, the transfer of sugars and amino acids occurs in opposite directions, just like a barter transaction between the symbiotic alga and the host. Sugars are a common nutrient and can also act as signaling substances in many organisms: sugar sensing and signaling pathways have been found in yeasts, plants, and mammals (Rolland et al. 2001). In free-living *C. kessleri* glucose and its analogues cause induction of a hexose and of the amino acid transport system (Cho et al. 1981; Tanner 1969). This suggests that green algae also have a sugar sensing and signaling system. Symbiotic *Chlorella* cells release sugars (Brown and Nielsen 1974; Pardy et al. 1989; Reisser and Windowski 1992) and each *Chlorella* cell is enclosed in a vacuolar membrane in the host cell (Meier et al. 1984). Therefore, sugars should exist around the algae, implying that they import amino acids in the presence of sugar in their host. Thus, the effect of sugars on algal amino acid uptake was studied. Glucose is the typical sugar known to play a signal role and is the first metabolite of maltose in *P. bursaria*, which is produced and supplied by the symbiont *Chlorella*. Hence, studies were first performed using glucose (Kato and Imamura 2008b).

The rate of L-serine uptake by F36-ZK increased in the presence of glucose. A pulse treatment with glucose for 1 min also increased the rate to nearly the same level as the treatment for 30 min. L-Serine uptake increased with increasing glucose concentration and reached a maximum value approximately double that observed without glucose. The $EC_{50}$ was 3 μM, assuming that the effect observed with 27.8 mM glucose was maximal, although glucose treatment in the range 0.05–27.8 mM showed almost the same effect. The rate of uptake of cells treated with glucose for 1 h was $3.06 \times 10^{-8}$ nmol $min^{-1}$ $cell^{-1}$, that of cells treated with glucose for 1 h and then left standing for 1 h without glucose was $2.40 \times 10^{-8}$nmol $min^{-1}$ $cell^{-1}$, and that of untreated control cells was $1.40 \times 10^{-8}$ nmol $min^{-1}$ $cell^{-1}$. Although some reduction in L-serine uptake was observed after the removal of glucose, the higher L-serine uptake rate was sustained for at least 1 h after cells had been treated with glucose. The effect showed pH dependence; L-serine uptake increased sixfold at pH 8.0 and an external acidic pH caused a decrease in the effect. At pH 5.5, no change was observed in L-serine uptake relative to that of cells without glucose treatment. This observation is in contrast to the pH dependence of L-serine uptake, which was optimal near pH 5.0. The transport of L-serine was more affected by $Ca^{2+}$and glucose under alkaline than acid conditions; both were pH-dependent. The uptake of other amino acids was measured using cells treated with glucose. The general amino acid transport system proved more sensitive than the basic amino acid transport system. Kinetic analysis of L-serine uptake with glucose treatment was performed. The $V_{max}$ value of L-serine uptake was doubled by treatment with glucose, but no change occurred in the $K_m$ value, suggesting that glucose treatment produced an increase in the amount of the transporter. However, cycloheximide, an inhibitor of protein synthesis, had no effect on accelerated L-serine uptake by glucose, indicating that post-translational modification and transcript levels were unchanged by glucose. The effect of glucose on the incorporation of L-serine into protein was also evaluated, but no effect was observed under either light or dark conditions. Furthermore, the amount of protein in the cells did not differ with or without glucose treatment.

Symbiotic *Chlorella* release sugars, but little is known about sugar uptake by symbionts. The uptake of glucose and other sugars by *Chlorella* spp. was measured to determine whether sugar is actually imported into cells. Free-living *C. vulgaris* imported glucose, sucrose, and maltose at uptake rates of 8.44, 18.92, and $0.68 \times 10^{-8}$ nmol $cell^{-1}$, respectively, over 15 min; fructose was not imported. In contrast, F36-ZK did not import any sugars at pH 6.0 or 7.5 under either light or dark conditions. No uptake of glucose confirmed that F36-ZK could not utilize glucose as a nutrient although glucose obviously stimulated its growth, as mentioned above. Because the increase in the rate of L-serine uptake resulted from glucose treatment even at a low concentration ($EC_{50}$ = 3 μM) and with a pulse treatment lasting only 1 min, a glucose sensing and signaling pathway seems to be involved in this response. In a glucose sensing system, some nonmetabolizable analogues of glucose are known to act as agonists (Rolland et al. 2001). To determine the presence of a glucose sensing and signaling pathway, the response of L-serine uptake to glucose-related compounds was measured. The uptake of L-serine increased following treatment with almost all of the monosaccharides tested. However, treatment with L-glucose and 6-amino-6-deoxyglucose did not increase the uptake of serine; application of D-glucuronic acid inhibited the uptake of L-serine. Cells treated with the disaccharides maltose, cellobiose, trehalose, and gentiobiose, which contain only glucose as a building block, responded to treatment. Nonmetabolizable glucose analogues such as 2-deoxyglucose and 3-*O*-methylglucose also activated L-serine uptake, indicating that the message is transmitted via a glucose sensing and signaling pathway. These sugars were thought to be sensed at the same site on the plasma membrane, because treatment with other sugars, such as maltose, xylose, and 3-*O*-methylglucose, in addition to glucose, produced no additive effect on the rate of L-serine uptake.

Mechanisms by which glucose transduces its effects via the signaling pathway could be (1) the activation of amino acid transporters and (2) the translocation of amino acid transporters. Given that $H^+$-ATPase creates a $H^+$ gradient for amino acid transport, activation of $H^+$-ATPase by sugars is most probable (Camoni et al. 2006). To examine this, the effects of inhibitors on a protein kinase (*N*-6-dimethylaminopurine and chelerythrine), inhibitors of a protein phosphatase (cyclosporine A and okadaic acid), and a calmodulin antagonist (W-7), were measured. Inhibitors of the protein kinase and the phosphatase showed no effect; the calmodulin antagonist exhibited a strong inhibitory effect, supporting the involvement of calmodulin in the signaling pathway, but activation of $H^+$-ATPase was not clear.

Because the uptake of L-serine by F36-ZK was inhibited by $Ca^{2+}$ (Kato and Imamura 2008a), whether glucose could counter the inhibition caused by $Ca^{2+}$ was investigated next. Glucose restored L-serine uptake in $Ca^{2+}$-treated cells. Sugars such as fructose and sucrose contributed to the restitution of serine uptake in $Ca^{2+}$-treated cells, although they had only a small effect on the rate of serine uptake. These findings, regarding the regulatory effects of $Ca^{2+}$ and sugars on the amino acid transport system of symbiotic *Chlorella* cells, imply the presence of a control system using these materials in the perialgal vacuole enclosing the symbiotic algae. Because the symbionts released sugars, calcium ions are likely to be transported from the host to the symbiont. Calcium has indeed been observed using electron microscopy in *Paramecium* cell symbionts (Kato and Imamura 2008a).

## 4 Photosynthesis of Symbiotic *Chlorella* and Its Regulatory Factors

Many examples of symbiosis between invertebrates and algae have been reported in marine organisms. For example, the dinoflagellate *Symbiodium* spp. has established symbiotic relationships with the anemone (Trench 1971), coral (Schlichter et al. 1983), and the giant clam (Ishikura et al. 1999; Streamer et al. 1988) and supplies glycerol, amino acids, or sugars to the host (Cook 1983; Hinde 1988; Trench 1971, 1979, 1993). Muscatine (1967) first reported that host tissue homogenates of symbiotic coral and clam activated excretion of fixed carbon by its symbiotic algae.

Further studies on anemones (Whitehead and Douglas 2003), corals (Gates et al. 1995), and giant clams (Masuda et al. 1994) also revealed that host homogenates stimulate release of photosynthetic products from symbiotic algae. In the case of the sea anemone, a low molecular mass fraction of host homogenate accelerated glycerol release from algae severalfold (Grant et al. 1997). Thus, it is thought that excretion from symbiotic algae could be stimulated by some specific compound(s) in the host homogenate, referred to as a host release factor. This factor has not been identified owing to its lability and given inherent difficulties in cultivating experimental organisms (Grant et al. 1998). An inhibitory effect on photosynthesis by a symbiotic coral homogenate was also observed (Sutton and Hoegh-Guldberg 1990) and the photosynthesis inhibiting factor was also suggested to be a low molecular mass compound (Grant et al. 2001).

On the basis of these symbiotic relationships in marine invertebrates, it is conceivable that there is a similar relationship, established through a host factor, in the green freshwater paramecium. In cell-free extracts of Japanese *P. bursaria*, the host factor that affects carbon fixation by symbiotic algae will be described.

## 4.1 Features of Photosynthesis in Symbiotic *Chlorella*

To obtain fundamental knowledge about the features of carbon fixation by the symbiont F36-ZK, the pH dependence of F36-ZK's carbon fixation was compared with that of free-living *C. vulgaris*. Radioactive carbon dioxide was used and the photosynthetic products of F36-ZK were studied. Surprisingly, the pH dependence of gross carbon fixation by these algae showed opposing tendencies, i.e., the gross carbon fixation by F36-ZK was increased at alkaline pH, while that of *C. vulgaris* was decreased, as shown in Fig. 3 (Kamako and Imamura 2006). In *Chlorella* spp., a decrease in carbon fixation at alkaline pH has been previously reported (Beardall 1981; Beardall and Raven 1981; Gehl et al. 1990; Shelp and Canvin 1985). Thus, the enhancement of carbon fixation at alkaline pH seems to be a notable feature of the Japanese symbiont F36-ZK. However, the release of photosynthetic products by F36-ZK was increased at low pH, as observed in an endosymbiotic *Chlorella* isolated from a green hydra (McAuley et al. 1996). The photosynthate of F36-ZK released at low pH was analyzed by silica gel thin-layer chromatography and the major photosynthetic product was identified as maltose. This observation strongly suggests that the Japanese symbiotic algae F36-ZK provides maltose to its host, like the European and American symbionts (Brown and Nielsen 1974; Reisser 1976, 1986).

## 4.2 Effects of Host Extracts on Carbon Dioxide Fixation by Symbiotic *Chlorella*

The effects of cell-free extracts of *P. bursaria* F36 on carbon fixation and the photosynthate release of symbiotic *Chlorella* F36-ZK were investigated (Kamako

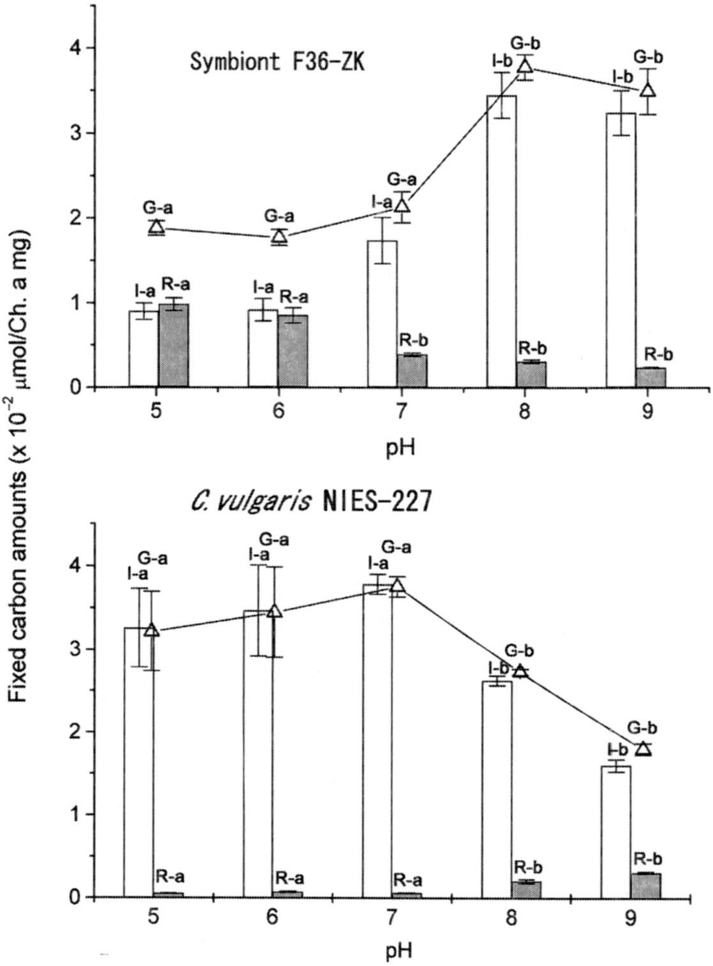


**Fig. 3** pH dependence of carbon fixation of F36-ZK and *C. vulgaris*. (Kamako and Imamura 2006). The *white bar* shows intracellular carbon fixation (*I*), the *shaded bar* represents the released photosynthate (*R*), and the *line chart* depicts gross carbon fixation (*G*). Mean values ± the standard error (SE; $n = 3$). Means showing statistical differences, based on one-way analysis of variance (ANOVA) followed by a Tukey test ($p < 0.05$), are indicated

and Imamura 2006). Intracellular carbon fixation by F36-ZK in sodium phosphate buffer (pH 7.0) increased with increasing host extract concentration and reached a maximum value about threefold that without the host extract, but the release of photosynthetic products hardly changed. Well-known factors limiting photosynthesis are light intensity and carbon dioxide concentration (Taiz and Zeiger 2002). Light intensity was kept constant, and carbon dioxide and bicarbonate were excluded as enhancing factors, because carbon fixation increased with host extract from which carbon dioxide was ready eliminated by acid treatment. From these experiments, proof was obtained for a host factor that stimulates algal carbon fixation.

The pH dependence of the effects of the host factor was studied at a pH of 5.0, 7.0, and 9.0. Intracellular carbon fixation by F36-ZK was enhanced with increasing host extract concentration at pH 5.0 and 7.0, although it remained high at pH 9.0 with or without host extract. Carbon fixation by F36-ZK is optimal at alkaline pH as mentioned above and thus it might be impossible to enhance the carbon fixation at pH 9.0. While the amount of photosynthate released was affected only at pH 5.0, about 45% of the total carbon fixed was released (a relatively high percentage compared with results at pH 7.0 or 9.0), irrespective of host extract concentration (Fig. 4). Therefore, photosynthate release depended on pH but not on the concentration of the host extract at any pH. In *P. bursaria* cells, a perialgal vacuole encloses the symbiotic algae and maintains acidic conditions (Schüßler and Schnept 1992). Inside the vacuole, the host factor could enhance symbiotic algal carbon fixation and the acidic environment could favor release of the photosynthate, maltose, from the symbiont. The enhancement of carbon fixation by an agent is still an unusual phenomenon, but evidence for increased symbiotic algal carbon fixation by a host cell homogenate was reported for coelenterates (Trench 1971) and giant clams (Gates et al. 1995). Dinoflagellates were the symbiotic algae in both cases, and several free amino acids were shown to enhance carbon fixation in the symbionts.

## 4.3 Factors Affecting Photosynthesis of Symbiotic *Chlorella*

Further studies on the host factor in *P. bursaria* extract were carried out and the host factor in Japanese *P. bursaria* F36 extract was characterized as described below. In many marine organisms, the host factors were known to be of low molecular mass. Experiments using ultrafiltration showed that the host factor in *P. bursaria* F36 extract was also a low molecular mass substance of less than 5 kDa (Kamako and Imamura 2006). Heat stability was measured, because some host factors in marine organisms were reported to be too labile to be isolated (Grant et al. 1998), and it was found that the activity remained even when the extract was autoclaved at 121°C for 20 min. Considering the role of minerals as potential host factors, the activity of the inorganic fraction of the extract was examined. After removal of the organic compound by burning at 700°C, the remaining ash solution and the extract enhanced algal carbon fixation, indicating that the active ingredient is inorganic material.

The major cellular inorganic materials are cations, such as $K^+$, $Ca^{2+}$ and $Mg^{2+}$, and the mixture of these cations were considered to be the host factor. Every cation is important for photosynthesis; $K^+$ is present in the chloroplast at approximately 100 mM (Wu and Berkowitz 1992a), and contributes to the stability of enzymes for carbon fixation and for the regulation of stromal pH (Berkowitz and Wu 1993). Activation of ribulose bisphosphate carboxylase/oxygenase and fructose 1,6-bisphosphatase requires $Mg^{2+}$ (Ishijima and Ohnishi 2002). Furthermore, the activity of a chloroplastic ATPase, which also requires $Mg^{2+}$ for function, drastically increases

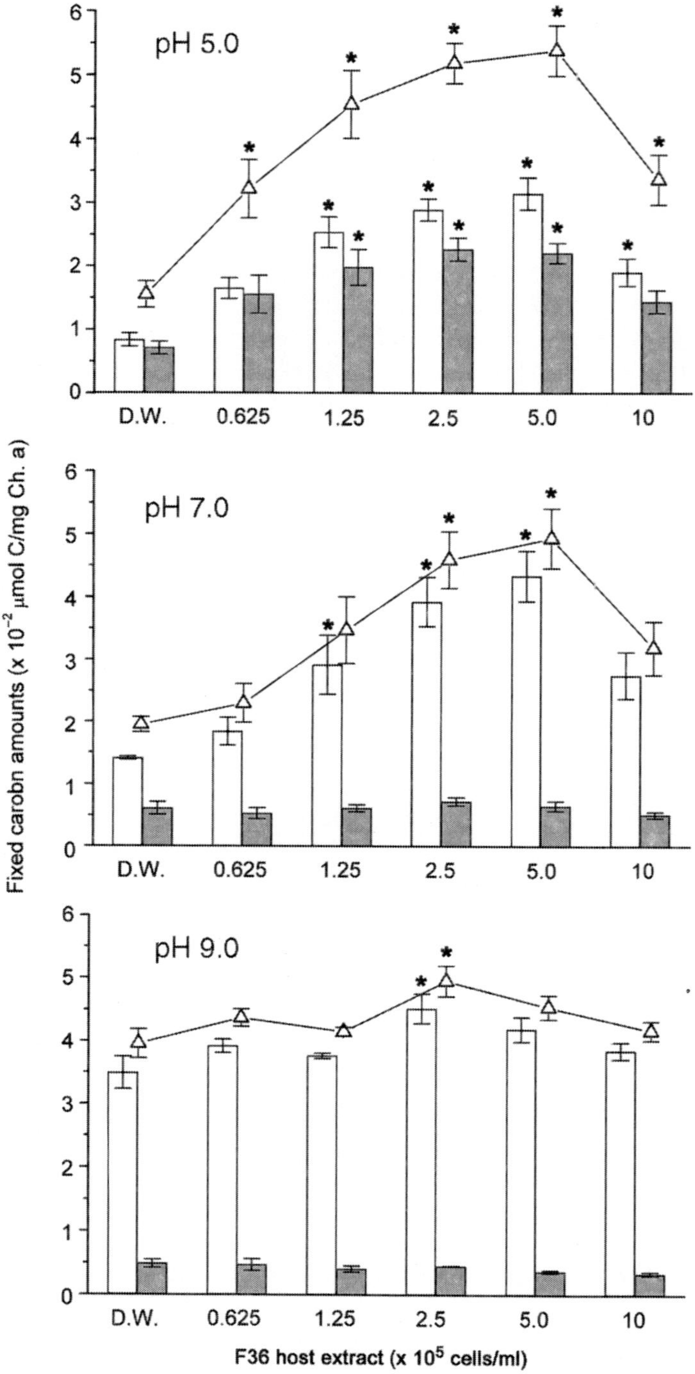


**Fig. 4** The pH dependence effect of the cell-free extract on carbon fixation of F36-ZK. (Kamako and Imamura 2006). The *top*, *middle*, and *lower* charts show carbon fixation at a pH of 5.0, 7.0, and 9.0, respectively. The *white bar* shows intracellular carbon fixation, the *shaded bar* shows released photosynthate, and the *line chart* shows gross carbon fixation. Mean values ± SE ($n = 3$). *Asterisks* indicate a statistically significant difference from the control (distilled water, *D.W.*) based on one-way ANOVA followed by a Dunett test ($p < 0.05$). The concentrations of the host extract were expressed as the cell density before homogenization

in the presence of $K^+$ and $Mg^{2+}$ (Wu and Berkowitz 1992b). Calcium ions also seem to play a role in oxygen evolution (Yocum 1991), although the effect of $Ca^{2+}$ on photosynthesis remains unclear.

In marine organisms, organic molecules have been postulated to be the host factors (Gates et al. 1995; Ritchie et al. 1997). However, the inorganic materials $K^+$, $Ca^{2+}$, and $Mg^{2+}$ were seemed to play important roles in the symbiosis of Japanese *P. bursaria* F36.

## 5 Conclusions

Many novel features of the Japanese symbiont F36-ZK were demonstrated. The lack of NR activity in F36-ZK indicated its irreversible adaptation to symbiotic circumstances. Information gathered about the transfer of materials and metabolic control between the host and the symbiont led to the following conclusions. The ability of F36-ZK to utilize nitrogen provided in the form of ammonium and amino acids implied that F36-ZK lives in environments in which several free amino acids are available but where ammonium is not. The utilizable amino acids seem to easily generate ammonium ions in the algal cells. Extracellular ammonium is not a good growth stimulant; therefore, these amino acids could be carriers of nitrogen in symbiosis. Although it remains unclear which amino acid is used as the nitrogen carrier, multiple amino acids are candidates, because F36-ZK possesses three developed amino acid transport systems. However, the presence of sugars in the symbiotic circumstances complicates this situation. Sugars are known to induce amino acid transport systems in free-living *C. kessleri* and *C. vulgaris* and *P. bursaria* symbionts release maltose in a perialgal vacuole during the entire day (Ziesenisz et al. 1981). Therefore, amino acid transport systems of symbiotic *Chlorella* evolved in a perialgal vacuole, and might become a constitutional system. Because a number of amino acid transport systems can be induced by sugars, it is doubtful that the three amino acid transport systems directly support the theory that the host supplies multiple amino acids. An interesting $Ca^{2+}$-dependent regulatory phenomenon in the F36-ZK amino acid transport systems was unveiled. The inhibition of L-serine uptake by $Ca^{2+}$ was a common feature with the transport system induced by glucose in *C. kessleri*. The general amino acid transport system of F36-ZK resembles that induced in free-living *C. kessleri*, indicating that these amino acid transport systems are common features of *Chlorella* spp. regardless of their expression. Therefore, the amino acid transport systems of F36-ZK do not appear to acquired characters but seem to be genetic characters of *Chlorella* spp. Another regulatory phenomenon involving glucose was also found; glucose restored the inhibition of L-serine uptake by $Ca^{2+}$. $Ca^{2+}$ and glucose appear to be an inhibitor-activator pair for the regulation of the amino acid uptake. The transfer of sugars and amino acids between the symbiont and the host occurs in opposite directions in the *P. bursaria* cell. When the $Ca^{2+}$ concentration in the perialgal vacuole is constant, the symbiont releases more sugar, leading to an import of more amino acid from the host, analogous to a barter transaction between the symbiotic alga and the host.

From the results of the study on host factors, the pH and cation concentrations of the extracellular fluid triggered drastic changes in carbon fixation by the Japanese symbiont. Quite unexpectedly, the photosynthetic ability of F36-ZK showed the opposite pH dependence compared with that of free-living *C. vulgaris*, because carbon dioxide is converted to bicarbonate under alkaline conditions and the impermeability of the plasma membrane to bicarbonate causes a general decrease in the rate of photosynthesis. The dependency of F36-ZK's photosynthetic ability on extracellular cation concentration is also unusual. However, the possibility remains that the host controls the cation concentration in the perialgal vacuole for optimal algal photosynthesis. Although further analyses of the host factor are still ongoing, it is assumed that several signals exchanged between the host and its symbiotic cells control each other to maintain the symbiotic relationship.

Finally, a schematic of the above-mentioned metabolic control mechanisms is illustrated in Fig. 5. It is important to remember that the symbionts are wrapped with a lipid bilayer in the host cell and that the perialgal vacuole membrane is thought to be one of the barriers to overcome for the transportation of materials between the host and the symbiont. The permeability of this membrane regulates material transfer between the host and the symbiont and the surroundings of the symbionts in the vacuole. Studies on

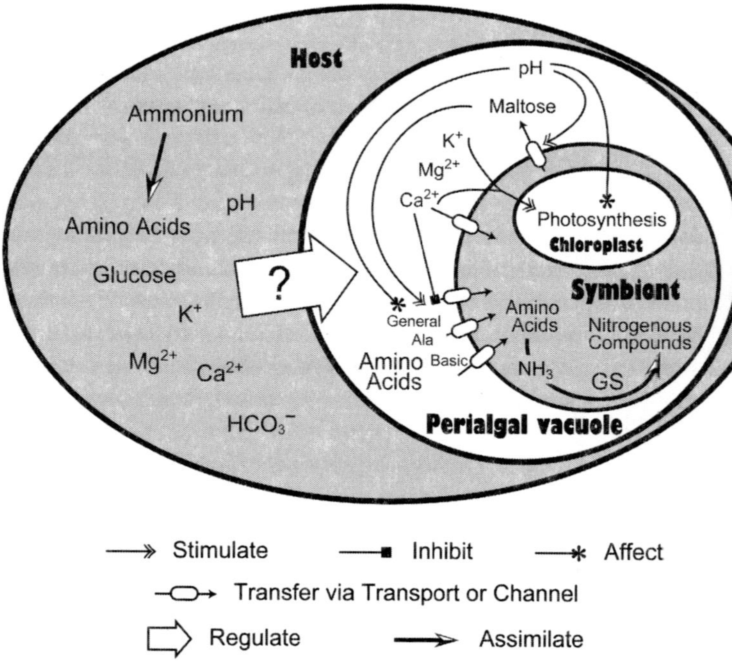


**Fig. 5** Material supposedly transported, and metabolic control mechanisms, between the host and the symbiont

the perialgal vacuole membrane are at early stages and little is known about the dynamics of membrane permeability. Recently, Kodama and Fujishima (2008) found a novel effect caused by the protein synthesis inhibitor cycloheximide on the perialgal vacuole. They speculated that a protein produced and released by the symbiont during photosynthesis plays an important role in the transfer of maltose on the perialgal vacuole membrane. Interestingly, proteins permitting the transportation through the perialgal vacuole membrane are thought to be produced both by the host and by the symbiont. Certainly, the surroundings in the perialgal vacuole are also important, e.g., the pH and cation concentrations are important regulatory factors for the physiological features of the symbiont. The Japanese symbiont F36-ZK possesses many unique features that differ from previously reported symbiotic algal characters, and thus seems an interesting symbiotic alga to study further. However, the surroundings of the perialgal vacuole of Japanese *P. bursaria* could be very similar to those of *P. bursaria* worldwide. Regardless of adaptation to the symbiotic milieu, material transport between the host and the symbiont could be almost the same. Details regarding the role of the perialgal vacuole in the symbiotic relationship should be investigated in the future.

**Acknowledgements** We thank the editor, M. Fujishima, and the series editor, A. Steinbüchel, for providing us with an opportunity to write on this theme. Grateful acknowledgement is made to the following sources for permission to reproduce material in this chapter: Kamako and Imamura 2006; Kato et al. 2006; Kato and Imamura 2008b.

## References

Albers D, Reisser W, Wiessner W (1982) Studies on the nitrogen supply of endosymbiotic *Chlorellae* in green *Paramecium bursaria*. Plant Sci Lett 25:85–90

BD Technical Center (2003) Typical analysis – BactoTM casamino acids. http://www.bd.com/ds/technicalCenter/typicalAnalysis/typ-casamino_acids.pdf

Beardall J (1981) $CO_2$ accumulation by *Chlorella saccharophila* (Chlorophyceae) at low external pH: evidence for active transport of inorganic carbon at the chloroplast envelope. J Phycol 17:371–375

Beardall J, Raven JA (1981) Transport of inorganic carbon and the '$CO_2$ concentrating mechanism' in *Chlorella emersonii* (Chlorophyceae). J Phycol 17:134–141

Berkowitz GA, Wu W (1993) Magnesium, potassium flux and photosynthesis. Magnes Res 6:257–265

Brown JA, Nielsen PJ (1974) Transfer of photosynthetically produced carbohydrate from endosymbiotic *Chlorella* to *Paramecium bursaria*. J Protozool 21:569–570

Bush DR (1993) Proton-coupled sugar and amino acid transporters in plants. Annu Rev Plant Physiol Plant Mol Biol 44:513–542

Cameron LE, Lejohn HB (1972) On the involvement of calcium in amino acid transport and growth of the fungus *Achlya*. J Biol Chem 247:4729–4739

Camoni L, Marra M, Garufi A, Visconti S, Aducci P (2006) The maize root plasma membrane $H^+$-ATPase is regulated by a sugar-induced transduction pathway. Plant Cell Physiol 47:743–747

Cho BH, Komor E (1983) Mechanism of proline uptake by *Chlorella vulgaris*. Biochim Biophys Acta 735:361–366

Cho BH, Komor E (1985) The amino acid transport systems of the autotrophically grown green alga *Chlorella*. Biochim Biophys Acta 821:384–392

Cho BH, Sauer N, Komor E, Tanner W (1981) Glucose induces two amino acid transport systems in *Chlorella*. Pro Natl Acad Sci U S A 78:3591–3594

Cook CB (1983) Metabolic interchange in algae-invertebrate symbioses. Int Rev Cytol (Suppl 14):177–210

Gates RD, Hoegh-Guldberg O, McFall-Ngai MJ, Bil' KY, Muscatine L (1995) Free amino acids exhibit anthozoan "host factor" activity: they induce the release of photosynthate from symbiotic dinoflagellates *in vitro*. Proc Natl. Acad Sci U S A 92:7430–7434

Gehl KA, Colman B, Sposato LM (1990) Mecanism of inorganic carbon uptake in *Chlorella soccharophila*: the lake of involvement of carbonic anhydrase. J Exp Bot 41:1385–1391

Grant AJ, Rémond M, People J, Hinde R (1997) Effects of host-tissue homogenate of the scleractinian coral *Plesiastrea versipora* on glycerol metabolism in isolated symbiotic dinoflagellates. Mar Biol 128:665–670

Grant AJ, Rémond M, Hinde R (1998) Low molecular-weight factor from *Plesiastrea versipora* (Scleractinia) that modifies release and glycerol metabolism of isolated symbiotic algae. Mar Biol 130:553–557

Grant AJ, Rémond M, Withers KJT, Hinde R (2001) Inhibition of algal photosynthesis by a symbiotic coral. Hydrobiologia 461:63–69

Grenson M, Hou C, Crabeel M (1970) Multiplicity of the amino acid permeases in *Saccharomyces cerevisiae*. IV. Evidence for a general amino acid permease. J Bacteriol 103:770–777

*Harrington* HM, Berry SL, Henke RR (1981) Amino acid transport into cultured tobacco cells. II. Effect of calcium. Plant Physiol 67:379–384

Hinde R (1988) Factors produced by symbiotic marine invertebrates which affect translocation between the symbionts. In: Scannerini S, Smith D, Bonfante-Fasolo P, Gianinazzi-Pearson V (eds) Cell to cell signals in plant, animal and microbial symbiosis. Springer, Berlin, pp 311–324

Ichimura T (1971) Sexual cell division and conjugation-papilla formation in sexual reproduction of *Closterium strigosum*. In: Proceedings of the Seventh International Seaweed Symposium. University of Tokyo Press, Tokyo, pp 208–214

Ishikura M, Adachi K, Maruyama T (1999) Zooxanthellae release glucose in the tissue of a giant clam, *Tridacna crocea*. Mar Biol 133:665–673

Ishijima S, Ohnishi M (2002) Regulation of enzyme activities by free $Mg^{2+}$ concentration. Regulation of stromal fructose-1,6-bisphosphatase and ribulose 1,5-bisphosphate carboxylase activities. J Appl Glycosci 49:199–203

Kamako SI, Imamura N (2006) Effect of Japanese *Paramecium bursaria* extract on photosynthetic carbon fixation of symbiotic algae. J Eukaryot Microbiol 5:136–141

Kamako SI, Hoshina R, Ueno S, Imamura N (2005) Establishment of axenic endosymbiotic strains of Japanese *Paramecium bursaria* and their utilization of carbohydrate and nitrogen compounds. Eur J Protistol 41:193–202

Kato Y, Imamura N (2008a) Effect of calcium ion on uptake of amino acids by symbiotic *Chlorella* F36-ZK isolated from Japanese *Paramecium bursaria*. Plant Sci 174:88–96

Kato Y, Imamura N (2008b) Effect of sugars on amino acid transport by symbiotic *Chlorella*. Plant Physiol Biochem 46:911–917

Kato Y, Imamura N (2009) Amino acid transport systems of Japanese *Paramecium* symbiont F36-ZK. Symbiosis

Kato Y, Ueno S, Imamura N (2006) Studies on the nitrogen utilization of endosymbiotic algae isolated from Japanese *Paramecium bursaria*. Plant Sci 170:481–486

Kessler E, Huss VAR (1990) Biochemical Taxonomy of symbiotic *Chlorella* strains from *Paramecium* and *Acanthocystis*. Bot Acta 103:140–142

Kirk DL, Kirk MM (1978) Carrier-mediated uptake of arginine and urea by *Chlamydomonas reinhardtii*. Plant Physiol 61:556–560

Kodama Y, Fujishima M (2008) Cycloheximide induced synchronous swelling of perialgal vacuoles enclosing symbiotic *Chlorella vulgaris* and digestion of the algae in the ciliate *Paramecium bursaria*. Protist 159:483–494

Lew RR (1989) Calcium activates an electrogenic proton pump in *Neurospora* plasma membrane. Plant Physiol 91:213–216

Loefer JB (1936) Bacteria-free culture of *Paramecium bursaria* and concentration of the medium as a factor in growth. J Exp Zool 72:387–407

Masuda M, Miyachi S, Maruyama T (1994) Sensitivity of zooxanthellae and non-symbiotic microalgae to stimulation of photosynthate excretion by giant clam tissue homogenate. Mar Biol 118:687–693

McAuley PJ (1986) Uptake of amino-acids by cultured and freshly isolated symbiotic *Chlorella*. New Phytol 104:415–428

McAuley PJ (1989) The effect of arginine on rates of internalization of other amino acids by symbiotic *Chlorella* cells. New Phytol 112:553–559

McAuley JP, Dorling M, Hodge H (1996) Effect of maltose release on uptake and assimilation of ammonium by symbiotic *Chlorella* (Chlorophyta). J Phycol 32:839–846

Meier R, Lefort-Tran M, Pouphile M, Reisser W, Wiessner W (1984) Comparative freeze-fracture study of perialgal and digestive vacuoles in *Paramecium bursaria*. J Cell Sci 71:121–140

Muscatine L (1967) Glycerol excretion by symbiotic algae from corals and *Tridacna* and its control by the host. Science 156:516–519

Nakahara M, Handa S, Nakano T, Deguchi H (2003) Culture and pyrenoid structure of a symbiotic *Chlorella* species isolated from *Paramecium bursaria*. Symbiosis 34:203–214

Nichols HW, Bold MC (1965) *Trichosarcina polymorpha gen.* et sp. nov. J Phycol 1:34–38

Nishihara N, Horiike S, Takahashi T, Kosaka T, Shigenaka Y, Hosoya H (1998) Cloning and characterization of endosymbiotic algae isolated from *Paramecium bursaria*. Protoplasma 203:91–99

Ogawa K, Shiraishi N, Mii M, Ida S, Komamine A, Nakagawa H (1994) Isolation and characterization of nitrate reductase – deficient mutants of cultured spinach cells: Biochemical, immunological and mRNA analysis. J Plant Physiol 143:279–285

Pardy RL, Spargo B, Crowe JH (1989) Release of trehalose by symbiotic algae. Symbiosis 7:149–158

Plakunov VK, Seifullina NKh, Voronia NA (1995) Specificity of induction of the "proline" transport system of neutral amino acids in *Chlorella vulgaris*. Microbiol 64:628–631

Reisser W (1976) Die stoffwechselphysiologischen Beziehungen zwischen *Paramecium bursaria* Ehrbg. und *Chlorella* spec. in der *Paramecium bursaria*-Symbiose. II. Symbiose-spezifische Merkmale der Stoffwechselphysiologie und der Cytologie des Symbioseverbandes und ihre Regulation. Arch Microbiol 111:161–170

Reisser W (1984) The taxonomy of green algae endosymbiotic in ciliates and a sponge. Br Phycol J 19:309–318

Reisser W (1986) Endosymbiotic associations of freshwater protozoa and algae. In: Corliss, JO, Patterson DJ (eds) Progress in protistology, vol 1. Biopress, Bristol, pp 195–214

Reisser W (1988) Signals in the Paramecium bursaria – *Chlorella* sp. - association. In: Scannerini S, Smith D, Bonfante-Fasolo P, Gianinazzi-Pearson V (eds) Cell to cell signals in plant, animal and microbial symbiosis. Springer, Berlin, pp 281–296

Reisser W, Widowski M (1992) Taxonomy of eukaryotic algae endosymbiotic in freshwater associations. In: Reisser W (ed) Algae and symbioses. Biopress, Bristol, pp 21–40

Reisser W, Radunz A, Wiessner W (1982) Participation of algal surface structures in the cell recognition process during infection of aposymbiotic *Paramecium bursaria* with symbiotic chlorellae. Cytobios 33:39–50

Reisser W, Vietze S, Widowski M (1988) Taxonomic studies on endocytobiotic chlorophycean algae isolated from different American and European strains of *Paramecium bursaria*. Symbiosis 6:253–270

Reisser W, Burbank DE, Meints RH, Becker B, Van Etten JL (1991) Viruses distinguish symbiotic *Chlorella* spp. of *Paramecium bursaria*. Endocytobiosis Cell Res 7:245–251

Rickauer M, Tanner W (1986) Effect of $Ca^{2+}$ on amino acid transportaccumulation in roots of *Phaseolus vulgaris*. Plant Physiol 82:41–46

Ritchie RJ, Grant AJ, Eltringham K, Hinde R (1997) Clotrimazole, a model compound for the host release factor of the coral *Plesiastrea versipora*. Aust J Plant Physiol 24:283–290

Rolland F, Winderickx J, Thevelein JM (2001) Glucose-sensing mechanisms in eukaryotic cells. Trends Biochem Sci 26:310–317

Sauer N (1984) A general amino-acid permease is inducible in *Chlorella vulgaris*. Planta 161:425–431

Schlichter D, Svoboda A, Kremer BP (1983) Functional autotrophy of *Heteroxenia fuscescens* (Anthozoa: Alcyonaria): carbon assimilation and translocation of photosynthates from symbionts to host. Mar Biol 78:29–38

Schüßler A, Schnept E (1992) Photosynthesis dependent acidification of perialgal vacuoles in the *Paramecium bursaria / Chlorella* symbiosis: visualization by monensin. Protoplasma 166:218–222

Seifullina NKh, Voronia NA, Plakunov VK (1995) The defferent nature of neutral amino acid transport systems induced by glucose and glycine in *Chlorella vulgaris*. Microbiology 64:501–503

Shelp BT, Canvin DT (1985) Inorganic carbon accumulation and photosynthesis by *Chlorella pyrenoidosa*. Can J Bot 63:1249–1254

Smith IK (1978) Role of calcium in serine transport into tobacco cells. Plant Physiol 62:941–948

Streamer M, Griffiths DJ, Luong-Van T (1988) The products of photosynthesis by zooxanthellae (*Symbiodinium microadriacticum*) of *Tridacna gigas* and their transfer to the host. Symbiosis 6:237–252

Sutton DC, Hoegh-Guldberg O (1990) Host-zooxanthellae interactions in four temperate marine invertebrate symbioses: assessment of effect of host extracts on symbionts. Biol Bull 178:175–186

Taiz L, Zeiger E (2002) Plant physiology, 3rd edn. Sinauer, Sunderland pp 117, 186

Takeda H (1995) Cell wall composition and taxonomy of symbiotic *Chlorella* from *Paramecium* and *Acanthocysts*. Phytochemistry 40:457–459

Tanner W (1969) Light-driven active uptake of 3-O-methylglucose via an inducible hexose uptake system in *Chlorella*. Biochem Biophys Res Commun 36:278–283

Trench RK (1971) The physiology and biochemistry of zooxanthrellae symbiotic with marine coelenterates. III. The effect of homogenates of host tissues on the excretion of photosynthetic products in vitro by zooxanthellae from two marine coelenterates. Proc R Soc Lond B 177:251–264

Trench RK (1979) The cell biology of plant-animal symbiosis. Annu Rev Plant Physiol 30:485–531

Trench RK (1993) Microalgal-invertebrate symbioses: a review. Endocytobiosis Cell Res 9:135–175

Van Etten JL, Meints HR, Kuczmarski D, Meints RH (1983) Virus infection for culturable Chlorella-like algae and development of a plaque assay. Science 219:994–996

Whitehead LF, Douglas AE (2003) Metabolite comparisons and the identity of nutrients translocated from symbiotic algae to an animal host. J Exp Biol 206:3149–3157

Willis RC, Iwata KK, Furiong CE (1975) Regulation of glutamine transport in *Escherichia coli*. J Bacteriol 122:1032–1037

Wu W, Berkowitz GA (1992a) $K^+$ stimulation of ATPase activity associated with the chloroplast inner envelope. Plant Physiol 99:553–560

Wu W, Berkowitz GA (1992b) Stromal pH and photosynthesis are affected by electroneutral $K^+$ and $H^+$ exchange through chloroplast envelope ion channels. Plant Physiol 98:666–672

Yocum CF (1991) Calcium activation of photosynthetic water oxidation. Biochim Biophys Acta 1059:1–15

Young K, Seale RB, Olsson K, Aislabie J, Cook GM (2003) Amino acid transport by *Sphingomonas* sp. strain Ant 17 isolated from oil-contaminated Antarctic soil. Polar Biol 26:560–566

Ziesenisz E, Reisser W, Wiessner W (1981) Evidence of de novo synthesis of maltose excreted by the endosymbiotic *Chlorella* from *Paramecium bursaria*. Planta 153:481–485

# Regulation of Circadian Rhythms of *Paramecium bursaria* by Symbiotic *Chlorella* Species

**Isoji Miwa**

**Contents**



**Abstract** Cells of *Paramecium bursaria* show many kinds of circadian rhythms, including mating reactivity and photoaccumulation. The period length and the phase of circadian rhythms are regulated by symbiotic *Chlorella* sp. under constant light. *Chlorella*-free white cells of an arrhythmic mutant are rescued from the aberrant mating rhythm by reinfecting them with *Chlorella* sp. isolated from wild-type *Chlorella*-containing green cells. Photosynthetic products of symbiotic *Chlorella* sp. are also effective for rescuing arrhythmic mutant white cells. Green cells show a higher tolerance to high-temperature stress and chemicals than white cells.

I. Miwa (✉)
Department of Biology, Faculty of Science, Ibaraki University, Mito, Ibaraki, 310-8512, Japan
e-mail: miwa@mx.ibaraki.ac.jp

M. Fujishima (ed.), *Endosymbionts in Paramecium*, Microbiology Monographs 12,
DOI: 10.1007/978-3-540-92677-1_4, 

The immaturity length of progeny from green cells is shorter than that from white cells. The functional symbiosis of *Chlorella* sp. in *P. bursaria* serves as a good model for the biological coevolution.

## 1 Introduction

Almost all levels of organisms living on the earth, from prokaryotes to humans and plants, exhibit circadian rhythms as a basic adaptive behavior to the earth's daily fluctuation in light and temperature (Johnson and Hastings 1986; Kondo et al. 1993). Circadian rhythms are controlled by one or more endogenous oscillators, often called the "circadian clock." Although almost all mechanisms of the circadian clock have recently been identified by molecular genetic studies, the essential nature of these mechanisms has not yet been elucidated (Dunlap et al. 2004). The first genetic experiment concerning circadian rhythms was performed in *Drosophila melanogaster* (Konopka and Benzer 1971). The researchers identified the *per* gene, which determines the length of the period in a circadian rhythm. The cloning of the *per* gene was performed 13 years after it was first found (Zehring et al. 1984). Subsequently, many kinds of genes related to circadian rhythms were discovered in *Drosophila* (Price et al. 1998) and also in mammals (King et al. 1997; Reppert and Weaver 2002). The basic molecular mechanism of the circadian rhythm consists of negative feedback in these clock gene products. However, it was reported recently that transcription-translation feedback is not necessary for the circadian rhythm of KaiC phosphorylation in *Cyanobacteria* (Tomita et al. 2005), which has been reconstituted in vitro (Nakajima et al. 2005). The whole picture of molecular mechanisms in a circadian rhythm should be clarified in the near future.

The unicellular ciliate *Paramecium bursaria*, as normally found in nature, contains several hundred *Chlorella* sp. (*Chlorella*) established in the cytoplasm as endosymbionts (Loefer 1936). *Chlorella*-free white cells can be easily derived from natural *Chlorella*-containing green cells by rapid growth in the dark. White cells can be restored quickly to green ones by reinfection with *Chlorella* isolated from green cells (Meier and Wiessner 1988). The detailed process by which reinfection of *Chlorella* establishes symbiosis is described by Kodama and Fujishima in this volume.

*P. bursaria* is an interesting model of a coexisting plant cell in a single animal cell. Symbiotic *Chlorella* release their photosynthetic products, 90% of which are maltose and oxygen, and the host cells use them for energy. Free-living *Chlorella* do not release their photosynthetic sugar products, and therefore maltose release could be one of the essential factors in establishing symbiosis (Weis 1979).

*P. bursaria* has many circadian rhythms (Johnson et al. 1989; Miwa et al. 1987), and symbiotic *Chlorella* also has a circadian photosynthetic rhythm (Butko 1988). They each have separated their own clocks. Since there are two different clocks operating in a *P. bursaria* cell, the means by which the correlation of two clocks occurs is an interesting problem. It is important to consider this issue from the point of view of cellular coevolution.

## 2 Features of Circadian Rhythms

The salient characters of circadian rhythm are as follows: (1) persistence under constant conditions, such as constant light (LL) or constant darkness (DD) (the free-running period is about 24 h); (2) temperature compensation (the free-running period is not altered with every 10°C change in temperature); and (3) resetting the phase of the rhythm by pulses of light or darkness. The circadian clock system can be considered to consist of (1) the input of external signals such as illumination, (2) the "pacemaker" that performs autonomous oscillations with a circadian period, and (3) the output system, the physiological activities of which oscillate within a circadian period driven by the pacemaker (Edmunds 1988). Though these three systems are found in different organs in multicellular organisms, all circadian systems could be included basically in one cell, since circadian rhythms have been found in populations of unicellular organisms (Johnson and Hastings 1986; Mergenhagen 1980; Schweiger and Hartwig 1986). The pacemaker, however, exists in every individual cell even in multicellular organisms (Plautz et al. 1997).

The first description of the circadian rhythm in *Paramecium* was Sonneborn's observation that cells of *P. triaurelia* were able to mate only at certain times of the day (Sonneborn 1938). Since then, the rhythms of mating reactivity in *P. bursaria* have been reported by Jennings (1939), Chen (1946), and Ehret (1953). A reversal of the mating-type rhythm in which mating types III and IV were changed during the day has been described in *P. multimicronucleatum*, syngen 2 (Sonneborn 1957). These rhythms are determined by one dominant gene and continue in DD for several days (Barnett 1966). Through extensive studies of circadian mating reactivity rhythms in *P. bursaria*, Ehret proposed the "chronon model" for the circadian clock mechanism based upon hundreds of genes whose purpose is time-keeping (Ehret and Truco 1967). The "chronon" is defined as the sum of these genes. The first gene is transcribed, the messenger RNA translated, and the protein product helps initiate transcription of the next element in the chronon, with the last element initiating the first one to restart the cycle. This is one of the first and most widely known models.

After that, the study of circadian rhythms in *Paramecium* was avoided for about 20 years until Japanese groups succeeded in measuring the circadian rhythms of locomotor activity (Hasegawa et al. 1984), mating reactivity (Miwa et al. 1987), and resting membrane potential (Nakajima and Nakaoka 1989). The circadian locomotor behavior in a population of *P. multimicronucleatum* was first measured automatically with a computerized video system. Individual cells swam fast and unidirectionally during the day, and they swam slowly and frequently turned at night. This oscillation pattern was sustained in DD (Hasegawa and Tanakadate 1984). Although molecular studies of circadian rhythms are not adequate in *Paramecium*, so far we have been able to record many kinds of circadian rhythms in *P. bursaria*, including mating reactivity, photoaccumulation, locomotor activity, negative gravitaxis, and cell division. The two rhythms of mating reactivity and photoaccumulation are described in detail in this chapter.

## 2.1 Mating Reactivity Rhythm

Sexual interaction in *Paramecium* begins with the coming together of cells, which occurs immediately upon mixing cells of complementary mating types. This phenomenon is called the "mating reaction." Many species of *Paramecium* exhibit mating reactivity continuously when they are in the stationary phase and are sexually mature (Hiwatashi 1981). However, mature cells of *P. bursaria* show a definite circadian rhythm of mating reactivity in the stationary phase (Cohen 1965; Miwa et al. 1989). The mating reactivity of single white cells can be clearly identified, since white cells are easily distinguished from green tester cells in mating clumps.

To assay the mating reactivity rhythm of single white cells, we tested each cell every 3 h by mixing it with about 100 highly reactive green tester cells. After every test, the white cell was reisolated from the green tester cells and incubated separately in K-DS solution (0.6 mM $KH_2PO_4$, 1.4 mM $Na_2HPO_4$, 2 mM $Na_3C_6H_5O_7$, 1.5 mM $CaCl_2$, pH 7.0) until the next test. Mating reactivity in cell populations was measured every 3 h as follows. Ten white cells were placed in each of six different wells of a depression glass plate, and about 100 highly reactive green tester cells were added to each well. After 5 min, the percentage of mating reactive white cells clumping with green tester cells was assayed. When the mating reactivity rhythm of green cells was assayed, white cells were used for tester cells. To prepare tester cells with high reactivity, four groups of green cells were entrained to four light-dark cycles (LD 12:12 h, staggered by 6 h). Each group of tester cells was used twice in consecutive 3-h testing intervals, such that the tester cells were always similarly highly reactive (Miwa et al. 1987).

Cell populations of *P. bursaria* show mating reactivity in the light period, but not in the dark period, when exposed to a LD 12:12 h cycle (12 h of light followed by 12 h of darkness). After they are transferred to LL (0.5 mW $cm^{-2}$), cells continue to show a clear circadian rhythm of mating reactivity. The rhythm, however, gradually dampens in LL such that mating reactivity in populations becomes arrhythmic in LL within 2 weeks. We wanted to know whether the arrhythmicity of the population was due to the absence of circadian rhythmicity within each individual cell, or merely due to asynchrony of a population of individually rhythmic cells. Therefore, single cells were isolated randomly from an arrhythmic population that had been exposed to LL for a long time. Then these single cells were tested individually for mating reactivity every 3 h for 2 days. Each single cell showed a circadian mating rhythm in LL. This shows that the disappearance of the mating rhythm in cell populations under LL is not caused by the disappearance of circadian rhythmicity within individual cells, but is due to desynchronization among cells in a population. When an arrhythmic population in LL is darkened for 9 h, the mating reactivity rhythm of the cell population reappears. This occurs by resynchronization of the rhythms among individual cells, as can be shown by exposing single cells to pulses of 9 h of darkness (Miwa et al. 1987). When the arrhythmic population in LL is exposed to dark pulses of various durations, the first peak of the recovered mating reactivity rhythm appears 6 h after the end of the dark pulse. Thus, in the case of dark pulses to cells in LL, the transition from dark to light sets the phase of the subsequent mating reactivity rhythm. When an arrhythmic population in LL is transferred to DD, a rhythm of mating reactivity also appears, and in this case the first peak of the rhythm occurs 18 h after the LL to DD transition. Therefore, arrhythmic populations in LL can be synchronized by either a dark pulse or transition to continuous darkness. The synchronization of mating reactivity rhythms in arrhythmic populations can be induced by both lights-on ("dawn") and lights-off ("dusk") signals. However, if cells are exposed to both lights-on and lights-off signals, the subsequent phase is determined by the lights-on signal (Miwa et al. 1989).

Cells of an arrhythmic population cultured in LL were treated with the protein synthesis inhibitor puromycin (400 μg $ml^{-1}$) for 18, 12, or 6 h. The cells completely lost mating reactivity 6 h after the beginning of treatment. After the puromycin had been removed by washing with K-DS solution, the cell population showed mating reactivity rhythms in which the peaks of the rhythms appeared at the same time. This suggested that the phases of individual cells of an arrhythmic population were synchronized by puromycin treatment. If puromycin inhibited only the expression of mating reactivity, the cells should show arrhythmic mating reactivity after treatment. Since the phases of individual cells were shifted by puromycin, this suggested that protein synthesis is concerned with the biological clock systems of *P. bursaria* (Miwa et al. 1989).

The period of circadian rhythm is inexact, and it is generally changeable at certain limits by exposing the organisms to LD cycles whose period deviates from 24 h. Such adaptability to new environmental cycles is called "entrainment" (Aschoff 1960). The range of entrainment of circadian rhythms to various periods of LD cycles has been examined in many organisms (Aschoff 1981). Imafuku (1975) reported that the mating reactivity rhythms in *P. bursaria* are adaptable to a reversal of the LD cycle and to LD cycles of various periods. His results indicate that the rhythm is reentrained within 2–3 days to a LD cycle which has been reversed (180°) and that the range of entrainment to various LD cycles is between 12 and 36 h. Though our methods are different from those of Imafuku, we have shown that when arrhythmic populations in LL are transferred to various LD cycles, the mating reactivity rhythms are entrained to a LD cycle of 18–30 h in duration (Miwa et al 1989).

## 2.2 Photoaccumulation Rhythm

Cells of *P. bursaria* show phototaxis by accumulating in a lighted area. One mechanism of photoaccumulation is an avoidance response when coming to a shaded area (Saji and Oosawa 1974). Another mechanism is low swimming velocity: cells which accumulate in a light spot decrease their swimming velocity or stay in the same place. As the light intensity increases from full darkness, swimming velocity decreases and reaches zero at the intensity at which cells show the most noticeable photoaccumulation. The degree of photoaccumulation is therefore inversely proportional to swimming velocity (Matsuoka and Nakaoka 1988).

Matsuoka and Nakaoka recorded the membrane potential of both green and white cells after photostimulation. They reported that a stepwise increase in light intensity induces steady depolarization and causes a decrease in swimming velocity, resulting in accumulation of cells in the lighted region. A stepwise decrease in light intensity induces hyperpolarization of the membrane potential, and induces an avoidance reaction in the swimming cells (Matsuoka and Nakaoka 1988). They also reported that the resting membrane potential of *P. bursaria* oscillates during the daily cycle (Nakajima and Nakaoka 1989).

We examined the rhythmicity of photoaccumulation in the green and white cells of *P. bursaria*. Assaying circadian photoaccumulation in *P. bursaria* is based on the same kind of assay as used by Bruce and Pittendrigh (1956) for *Euglena*. Aliquots (10 ml) of *Paramecium* culture were placed in 30-ml disposable tissue culture flasks (Costar 3025). The flasks were placed in a vertical position so that a focused test light beam was directed through the cell suspension. The beam was placed in the middle to upper third of the culture. The photoaccumulation rate was tested at hourly intervals as follows. At the beginning of each assay, the test light bulb (General Electric 75X filtered through an infrared reflecting filter) was turned on, and the beam was passed through a 3-mm aperture into the culture (in the case of LL experiments, the LL background illumination was turned off during the photoaccumulation assay). A phototransistor (Hamamatsu S2041) positioned on the opposite side of the culture monitored the transmittance of the test light.

Photoaccumulation of *Paramecium* cells causes a decrease in the transmittance of the test light beam through the culture. This decrease was measured by the phototransistor, and the analog signal was converted to a digital signal by an analog-to-digital converter (Keithley series 500) controlled by a microcomputer. Transmittance was measured when the test light was first turned on ("initial" transmittance) and again just before the test light was turned off ("final" transmittance). The test light remained on between these measurements for 12 min. The difference between the initial transmittance and the final transmittance was taken as the photoaccumulation for that assay, which was stored on a microcomputer disk for later retrieval and data analysis (Fig. 1b). The test was repeated at hourly intervals (Johnson et al. 1989).

Cells of *P. bursaria* in the light period accumulate more in lighted areas than cells in the dark period in the LD cycle (Fig. 1a). *Chlorella*-containing green cells exhibit excellent circadian rhythms in either LL or DD for several days. *Chlorella*-free white cells also express a clear rhythm of photoaccumulation, albeit at lower amplitude than that of green cells. Therefore, the rhythm of photoaccumulation is intrinsic to the *Paramecium* cells, and is not an artifact associated with the algal symbionts. Moreover, sexually immature cells also show a robust circadian rhythm of photoaccumulation. This is significant, because even though these immature cells cannot express the rhythm of mating reactivity, they contain an operative circadian pacemaker, as expressed though the photoaccumulation rhythm. Green and white cells in various strains of *P. bursaria* also show good rhythms. On the other hand, certain other species of *Paramecium* do not exhibit the circadian rhythm of photoaccumulation under conditions that allow rhythmic accumulation in *P. bursaria*. We have tested photoaccumulation in the following species and found

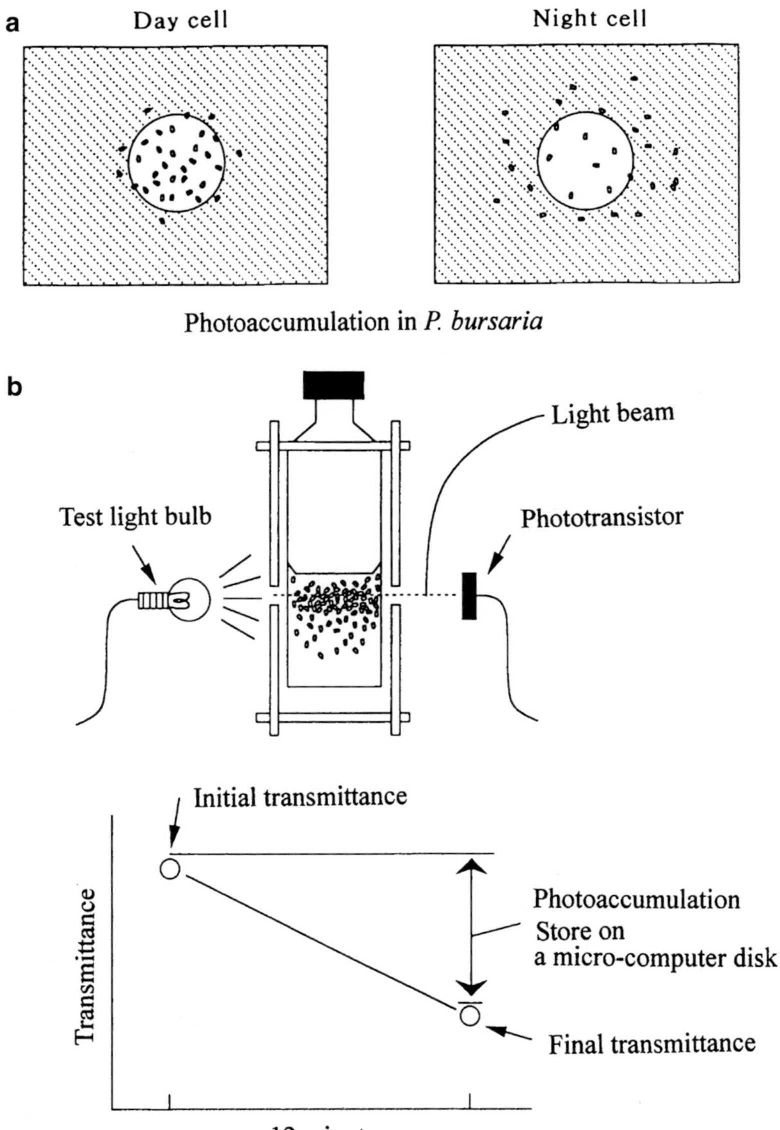


**Fig. 1** Photoaccumulation in *Paramecium bursaria*. **a** Cells of *P. bursaria* in the light period (day cell) accumulate more in lighted areas than cells in the dark period (night cell) in the light-dark (LD) cycle. **b** Assay of circadian photoaccumulation rhythm. Photoaccumulation of cells causes a decrease in the transmittance of the test light beam through the culture. The difference between the initial transmittance and the final transmittance for 12 min is taken as the photoaccumulation, which is stored on a microcomputer disk. The test is repeated at hourly intervals

it to be nonrhythmic: *P. tetraurelia*, *P. caudatum*, and *P. multimicronucleatum*. In addition, another protozoan, *Blepharisma japonicum*, has been tested; it is also nonrhythmic under these conditions (Johnson et al. 1989).

In some neural models of circadian rhythmicity, membrane potential and transmembrane flux of potassium and calcium ions appear to play an important role in the entrainment and central mechanisms of the biological clock (Block et al. 1993). What is the correlation between the magnitude of membrane potential change and the magnitude of phase resetting of the photoaccumulation rhythm in *P. bursaria*? The membrane potential response to changes in extracellular potassium levels is well documented in *P. caudatum* (Naitoh and Eckert 1968; Oka et al. 1986; Machemer 1989) and in *P. bursaria* (Johnson et al. 1994). And the magnitude of light-induced membrane depolarization is dependent upon the extracellular concentration of calcium in *P. bursaria* (Nakaoka et al. 1987). We found that pulses or step changes in extracellular potassium concentrations did not phase-shift the circadian clock of *P. bursaria*. Furthermore, modifying the extracellular concentration of calcium did not affect the magnitude of light-induced phase resetting. Therefore, while membrane potential and calcium fluxes may be crucial components of the circadian clock system in some organisms, especially in neural systems that involve intercellular communication, the *P. bursaria* data indicate that membrane potential changes are not necessarily an intrinsic component of circadian organization at the cellular level (Johnson et al. 1994).

## 2.3 Phase Shift by Light Pulses

When white cells of *P. bursaria* that have been entrained to LD 12:12 h are put into DD, they show a persistent free-running rhythm of mating reactivity. Cells were exposed to light pulses (0.5 mW cm$^{-2}$) of various durations 27 h after the onset of DD, when the cells showed no mating reactivity. After the light pulses, the cells were transferred to DD again and the mating reactivity of the cells was examined by mixing them with green tester cells at 30-min intervals. No cells showed mating reactivity during the first hour after the light pulse, but mating ability began to appear after 1.5 h. The extent of mating reactivity 3 h after the light pulse was correlated with the duration of the light pulse. Next, the duration of the light pulse that induces a shift in phase of the mating reactivity rhythm was examined. As in the experiment described above, the cells were exposed to light pulses of various durations (5 s to 6 h, 0.5 mW cm$^{-2}$) 27 h after the onset of DD. After exposure, the cells were transferred to DD and mating reactivity was tested at 3-h intervals for 2 days. Phase shifts did not occur in cells exposed to light for less than 1 h, but a large phase shift occurred after a light pulse of 2–3 h. Maximum phase shifts occurred after a 6-h light pulse as well as upon transfer to LL. Light pulses of less than 1 h induced mating reactivity only transiently; the clock was not influenced by such stimulation. Even though phase shifts can be induced by very short pulses of light in some organisms, 2 h or more is necessary for the induction of a phase shift in the mating-reactivity rhythm in *P. bursaria* (Miwa and Wada 1995).

To determine the wavelengths of light that are most effective in inducing mating reactivity, cells placed in ten quartz-glass cuvettes (4.5 ml) were irradiated with ten different wavelengths of monochromatic light (12.5-nm half-bandwidth, 10 mW $cm^{-2}$) for 5 s by a spectra irradiator (CRM-FA; Nihonbunkoh, Japan) at the same phase as in the experiment described above. The cells showed maximum mating reactivity when they had been exposed to light pulses at 416, 547, and 626 nm, with less reactivity after exposure to light at other wavelengths. To date, a few action spectra have been reported for *P. bursaria*. Pado (1972) reported that the peak of the action spectrum for photoaccumulation in *Chlorella*-containing green cells is around 420 nm, and Iwatsuki and Naitoh (1988) reported that the peak wavelength of action spectra differs between two photophobic responses (560 and 680 nm for step-up, 520 nm for step-down) in *Chlorella*-free white cells. Moreover, Matsuoka and Nakaoka (1988) reported that the action spectrum of potential change (the peak wavelength at which cells are most sensitive to changes in light intensity) has peaks at 420 nm and 560 nm for both green and white cells. These results indicate that cells of *P. bursaria* are sensitive to light at three wavelengths (around 420, 560, and 680 nm). In our experiments, we found three similar peaks in the action spectrum for induction of mating ability (Miwa and Wada 1995). It is difficult to identify the photoreceptor pigment from our action spectrum. However, the photosensitive site in *P. bursaria* is located on the anterioventral surface, particularly within the oral groove of the cell (Nakaoka 1989). Retinal can be extracted from the cells and an antibody against rhodopsin binds to the cell surface of *P. bursaria*, so therefore it seems likely that the photoreceptor of *P. bursaria* is a rhodopsin-like protein (Nakaoka et al. 1991).

It seems to take about 1.5 h for transduction of the light signal from the photoreceptor on the cell surface to the expression of mating reactivity. This raises the question of what second-messenger systems are involved in intracellular transmission of light stimulation in *Paramecium*. Inositol phosphate is related to an increase in cytosolic levels of free $Ca^{2+}$ ions, which triggers a variety of physiological events (Berridge and Irvine 1989). Some inositol phosphates have been found in *Paramecium* cells (Freund et al. 1992). To investigate the intracellular light-transduction pathway, we examined whether inositol triphosphate ($IP_3$) could induce mating activity in DD. Ten picoliters of a solution of $IP_3$ (120 μM) was injected into nine cells in the nonreactive phase of DD under dim green light (480 nm, 8 μW $cm^{-2}$), which did not induce mating ability. As a control, the same volume of water was injected into another nine nonreactive cells under the same conditions. In addition, another nine single cells were exposed to a 3-h light pulse (0.5 mW $cm^{-2}$). The nine single cells subjected to each treatment were tested individually for mating reactivity under nonsensitive light (480 nm, 8 μW $cm^{-2}$) at 30-min intervals for 3 h. The cells into which $IP_3$ had been injected began to show mating ability 30 min after injection, and the extent of the activity was greater than that obtained by exposing cells to a 3-h light pulse. None of the nine single cells into which water had been injected showed any mating reactivity after 3 h. This result suggests that cells of *P. bursaria* might use $IP_3$ as a second messenger for intracellular transmission. However, the injection of $IP_3$ did not cause a shift in the phase of the mating reactivity rhythm. It is possible that the injection of $IP_3$ did not induce a phase shift because too little $IP_3$ was injected. Alternatively, the pathway that leads to a phase shift might be different from that which leads to mating reactivity (Miwa and Wada 1995).

# 3 Effects of Symbiotic *Chlorella* sp. on Circadian Rhythms

Circadian systems in higher organisms may involve several oscillators in a hierarchical combination of master and slave oscillators, or as coupled circadian oscillators, each of which derives different rhythms (Pittendrigh 1981). Unicellular organisms such as *Gonyaulax poryedra* also have two oscillators (clocks) in one cell (Roenneberg and Morse 1993). The complication of multiple circadian systems is therefore not a feature restricted to multicellular organisms. Cells of *P. bursaria* have a clock to control the many kinds of circadian rhythms in reproduction and physiological activities. Symbiotic *Chlorella* also has a clock to control the circadian photosynthetic rhythm. So which clock is dominant in the green cells of *P. bursaria*?

It is useful to measure the contributions of symbiotic *Chlorella* by comparing circadian rhythms between green and white cells of the same strain. Furthermore, since the white cells can be reinfected easily with isolated *Chlorella*, an informative experiment is to reinfect white cells with *Chlorella* during a different phase of the circadian rhythm. If the host cells exhibit rhythms in the same phase as that of *Chlorella*, *Chlorella* might work to regulate the rhythms of host cells. Thus, the relationship between the separate clocks of *P. bursaria* and symbiotic *Chlorella* makes for an interesting system to determine the role of symbiosis on the behavior of a host.

## 3.1 Period Length

The circadian photoaccumulation rhythm in *P. bursaria* was measured by a micro-computer-assisted data collection apparatus, as described earlier. The trough or peak of a rhythm was used as the phase reference point. The time of the phase reference point was calculated by a computer program created by Kondo et al. (1991), which fitted a parabola to the data of each peak and trough. To calculate the period, the time of a phase reference point was subjected to a least-squares regression. By this method, we can constantly obtain the exact period length of the photoaccumulation circadian rhythm. The period length of circadian rhythms is inherent in each stock.

The photoaccumulation rhythms of green and white cells in three natural stocks (Sj2, Kz1, and T316) were measured in LL (0.5 mW cm$^{-2}$) and DD. The period lengths of each stock are shown in Table 1. Green cells exhibited about a 3 h longer period than white cells in LL. There is statistical significance between green and white cells. On the other hand, no statistically significant difference in period length between green and white cells was recognized in DD, and these periods were almost the same length as those of white cells in LL. Stock T316 showed a longer period compared with the other two stocks (Miwa et al. 1996b).

To understand the influence of symbiotic *Chlorella* on the period length, the photoaccumulation rhythm was measured in light-green cells with reduced numbers of *Chlorella*. The light-green cells were obtained by growth in darkness for 2 weeks. Three strains of stock Kz1 containing different numbers of *Chlorella* were assayed for photoaccumulation rhythms in LL. As seen in Table 2, the period length was

**Table 1** The period lengths of photoaccumulation rhythm in constant light (LL) and constant darkness (DD)

| Stock (green cell) | Period (h) | Stock (white cell) | Period (h) |
|---|---|---|---|
| LL | | | |
| Sj2 | 25.2 ± 0.2 | Sj2w | 21.2 ± 0.2 |
| Kz1 | 25.4 ± 0.2 | Kz1w | 22.5 ± 0.1 |
| T316 | 27.3 ± 0.4 | T316w | 24.6 ± 0.4 |
| DD | | | |
| Sj2 | 21.8 ± 0.3 | Sj2w | 21.3 ± 0.2 |
| Kz1 | 23.0 ± 0.4 | Kz1w | 23.5 ± 0.3 |
| T316 | 24.5 ± 0.3 | T316w | 23.8 ± 0.4 |

The standard deviation (SD) was obtained from three or more experiments.

**Table 2** Elongation of the period length of photoaccumulation rhythm in stock Kz1 depending on the number of symbiotic *Chlorella* in the cell in LL

| | | Number of symbiotic *Chlorella*[a] | |
|---|---|---|---|
| Cells | Period (h) | Before assay | After assay |
| Green cells | 25.9 ± 0.2 | 562.4 | 798.6 |
| Light-green cells | 23.9 ± 0.5 | 174.0 | 222.2 |
| White cells | 22.6 ± 0.3 | 0 | 0 |

The SD was obtained from three or more experiments.
[a]Mean of five cells.

**Table 3** Effects of photosynthesis inhibitor (5 μM dichlorophenyl dimethylurea) on the period length of photoaccumulation rhythm in LL and in DD

| Stock (green cell) | Period (h) | Stock (white cell) | Period (h) |
|---|---|---|---|
| LL | | | |
| Sj2 | 22.2 ± 0.8 | Sj2w | 21.8 ± 0.5 |
| Kz1 | 22.5 ± 0.1 | Kz1w | 22.6 ± 0.0 |
| T316 | 24.2 ± 0.4 | T316w | 23.6 ± 0.2 |
| DD | | | |
| Sj2 | 22.9 ± 0.1 | Sj2w | 21.9 ± 0.2 |
| Kz1 | 22.7 ± 0.2 | Kz1w | 22.9 ± 0.3 |
| T316 | 23.9 ± 0.7 | T316w | 23.6 ± 0.7 |

The SD was obtained from three or more experiments.

different depending on the number of *Chlorella* in the cells. *Chlorella* numbers increased during the measurement period for 1 week. To further confirm the contribution of symbiotic *Chlorella*, the photoaccumulation rhythms of three stocks were measured after adding 5 μM dichlorophenyl dimethylurea (DCMU), a photosynthetic inhibitor. The period lengths of green cells in each stock were almost the same as those of white cells in LL, which were also statistically the same as those of green cells and white cells in DD (Table 3).

Thus, photosynthesis of symbiotic *Chlorella* should lengthen the circadian period of host cells. The products of *Chlorella* photosynthesis are maltose and oxygen. Which of these is effective in lengthening the period of photoaccumulation rhythm in the host cell? First, green and white cells of stock T316 were assayed for photoaccumulation rhythm in DD after adding 1 mM maltose. Neither green nor white cells showed a lengthened period. Next, green cells of stock T316 were assayed for photoaccumulation rhythm in DD by adding 10 ml oxygen every 6, 12, and 24 h. Contrary to our expectation, the period length was shortened relative to the volume of oxygen. Thus, a longer period was not seen when photosynthesis products were added to cells in DD. However, since these products were added to the cells from outside rather than being introduced intracellularly, the results are not conclusive (Miwa et al. 1996b).

## 3.2 Phase Shift

White cells can be restored to green ones quickly by reinfecting them with *Chlorella* isolated from green cells. To determine the contribution of symbiotic *Chlorella* on the phase of the photoaccumulation rhythm, white cells entrained to LD 12:12 h cycles were reinfected with *Chlorella* isolated from green cells entrained in the reverse phase or a 90° different phase of the LD cycle. These showed the same phase as the original white cells during the first 2 days, but began to shift depending on the phase of *Chlorella* 3 days after reinfection in LL. When assays were conducted in DD, the phase shift did not occur in reinfected cells. Thus, reinfection with *Chlorella* forced the host to shift the phase in LL, but not in DD (Miwa et al. 1996b).

The same assays as described above were performed for the mating reactivity rhythm of *P. bursaria*. Green cells in a 100-ml culture were concentrated by low-speed centrifugation. One milliliter of cell suspension was put into a micro test tube and then sonicated with an ultrasonicator for 10 s. *Chlorella* cells were collected through a 15-$\mu m^2$-opening nylon mesh and washed with K-DS solution. About $2 \times 10^7$*Chlorella* cells/ml were mixed with about $2 \times 10^3$ host cells/ml for 6 h. White cells were reinfected with *Chlorella* of a reverse phase of the LD cycle. They exhibited a phase shift of mating reactivity rhythm depending on the phase of *Chlorella* 3 days after reinfection. These phase shifts were not seen in DD or with 5 μM DCMU (Fig. 2). When white cells entrained to LD 12:12 h cycles were reinfected with *Chlorella* isolated from green cells entrained to the same LD cycle, the phase of the rhythms that appeared conformed exactly to the ones of original white cells.

Cells of *P. bursaria* exhibit mating reactivity in the warmer (25°C) period, but not in the colder (15°C) period, in the warmer and colder (WC) temperature cycle (WC 12:12 h) in LL and DD. To observe the effect of *Chlorella* entrained to the WC temperature cycle on the circadian rhythms of host cells, the following

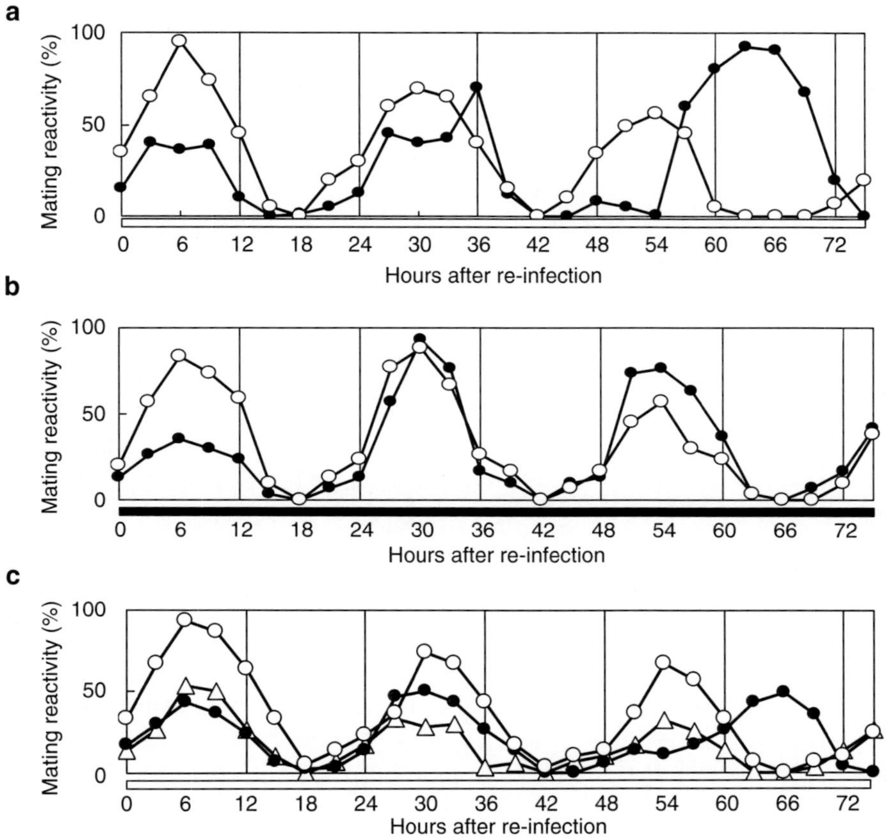


**Fig. 2** Phase shift of mating reactivity rhythms in reinfected cells. White cells entrained in a LD cycle (12:12 h) were reinfected with *Chlorella* isolated from green cells entrained in a reverse LD cycle. They were immediately assayed for mating reactivity rhythms in constant light (LL) (**a**, **c**) and in constant darkness (DD) (**b**). *White circles* represent control white cells, *black circles* represent green cells reinfected with reverse-phase *Chlorella*, and *triangles* represent cells reinfected with reverse-phase *Chlorella*, which were treated with the photosynthesis inhibitor dichlorophenyl dimethylurea (DCMU). They showed the same phase as the original white cells during the first 2 days, but began to shift depending on the phase of *Chlorella* 3 days after reinfection in LL; however, phase shifts were not seen in DD or with DCMU

experiments were performed. White cells entrained to the LD cycle at 25°C were reinfected with *Chlorella* isolated from green cells entrained to a reverse phase of the WC temperature cycle (WC 12:12 h). The reinfected white cells showed a phase shift in mating reactivity rhythms depending on the phase of *Chlorella* 4 days after reinfection in LL. This phase-shift phenomenon did not occur in DD. This indicates that the clock of symbiotic *Chlorella* can be entrained with the WC temperature cycle; however, it is necessary to confirm whether the photosynthetic activity of symbiotic *Chlorella* is entrained with the WC temperature cycle.

## 3.3 Expression of Mating Reactivity Rhythms

The rhythm of mating reactivity has been investigated with many stocks of *P. bursaria* collected in Japan and China. Most stocks exhibit normal mating reactivity rhythms in LD, LL, and DD. We found an interesting stock (Sj2) among them, which exhibited normal rhythmicity in LD and LL, but showed arrhythmicity in DD. This stock exhibited high mating reactivity continuously during the day in DD. *Chlorella*-free white cells (Sj2w) derived from Sj2 also showed rhythmicity in LD and LL, but arrhythmicity in DD. To determine the heredity of the character of arrhythmicity in DD, progenies from a cross between stock Sj2w and wild-type Kz1 were analyzed. Arrhythmic and rhythmic clones segregated close to a 1:1 ratio in the $F_1$ progeny. The $F_2$ progeny from a sibling cross between arrhythmic $F_1$ clones had low survivorship, but arrhythmicity and rhythmicity in DD segregated at a ratio close to 3:1. The results suggest that arrhythmicity in DD is a dominant character and is inherited by progeny in a Mendelian fashion. Next, log-phase cells of Sj2w were treated with 2 μg ml$^{-1}$ nitrosoguanidine for 5 h. After that, cells were tested for mating reactivity using green tester cells of Kz1 after entrainment in LD for 7 days. Five cells that showed high mating reactivity in spite of being in the dark phase in the LD cycle were isolated. Five clones established from the five cells were tested for their mating reactivity again in LD for a few days. Finally, we isolated one clone that continuosly expressed high mating reactivity in LL in addition to DD and weak rhythmicity in LD. We called this clone "mating reactivity continuing mutant" (MC1w). To determine the circadian rhythmicity of each individual cell in MC1w, the mating reactivity rhythms of 15 single cells isolated randomly from a population of MC1w were tested every 3 h in LL. Every single cell showed irregular expression of the mating reactivity rhythm, exhibiting long mating reactive phases and very short or no reactive phases. This feature causes the expression of continuous high mating reactivity in a population of MC1w. We genetically analyzed the characters of MC1w. The mating reactivity of progeny from a cross between white cells of mutant strain MC1w and green cells of wild-type stock Kz1 was examined in LL. The cytoplasmic parents of progeny cells were distinguished clearly by the presence or absence of symbiotic *Chlorella*. White progeny cells were derived from mutant MC1w, and green progeny cells were derived from wild-type Kz1. All white progeny clones showed arrhythmicity in LL, and all green progeny clones showed rhythmicity in LL. Therefore, the character of arrhythmicity in LL is most likely transferred to progeny cytoplasmically (Miwa et al. 1996a).

These white cells (MC1w) were reinfected with *Chlorella* isolated from stock T316 green cells, and the mating reactivity of the reinfected green cells (MCwT) was measured in LD, LL, and DD. MC1w had kept high mating reactivity continuosly under each condition, but green cells (MCwT) produced from MC1w showed rhythmic mating reactivity in LD and LL. However, they did not show rhythmicity in DD and their reactivity was less than 50%. It seems that the mating reactivity rhythms reverted to an arrhythmic state upon reinfection with *Chlorella* in LD and LL (Tanaka and Miwa 1996).

Although cells of MC1w are mutants that express arrhythmic mating reactivity continuously, they show a circadian rhythm of photoaccumulation. Therefore, their circadian system oscillates normally if these two rhythms are operated by the same oscillator. In addition, it is expected that the arrhythmicity of mating reactivity is attributed to a mitochondrial mutation because the progenies of MC1w inherit the trait cytoplasmically (Miwa et al. 1996a). If this is true, it suggests that mitochondria play an important role in the expression of circadian mating reactivity rhythms. Furthermore, symbiotic *Chlorella* might interact with host cells through the mitochondria. Further experiments and consideration of this point of view are needed.

## 4 Effects of Photosynthetic Products

As seen already, symbiotic *Chlorella* forces *Paramecium* cells to lengthen the period and shift the phase of photoaccumulation rhythms in LL. Furthermore, symbiotic *Chlorella* rescues the mating reactivity rhythms of arrhythmic mutant cells in LL. These events occur only in LL, not in DD. This fact suggests that photosynthesis in symbiotic *Chlorella* is correlated with the regulation of circadian rhythms in *P. bursaria*. Symbiotic *Chlorella* cells provide their photosynthetic products, maltose and oxygen, to host cells for energy (Weis 1979). We investigated the effects of the photosynthetic products of symbiotic *Chlorella* on the mating reactivity circadian rhythm. Additionally, green cells exhibit a higher tolerance for chemicals and some forms of stress than white cells in LL. Symbiotic *Chlorella* might contribute to this increase in tolerance. We therefore also investigated the effects of the photosynthetic products on stress.

### 4.1 Rescue of Mutant

Stock Ok2 is an interesting strain collected from nature. *Chlorella*-containing Ok2 cells show normal mating reactivity rhythms in LL and DD. *Chlorella*-free Ok2w cells display a normal mating reactivity rhythm in DD, but they do not show any mating reactivity in LL except for the first cycle of the rhythm. This fact clearly suggests that Ok2w cells have a normal circadian oscillator, and that it ordinarily functions irrespective of the presence of symbiotic *Chlorella*. Constant light may inhibit the output route from a circadian oscillator to the mating reactivity rhythm. On the other hand, Ok2w cells display a normal mating reactivity rhythm in LL after reinfection with *Chlorella* isolated from Sj2. This suggests that reinfection with *Chlorella* could affect the output process in the circadian systems of Ok2w cells, making them suitable material for investigating the output systems from a circadian oscillator.

To determine whether the photosynthetic activity of *Chlorella* contributes to the expression of the mating reactivity rhythm, Ok2 green cells were treated with a photosynthetic inhibitor. In the experiments described above, DCMU was used to inhibit the effects of *Chlorella* on the period length of the photoaccumulation rhythm. However, it was found that DCMU is toxic for the expression of mating reactivity in *P. bursaria*; therefore, paraquat was used as a photosynthetic inhibitor in this experiment. Paraquat forms free radicals in plant cells and destroys chloroplasts, but has no effect on mating reactivity in *P. bursaria*. When Ok2 green cells were treated with 5 μg $ml^{-1}$ paraquat in LL, their mating reactivity rhythms disappeared, as with Ok2w cells. It has been reported that symbiotic *Chlorella* are destroyed and that green cells change into white cells when treated with paraquat for a period of 4–10 days (Hosoya et al. 1995). However, in the present study, Ok2 cells treated with paraquat remained green during the period of measurement of mating reactivity (about 60 h). It is thought that the disappearance of mating reactivity in Ok2 cells is caused by interruption of the supply of photosynthetic products from symbiotic *Chlorella* (Tanaka and Miwa 2000).

To clarify that symbiotic *Chlorella* cells release photosynthetic products in the cytoplasm of Ok2 cells, sugar components in the cytosol were analyzed by high-precision liquid chromatography. For the analysis, the following four kinds of samples were prepared: (1) Ok2 green cells in the light period of the LD cycle (day cells); (2) Ok2 green cells in the dark phase (night cells); (3) Ok2w white cells in the light phase (day cells); and (4) cultured symbiotic *Chlorella* isolated from green cells. Each sample was centrifuged at 14,000 rpm, 4°C for 20 min to separate free *Chlorella* and cell fragments. The supernatant was dried under a vacuum at 40°C, and then the dried powder was dissolved in a small amount of ultrapure water. After centrifugation again at 14,000 rpm, 4°C for 20 min, the extract was separated from nucleic acids, polysaccharides, and free peptides by ultrafiltration. Finally, it was lyophilized and analyzed for sugar components by high-precision liquid chromatography using an aminosilica column. As shown in Fig. 3, glucose, maltose, and an unidentified trisaccharide and tetrasaccharide were detected in the cytosol of Ok2 day cells, while only glucose was detected in Ok2 night cells, Ok2w day cells, and *Chlorella* in the light period of the LD cycle cultured for a long time. The unidentified trisaccharide and tetrasaccharide were strongly suspected to be maltose-type sugars, maltotriose and maltotetraose, respectively. These sugars might be released in Ok2 cells in the light phase.

We investigated the effects of the sugar components of Ok2 cells on the expression of mating reactivity rhythms. Ok2w cells in a 20-ml culture were given 0.5 ml of $10^{-2}$ M sugars every 3 h during 12 h of subjective day phases (0–12, 24–36, and 48–60 h), and their mating reactivity rhythms were assayed in LL. Mating reactivity rhythms of Ok2w cells were induced by the supply of both maltose and maltotetraose in LL. However, the other sugars, glucose and maltotetraose, had no effect on the induction of mating reactivity. Thus, the secretion of maltose and maltotriose from symbiotic *Chlorella* has a close connection with the expression of circadian rhythms in Ok2w cells. These results are possibly due to the recognition of a range of sugar receptors in *P. bursaria*. It is generally thought that the sugar receptor (lectin) on the cell surface recognizes disaccharides and trisaccharides, but that it is difficult for the receptor to recognize monosaccharides and tetrasaccharides. In the

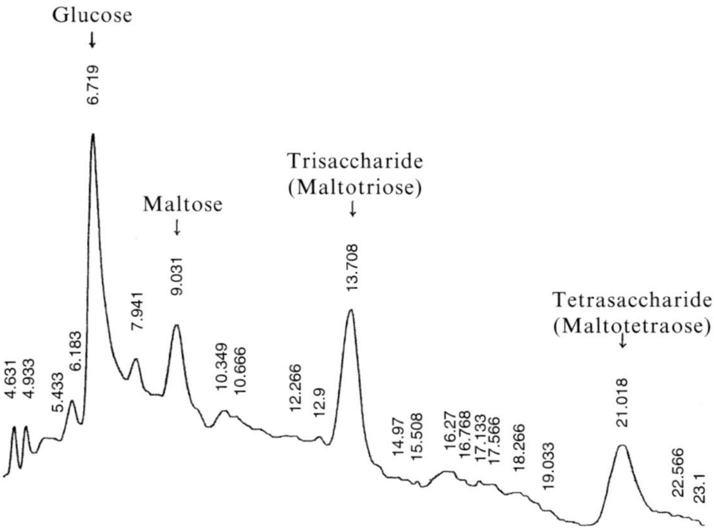


**Fig. 3** High precision liquid chromatography analysis pattern of saccharide in the cytosol of green cells in daytime in *P. bursaria*

present study, since Ok2w cells showed mating reactivity after reinfection of *Chlorella* in LL, it is thought that the cells also have a sugar receptor like lectin in the cytoplasm. It will be necessary to identify this sugar receptor in *P. bursaria* cells (Tanaka and Miwa 2000).

## 4.2 Acquisition of Various Stress Resistances

Symbiotic *Chlorella* might contribute to the increase in tolerance for many chemicals and types of stress because *P. bursaria* cells show different survival rates between green and white cells. Green cells show a higher survival rate (95%) to 0.5 mM nickel chloride ($NiCl_2$) than white cells (10%) 12 h after treatment in LL. This difference in survival rate can be seen only in LL, not in DD. These facts suggest that symbiotic *Chlorella* might contribute to the host's tolerance to environmental fluctuations.

Stationary-phase green (OS1) and white (OS1w) cells in the light phase of the LD cycle were treated with 150 μM hydrogen peroxide ($H_2O_2$), and 60 cells were isolated into each well of a microplate. After 12 h in LL, the percentage of surviving cells was counted under a binocular microscope. We judged the death of a cell on the basis of motionlessness and cell rupture. Green cells showed a higher survival rate (90%) than white cells (40%) 12 h after treatment with $H_2O_2$ in LL. Both green and white cells in the dark phase of the LD cycle showed almost the same survival rate 12 h after treatment with $H_2O_2$ in LL. Interestingly, green cells showed a higher survival rate when cells were kept in DD after treatment with $H_2O_2$ than in LL. However, white cells showed little difference in survival rate between in LL and

DD after treatment. It appears that symbiotic *Chlorella* enhances the tolerance of host cells in LL, and that this may be through the maltose photosynthetic product. However, oxygen might reduce the tolerance to $H_2O_2$ of host cells in LL, because the survival rate of green cells advanced substantially when cells were kept in DD after treatment with $H_2O_2$. Green cells of *P. bursaria* showed the highest tolerance to $H_2O_2$ compared with other species of *Paramecium* in DD, followed by white cells of *P. bursaria, P. tetraurelia, P. multimicronucleatum*, and *P. caudatum*, in that order.

We investigated the survival rate of both green and white cells after irradiation with ultraviolet (UV) light (257 and 282 nm wavelengths) for 1 min. Sixty cells in the light phase of the LD cycle were isolated into a microplate individually under dim red light (8 $\mu W\ cm^{-2}$) after UV irradiation, and were kept in DD to avoid photorecovery. Green cells showed a higher survival rate (83%) than white cells (46%) 24 h after irradiation with 257 nm UV light; survival was also higher for green cells (45%) than for white cells (10%) 24 h after irradiation with 282 nm UV light. However, both green and white cells in the dark phase of the LD cycle showed almost the same survival rate 24 h after UV irradiation (at both 257 and 282 nm). To determine the effects of photosynthetic products on the effects of UV irradiation, white cells were treated with 10 mM sucrose, glucose, maltose, and maltotriose 3 and 24 h before UV exposure. The treatment groups showed a higher survival rate than the controls with maltose and maltotriose, but not with sucrose and glucose. The positive effects of maltose and maltotriose increased depending on the length of treatment. When oxygen was added to a white cell culture 3 h before UV irradiation, a higher survival rate was detected 24 h after irradiation (at both 257 and 282 nm). This effect was dependent on the volume of added oxygen. Other gases such as carbon dioxide and nitrogen were not effective. This indicates that symbiotic *Chlorella* enhances tolerance to UV irradiation through its photosynthetic products.

The influence of symbiotic *Chlorella* can also be observed when cells of *P. bursaria* are kept at high temperature. Iwatsuki et al. (1998) reported that symbiotic *Chlorella* enhances the tolerance to high temperature (41°C) in *P. bursaria*; green cells show a higher survival rate than white cells in LL. The degree of tolerance in DCMU (an inhibitor of photosynthesis) treated green cells and green cells kept in DD for 24 h is as low as in white cells. White cells treated with maltose, glucose, fructose, and oxygen show as high a tolerance to high temperature as that of green cells.

We obtained almost the same results as in Iwatsuki's report. Stationary-phase green (OS1) and white (OS1w) cells were exposed to various temperatures from 37 to 46°C in LL. Sixty cells were isolated into each well of a microplate and kept at various temperatures. After 12 h, the percentage of surviving cells was counted under a binocular microscope (Fig. 4). We judged the death of cell according to motionlessness and cell rupture. The experiment was repeated three times. Green cells showed a higher survival rate than white cells 12 h after treatment with high temperature in LL. Then we assayed the tolerance of cells reinfected with *Chlorella* to high temperature (42°C) each day while *Chlorella* became established as a symbiont. About $2 \times 10^7$ *Chlorella* per milliliter were mixed with about $2 \times 10^3$ host cells per milliliter for 6 h. Host cells contained about 450 *Chlorella* per cell 6 h after reinfection, and after washing the number of *Chlorella* decreased to about 100 per cell 1 day after reinfection. After that, the number increased gradually for 3–6 days

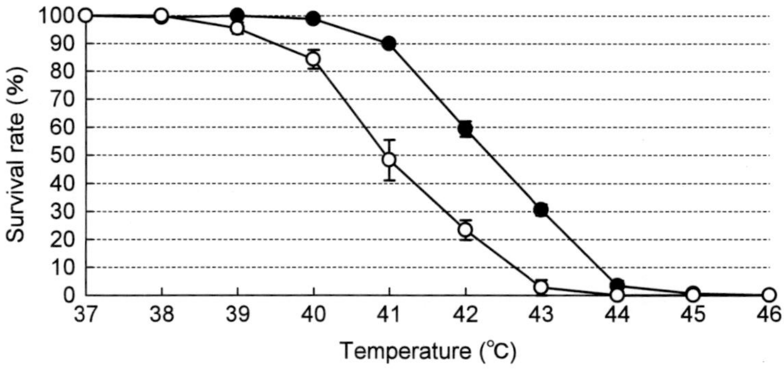


**Fig. 4** Survival rate of *P. bursaria* at various temperatures ranging from 37 to 46°C. *Black circles* represent *Chlorella* containing green cells and *white circles* represent *Chlorella*-free white cells. Sixty green and 60 white cells were isolated into separate wells of a microplate and kept at the various temperatures for 12 h. The percentage of surviving cells was counted under a binocular microscope. Motionless and ruptured cells were judged to be dead. The *bar* at each point indicates the standard deviation (SD) of results from three experiments

(Fig. 5a). The survival rate of reinfected cells at 42°C was assayed at the same time as the assay of *Chlorella* proliferation. Their survival rate was as low as that of white cells on the first day, in spite of the host cells containing many *Chlorella*. The survival rate increased gradually to the level of the original green cells 3–6 days after reinfection in relation to the increase of *Chlorella* in the host cells (Fig. 5b). This is almost same time as the phase shift of the circadian rhythm occurred when white cells were reinfected with *Chlorella* entrained to a reverse phase.

## 5 Effects of Symbiotic *Chlorella* sp. on the Life Cycle

An exconjugant clone of *Paramecium*, as with many other ciliates, has a well-defined life cycle, consisting of a sequence of immaturity, maturity, and senility (Jennings 1944; Siegel 1957; Sonneborn 1957). During the immature period, cells cannot mate even when they enter the stationary phase. During the mature period, cells show mating reactivity when cells of a complementary mating type are brought together under appropriate conditions. In the senility stage, cells exhibit a phenotypic instability and a progressive loss of reproductive function (Siegel 1967). The duration of the immaturity period is measured in terms of the number of fissions after conjugation, because even when environmental conditions cause the immature period to be shortened or lengthened, the number of fissions through which a given clone passes during the immature period is almost the same (Miwa and Hiwatashi 1970). The duration of each phase of the life cycle has been shown to be controlled genetically (Siegel 1961; Takagi et al. 1987). Some mutants with an altered length of immaturity also have been discovered in *Tetrahymena pyriformis* (Bleyman and Simon 1967) and in *P. caudatum* (Myohara and Hiwatashi 1978). We reported previously that cells of *P. caudatum* in the immature period contain immaturin, which represses the expression of mating

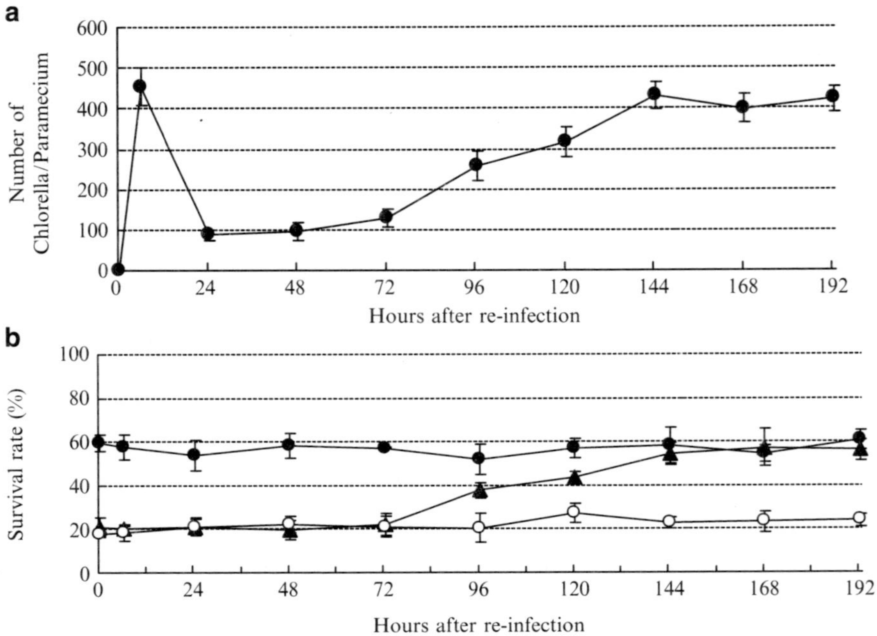


**Fig. 5** Numbers of *Chlorella* per host cell and the increase of survival rate of reinfected cells at 42°C. **a** About $2 \times 10^7$ *Chlorella* per milliliter were mixed with about $2 \times 10^3$ host cells per milliliter for 6 h. Host cells contained about 450 *Chlorella* per cell 6 h after reinfection, and after washing the number of *Chlorella* decreased to about 100 per cell 1 day after reinfection. After that the number increased gradually for 3–6 days. **b** The survival rate at 42°C was assayed in green cells (*black circles*), white cells (*white circles*), and white cells reinfected with *Chlorella* (*triangles*). The survival rate of reinfected cells was increased gradually to the level of the original green cells after 3–6 days. The *bar* at each point indicates the SD of results from three experiments

reactivity when it is injected into sexually mature cells (Miwa et al. 1975). Immaturin has been isolated and characterized; it is a heat-labile protein with a molecular mass of about 10,000 Da (Haga and Hiwatashi 1981) and is common to three different species complexes in *Paramecium* (Miwa 1979a, b).

How does symbiotic *Chlorella* affect to the life cycle of *P. bursaria*? We compared the length of immaturity of progeny from a cross between green cells and white cells of *P. bursaria*.

## 5.1 Length of Immaturity

Syngen 1 of *P. bursaria* has four complementary mating types which are determined by two kinds of genes. They have a maturation process consisting of a sequence of immaturity, adolescence, and maturity. In adolescence, progenies from a cross express one gene and are able to mate with cells from two of the four mating types, and then they grow to maturity, at which stage they can mate with cells from all three mating types besides their own (Siegel 1957).

The progenies from a cross between stocks STL4 (green cell) and MO1w (white cell), which belong to mating types I and III, respectively, in syngen 1 of *P. bursaria*, were used in this experiment. Each exconjugant line was grown at 25°C under the condition of daily isolation culture in LD 12:12 h. After daily isolation, the remaining cells in the depression slide in each line were allowed to starve for 2 days, during which the cells divided approximately six more times. The starved cells were divided into four small quantities and used to test mating reactivity by mixing them with tester cells of the respective four mating types. It is known that exconjugant clones of *P. bursaria* become adolescent after about 40 fissions and mature about 55 fissions after conjugation (Siegel 1957). In this experiment, exconjugant clones of green cells grew to adolescence after 30–50 fissions and to maturity by 40–60 fissions after conjugation. On the other hand, white cell clones grew to adolescence after 50–60 fissions and to maturity by 60–70 fissions after conjugation. The maturation of green cell clones appears about ten fissions earlier than that of white cells.

Why do the progenies from green cells mature earlier? It has been reported that the length of immaturity is shortened by about ten fissions by treatment with mitomycin C (Miwa and Hiwatashi 1970) and by UV irradiation (Takagi 1974) in *P. caudatum*. These two treatments affect DNA damage in a *Paramecium* cell. During maturation of *P. bursaria*, the photosynthetic product $O_2$ of symbiotic *Chlorella* changes to active oxygen, and this might affect the damage to DNA. If this idea is correct, green cells could have a shorter lifespan than white cells. In this experiment, the immature periods were compared within exconjugant clones that have different cytoplasm. Next, we have to compare the length of immaturity within sub-exconjugant clones derived from white cells, one of which is a white cell clone without reinfection, and the other of which is reinfected with *Chlorella*.

## 5.2 Circadian Clock and Developmental Clock

The oscillating mechanism which controls circadian rhythms is often called a "circadian clock". The period of the circadian clock is inherent in the whole-cell clock of each strain. The period lengths of the photoaccumulation rhythm are different within stocks in LL. Various physiological activities were compared in cells of four strains (Sj2w, Kz1w, T316w, and T316), and various cellular parameters changed in parallel to the changes of the circadian period (Table 4). Periodic contractions of the contractile vacuoles and the frequency of ciliary beating as estimated by swimming velocity became faster in strains having the shortest circadian periods, even though the time scales of these periods differed dramatically from one another. The resting membrane potential was more hyperpolarized in strains with shorter circadian periods. The membrane potential of *Paramecium* is closely associated with swimming behavior (Machemer 1988). When the *Paramecium* membrane is hyperpolarized, the ciliary

**Table 4** Correlation of circadian period with physiological activities in *Paramecium bursaria*

| | Sj2w | Kz1w | T316w | T316 |
|---|---|---|---|---|
| Circadian period (h) | 20.9 | 22.6 | 24.5 | 27.9 |
| Contraction period (s) | 10.6 | 12.7 | 14.7 | ND |
| Swimming velocity (mm $s^{-1}$) | 2.2 | 2.0 | 1.1 | 0.9 |
| Membrane potential (mV) | −24 | −23 | −22 | −21 |
| Membrane resistance (MΩ) | 117 | 105 | 91 | ND |
| K concentration (mM) | 9 | 12 | 16 | 22 |

These are values measured in LL. Stocks Sj2w, Kz1w, and T316w are *Chlorella*-free white cells; stock T316 is *Chlorella*-containing green cells.
*ND* means no data.

beating frequency rises and presumably increases the swimming velocity (Machemer and Eckert 1975; Nakaoka and Machemer 1990). It is possible that the factor which modulates the membrane potential may also influence or be involved in the circadian oscillation. Finally, the membrane resistance of the resting state is reduced in proportion to the increase in the circadian period. This correlation between cellular properties and the circadian period suggests that the circadian clock mechanism is associated with various physiological activities of the cell (Tokushima et al. 1994). Since there are differences in physiological activities between strains of green cells (T316) and white cells (T316w), symbiotic *Chlorella* may contribute to these differences.

Mating-type inheritance and determination in *P. bursaria* are controlled by pairs of alleles at two independently assorting loci. The mating-type loci are expressed sequentially; one locus is expressed in the adolescent phase and both loci are expressed when the cells become mature (Siegel and Larison 1960). Thus, the phenomenon of adolescence and immaturity in *P. bursaria* is a suitable model for studying biological timing measured by a "developmental clock."

Is there a relationship between the timekeeping mechanisms of the circadian clock and the developmental clock? The correlation between the period length of circadian rhythms and the length of sexual immaturity has been investigated. In genetic analysis of shorter-period and longer-period strains, the progeny from a cross between natural strains UK1 and T316 does not segregate with a Mendelian ratio; rather, the range of the period is spread widely. Therefore, the phenotypes of shorter and longer periods are probably controlled polygenetically or by an ecotype which is influenced by the ecosystem. It is thought that polygenetic control is much more likely than control by an ecotype, because stocks UK1 and T316 were collected near the same place.

To get a typical mutant clone of the circadian rhythm, we treated cells of Kz1 with nitrosoguanidine, and screened them with 9 mM $BaCl_2$ for 15 min. We could obtain 24 barium-resistant cells after performing this protocol five times. The photoaccumulation rhythms of these clones were measured. Almost all clones showed a period of about 26.0 h, which is nearly the same as the period of Kz1. One clone (E2), however, showed a short-period rhythm of 21.8 h, and we found no clones which exhibited a longer-period rhythm or an arrhythmic pattern.

In the case of *P. bursaria*, an exconjugant becomes adolescent after about 40 fissions and mature about 55 fissions after conjugation. Clones never show mating reactivity after fewer than 30 fissions. For genetic analysis of the early maturation character, E2 (mating-type III) was mated with Sj2w (mating-type I). Almost all progenies were able to mate about 25 fissions after conjugation and showed mating-type I or IV. The appearance of maturity of $F_2$ progeny clones (from SE11 × SE14, which are $F_1$ progenies from Sj2 × E2) segregated into three groups, about 15, 22, and 52 fissions after conjugation. The number of $F_2$ progeny was too small for an $x^2$ statistical test, but its ratio (3:10:4) is close to 1:2:1. It seems that strain E2 is an early mature mutant. The period length of the circadian photoaccumulation rhythm in LL of three groups of $F_2$ clones was, on average, 20.8 ± 0.5, 22.0 ± 0.9, and 25.6 ± 1.0 h, respectively. The early maturation groups showed a shortened period of circadian rhythm. Further, to investigate the correlation between the period length of circadian rhythms and the length of immaturity, E2 (early mature mutant) and Kz1 (wild-type) were mated with a short-period strain (UK1) and a long-period strain (T316) in the photoaccumulation rhythm, respectively. The progeny of short-period strains matured earlier than that of longer-period strains. The length of adolescence of the progeny from the early mature strain (E2) was shorter than that of progeny from the wild-type strain (Kz1). It seems that maturation in *P. bursaria* depends on both of the number of fissions after conjugation (a "developmental clock") and the period length of the circadian rhythm (a "circadian clock"). Therefore, it may be important to clarify the relationship between the rate of cell division and the period length of the circadian rhythm (Miwa and Yajima 1995).

## 6 Concluding Remarks and Future Perspectives

Symbiotic *Chlorella* of *P. bursaria* lengthens the period of the circadian photoaccumulation rhythm of host cells by about 3 h in LL. Symbiotic *Chlorella*, which reinfect host cells of a different phase, forces them to shift the phase of the photoaccumulation and mating reactivity rhythms in LL. These increases of period and shifts in phase do not occur in DD or with DCMU treatment in LL. Therefore, the photosynthetic products of symbiotic *Chlorella* are likely related to these events, although they are not effective when maltose and oxygen are added to the cells externally in DD. Furthermore, reinfected *Chlorella* rescues the mating reactivity rhythm of arrhythmic mutant cells in LL, but not in DD. It is certain that symbiotic *Chlorella* influences the circadian rhythms of *P. bursaria* through their photosynthetic products.

Circadian rhythms have been observed in almost all levels of organisms living on the earth, from prokaryotes to humans, and are a very important property associated with various physiological activities and lifespan. It has been reported that unicellular *G. poryedra* has two circadian clocks in one cell (Roenneberg and Morse 1993). As these clocks are sensitive to different wavelengths of light, one is thought to be the clock of the whole cell, and the other is in the chloroplast

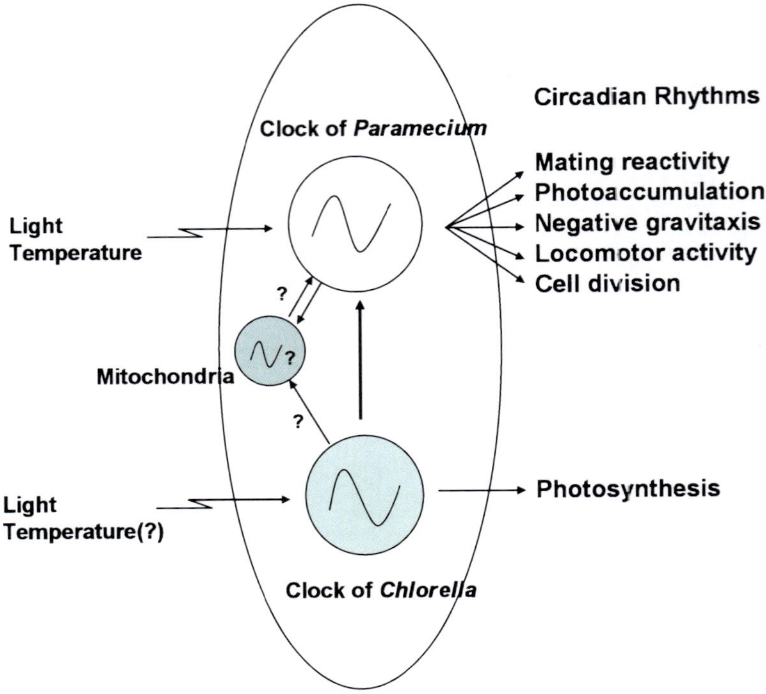


**Fig. 6** Circadian system of *P. bursaria*. Cells of *P. bursaria* exhibit many kinds of circadian rhythms and symbiotic *Chlorella* shows the circadian rhythm of photosynthesis. The clock of symbiotic *Chlorella* affects the clock of *P. bursaria* through photosynthetic products. In some cases, the clock of *Chlorella* might influence the mitochondria of the host cell. Mitochondria may have a circadian clock of their own

(Morse et al. 1994). It is possible that chloroplasts have their own circadian clocks since they evolved from *Cyanobacteria*, which are known to have a clock. Cells of *P. bursaria* have two circadian clocks per cell, one whole-cell clock and *Chlorella*'s clock (Fig. 6). As seen above, the clock of *Chlorella* has a great influence on the activities of *P. bursaria*. Currently, the relationship between host and symbiont in *P. bursaria* is not so intimate, since each cell can be free-living without the other. In the future, if their relationship evolves to become more interdependent, *Chlorella*'s clock might take the place of the host clock. On the other hand, *Chlorella*'s clock may influence the host cell through the mitochondria in some cases. The mitochondria themselves are derived from symbiosis of aerobic bacteria at some point in the distant past, so they may possibly have their own circadian clocks, although as of yet we do not have any evidence of this. The relationship of host and symbiont in *P. bursaria* is an interesting model for investigating biological coevolution.

**Acknowledgements** I am indebted to all coworkers and all students in my laboratory at Ibaraki University who have contributed to *Paramecium* research in the past.

## References

Aschoff J (1960) Exogenous and endogenous components in circadian rhythms. Cold Spring Harb Symp Quant Biol 25:11–28

Aschoff J (1981) Free-running and entrained circadian rhythms. In: Aschoff J (ed) Handbook of behavioral neurobiology. Plenum, New York, pp 81–93

Barnett A (1966) A circadian rhythm of mating type reversals in *Paramecium multimicronucleatum*, syngen 2, and its genetic control. J Cell Physiol 67:239–270

Berridge MJ, Irvine RF (1989) Inositol phosphates and cell signaling. Nature 341:197–205

Bleyman LK, Simon EM (1967) genetic control of maturity in *Tetrahymena pyriformis*. Genet Res 10:319–321

Block GD, Khalsa SBS, McMahon DG, Michel S, Geusz M (1993) Biological clocks in the retina: cellular mechanisms of biological timekeeping. Int Rev Cytol 146:83–144

Bruce VG, Pittendrigh CS (1956) Temperature independence in a unicellular 'clock'. Proc Natl Acad Sci U S A 42:676–682

Butko P (1988) Oxygen evolution of synchronous *Chlorella* under continuous and flash illumination. J Photochem Photobiol B 1:447–450

Chen TT (1946) Temporary pair formation in *Paramecium bursaria*. Biol Bull 91:274–300

Cohen LW (1965) The basis for the circadian rhythm of mating in *Paramecium bursaria*. Exp Cell Res 37:360–367

Dunlap JC, Loros JJ, DeCoursey PJ (2004) Chronobiology: biological timekeeping. Sinauer, Sunderland

Edmunds LN Jr (1988) Cellular and molecular bases of biological clock. Springer, New York

Ehret CF (1953) An analysis of the role of electromagnetic radiations in the mating reaction of *Paramecium bursaria*. Physiol Zool 26:274–300

Ehret CF, Truco E (1967) Molecular models for the circadian clock. I. The chronon concept. J Theor Biol 15:240–262

Freund WD, Mayr GW, Tietz C, Schultz J (1992) Metabolism of Inositol phosphates in the protozoan *Paramecium*. Eur J Biochem 207:359–367

Haga N, Hiwatashi K (1981) A protein called immaturin controlling sexual immaturity in *Paramecium*. Nature 289:177–179

Hasegawa K, Tanakadate A (1984) Circadian rhythm of locomotor behavior in a population of *Paramecium multimicronucleatum*: its characteristics as derived from circadian changes in the swimming speeds and the frequencies of avoiding response among individual cells. Photochem Photobiol 40:105–112

Hasegawa K, Katakura T, Tanakadate A (1984) Circadian rhythm in the locomotor behavior in a population of *Paramecium multimicronucleatum*. J Interdiscipl Cycle Res 15:45–56

Hiwatashi K (1981) Sexual interactions of the cell surface in *Paramecium*. In: O'Day DH, Horgen PA (eds) Sexual interaction in eukaryotic microbes. Academic, New York, pp 351–378

Hosoya H, Kimura K, Matsuda S, Kitaura M, Takahashi T, Kosaka T (1995) Symbiotic algal-free strains of the green paramecium *Paramecium bursaria* produced by herbicide paraquat. Zool Sci 12:807–810

Imafuku M (1975) Adaptation of the circadian rhythm of mating-reactivity to abnormal light-dark cycle in *Paramecium bursaria*. J Interdiscipl Cycle Res 6:141–151

Iwatsuki K, Naitoh Y (1988) Behavioral responses to light in *Paramecium bursaria* in relation to its symbiotic green alga *Chlorella*. J Exp Biol 134:43–60

Iwatsuki K, Nishidoi M, Suehiro K (1998) Symbiotic *Chlorella* enhances the thermal tolerance in *Paramecium bursaria*. Comp Biochem Physiol Part A 121:405–409

Jennings HS (1939) Genetics of Paramecium bursaria: I. Mating type and groups, their interrelations and distribution; mating behavior and self sterility. Genetics 24:202–233

Jennings HS (1944) *Paramecium bursaria*: Life history. I. Immaturity, maturity and age. Biol Bull Mar Biol Lab Woods Hole 86:131–145

Johnson CH, Hastings JW (1986) The elusive mechanism of circadian clock. Am Scient 74:29–36

Johnson CH, Miwa I, Kondo T, Hastings JW (1989) Circadian rhythm of photoaccumulation in *Paramecium bursaria*. J Biol Rhythms 4:405–415

Johnson CH, Nakaoka Y, Miwa I (1994) The effects of altering extracellular potassium ion concentration on the membrane potential and circadian clock of *Paramecium bursaria*. J Exp Biol 197:295–308

King DP, Zhao Y, Sangoram AM, Wilsbacher LD, Tanaka M, Antoch MP, Steeves TD, Vitaterna MH, Kornhauser JM, Lowrey PL, Turek FW, Takahashi JS (1997) Positional cloning of the mouse circadian clock gene. Cell 89:641–653

Kondo T, Johnson CH, Hastings JW (1991) Action spectrum for resetting the circadian phototaxis rhythm in the CW15 strain of *Chlamydomonas*. Plant Physiol 95:197–205

Kondo T, Strayer CA, Kulkarni RD, Taylor W, Ishiura M, Golden SS, Johnson CH (1993) Circadian rhythms in prokaryotes: luciferase as a reporter of circadian gene expression in cyanobacteria. Proc Natl Acad Sci U S A 90:5672–5676

Konopka RJ, Benzer S (1971) Clock mutants of *Drosophila melanogaster*. Proc Natl Acad Sci U S A 68:2112–2116

Loefer JB (1936) Isolation and growth characteristics of the 'zoochlorella' of *Paramecium bursaria*. Am Midl Nat 70:184–188

Machemer H (1988) Motor control of cilia. In: Görtz H-D (ed) *Paramecium*. Springer, Berlin, pp 216–235

Machemer H (1989) Cellular behaviour modulated by ions: electrophysiological implications. J Protozool 36:463–487

Machemer H, Eckert R (1975) Ciliary frequency and orientational responses to clamped voltage steps in Paramecium. J Comp Physiol 104:247–260

Matsuoka K, Nakaoka Y (1988) Photoreceptor potential causing phototaxis of *Paramecium bursaria*. J Exp Biol 137:477–485

Meier R, Wiessner W (1988) Infection of algae-free *Paramecium bursaria* with symbiotic *Chlorella* sp isolated from green paramecia. I. Effect of the incubation period. Europ J Protistol 24:69–74

Mergenhagen D (1980) Circadian rhythms in unicellular organisms. Curr Top Microbiol Immunol 90:123–147

Miwa I (1979a) Specificity of the immaturity substances in *Paramecium*. J Cell Sci 36:253–260

Miwa I (1979b) Immaturity substances in *Paramecium primaurelia* and their specificity. J Cell Sci 38:193–199

Miwa I, Hiwatashi K (1970) Effect of mitomycin C on the expression of mating ability in *Paramecium caudatum*. Jpn J Genet 45:269–275

Miwa I, Wada T (1995) Light pulses and injection of $IP_3$ induce mating ability in *Paramecium bursaria*. J Exp Zool 272:338–344

Miwa I, Yajima H (1995) Correlation of the period length of circadian rhythms with the length of immaturity in *Paramecium bursaria*. Zool Sci 12:53–59

Miwa I, Haga N, Hiwatashi K (1975) Immaturity substances: Material basis for immaturity in *Paramecium*. J Cell Sci 19:369–378

Miwa I, Nagatoshi H, Horie T (1987) Circadian rhythmicity within single cell of *Paramecium bursaria*. J Biol Rhythms 2:57–64

Miwa I, Kanazawa Y, Ishikawa K, Hirose M (1989) Synchronization of mating reactivity rhythms in populations of *Paramecium bursaria*. J Protozool 36:24–28

Miwa I, Izumo T, Sonoda T (1996a) Cytoplasm rescues an arrhythmic mutant on the circadian rhythm of mating reactivity in *Paramecium bursaria*. J Eukaryot Microbiol 43:231–236

Miwa I, Fujimori N, Tanaka M (1996b) Effects of symbiotic *Chlorella* on the period length and the phase shift of circadian rhythms in *Paramecium bursaria*. Europ J Protistol 32(Suppl 1):102–107

Morse J, Hastings JW, Roenneberg T (1994) Different phase responses of the two circadian oscillators in *Gonyaulax*. J Biol Rhythms 9:263–274

Myohara K, Hiwatashi K (1978) Mutants of sexual maturity in *Paramecium caudatum* selected by erythromycin resistance. Genetics 90:227–241

Naitoh Y, Eckert R (1968) Electrical properties of *Paramecium*: modification by bound and free cations. Z Vergl Physiol 61:427–452

Nakajima K, Nakaoka Y (1989) Circadian change of photosensitivity in *Paramecium bursaria*. J Exp Biol 144:43–51

Nakajima M, Imai K, Ito H, Nishiwaki T, Murayama Y, Iwasaki H, Oyama T, Kondo T (2005) Reconstitution of circadian oscillation of cyanobacterial KaiC phosphorylation in vitro. Science 308:414–415

Nakaoka Y (1989) Localization of photosensitivity in *Paramecium bursaria*. J Comp Physiol A 165:637–641

Nakaoka Y, Machemer H (1990) Effects of cyclic nucleotides and intracellular Ca on voltage-activated ciliary beating in *Paramecium*. J Comp Physiol A 166:401–406

Nakaoka Y, Kinugawa K, Kurotani T (1987) $Ca^{2+}$-dependent photoreceptor potential in *Paramecium bursaria*. J Exp Biol 131:107–115

Nakaoka Y, Tokioka R, Shinozawa T, Fujita J, Usukura J (1991) Photoreception of *Paramecium* cilia: localization of photosensitivity and binding with anti-frog-rhodopsin IgG. J Cell Sci 99:67–72

Oka T, Nakaoka Y, Oosawa F (1986) Changes in membrane potential during adaptation to external potassium ions in *Paramecium caudatum*. J Exp Biol 126:111–117

Pado R (1972) Spectral activity of light and phototaxis in *Paramecium bursaria*. Acta Protozool 11:387–393

Pittendrigh CS (1981) Circadian system: General perspective. In: Aschoff J (ed) Handbook of behavioral neurobiology, vol 4. Biological rhythms. Plenum, New York, pp 57–80

Plautz JD, Kaneko M, Hall JC, Kay SA (1997) Independent photoreceptive circadian clocks throughout *Drosophila*. Science 278:1632–1635

Price JL, Blau J, Rothenfluh A, Abodeely M, Kloss B, Yong MW (1998) double-time is a novel Drosophila clock gene that regulates PERIOD protein accumulation. Cell 94:83–95

Reppert SM, Weaver DR (2002) Coordination of circadian timing in mammals. Nature 418:935–941

Roenneberg BM, Morse D (1993) Two circadian oscillators in one cell. Nature 362:362–364

Saji M, Oosawa F (1974) Mechanism of photoaccumulation in *Paramecium bursaria*. J Protozool 21:556–561

Schweiger HG, Hartwig R (1986) Cellular aspects of circadian rhythms. J Cell Sci 4:181–200

Siegel RW (1957) An analysis of the transformation from immaturity to maturity in *Paramecium aurelia*. Genetics 42:394–395

Siegel RW (1961) Nuclear differentiation and transitional cellular phenotypes in the life cycle of *Paramecium*. Exp Cell Res 24:6–20

Siegel RW (1967) Genetics of ageing and the life cycle in ciliates. Symp Soc Exp Biol 21:127–148

Siegel RW, Larison LL (1960) The genetic control of mating type in *Paramecium bursaria*. Proc Natl Acad Sci U S A 46:344–349

Sonneborn TM (1938) Mating type in *Paramecium aurelia*: diverse conditions for mating in different stocks; occurrence, number and interrelations of the types. Proc Am Philos Soc 79:411–434

Sonneborn TM (1957) Diurnal change of mating type in *Paramecium*. Anat Rec 128:626

Takagi Y (1974) The effect of ultraviolet irradiation on the period of immaturity in *Paramecium caudatum*. Zool Mag 83:96–98

Takagi Y, Suzuki T, Shimada C (1987) Isolation of a *Paramecium tetraurelia* mutant with short clonal life-span and with novel life-cycle features. Zool Sci 4:73–80

Tanaka M, Miwa I (1996) Significance of photosynthetic products of symbiotic *Chlorella* to establish the endosymbiosis and to express the mating reactivity rhythm in *Paramecium bursaria*. Zool Sci 13:685–692

Tanaka M, Miwa I (2000) Correlation of photosynthetic products of symbiotic *Chlorella* with the mating reactivity rhythms in a mutant strain of *Paramecium bursaria*. Zool Sci 17:735–742

Tokushima H, Okamoto KI, Miwa I, Nakaoka Y (1994) Correlation between circadian periods and cellular activities in *Paramecium bursaria*. J Comp Physiol A 175:767–772

Tomita J, Nakajima M, Kondo T, Iwasaki H (2005) No transcription-translation feedback in circadian rhythm of KaiC phosphorylation. Science 307:251–254

Weis DS (1979) Correlation of sugar release and concanavalin A agglutinability with infectivity of symbiotic algae from *Paramecium bursaria* for aposymbiotic *P. bursaria*. J Protozool 26:117–119

Zehring WA, Wheeler DA, Reddy P, Konopka RJ, Kyriacou CP, Rosbash M, Hall JC (1984) P-element transformation with period locus DNA restores rhythmicity to mutant, arrhythmic *Drosophila melanogaster*. Cell 39:369–376

# Photobiological Aspects of the Mutualistic Association Between *Paramecium bursaria* and *Chlorella*

**Ruben Sommaruga and Bettina Sonntag**

**Contents**



**Abstract** The main benefit involved in the mutualistic relationship between *Paramecium bursaria* and *Chlorella* has been traditionally interpreted as an adaptation to the struggle against starvation and nutrient limitation. However, other benefits such as the minimization of mortality and protection against damage by sunlight have been proposed recently. Here, we explore the photobiological adaptations and responses of *P. bursaria* and its algal symbionts when exposed to photosynthetically active and ultraviolet radiation, as well as the role of (photo-) oxidative and antioxidative defenses in the symbiosis. We conclude that the benefits are multiple and should be considered as a whole when assessing the selective advantage of living in mutualistic symbiosis.

## 1 Introduction

The study of mutualistic interactions in natural communities has become an important research area to understand the environmental adaptation and coevolution of species (Doebeli and Knowlton 1998; Herre et al. 1999). Symbiotic relationships

R. Sommaruga (✉) and B. Sonntag
Laboratory of Aquatic Photobiology and Plankton Ecology, Institute of Ecology, University of Innsbruck, Technikerstr. 25, 6020, Innsbruck, Austria
e-mail: ruben.sommaruga@uibk.ac.at

M. Fujishima (ed.), *Endosymbionts in Paramecium*, Microbiology Monographs 12,
DOI: 10.1007/978-3-540-92677-1_5, 

where each species benefits are mutualistic, but depending on whether the species involved are able to survive or not in the absence of each other can be considered obligate or facultative. The symbiosis between the ciliate *Paramecium bursaria* and representatives of the green algae *Chlorella* is a good example of a facultative mutualistic trophic interaction and is one of the best studied. Indeed, the fact that it is possible to obtain aposymbiotic cell lines of *P. bursaria* in the laboratory and to infect them again with *Chlorella* (Niess et al. 1982; Meier and Wiessner 1988, 1989; Summerer et al. 2007) has largely contributed to the understanding of some basic principles and benefits in algae-ciliate mutualism.

Paramecia are ubiquitous freshwater organisms found in almost all kinds of freshwater habitats (Landis 1988). Whereas natural populations of aposymbiotic *P. bursaria* are known (Tonooka and Watanabe 2002), their *Chlorella*-bearing counterparts are commoner in natural habitats. This brings us to the central question of what kind of cost-benefit balance or selective advantage is involved in the *P. bursaria–Chlorella* mutualism and in general for other species of *Chlorella*-bearing ciliates. In the past, this question attracted the interest of many researchers (see Dolan 1992; Jones 1994; Stoecker 1998), but it is still relevant nowadays because we know little about the molecular, cellular, and organismal adaptations that made possible the establishment of this successful mutualism. Further, the recent evidence that ciliates may once have been photosynthetic (Reyes-Prieto et al. 2008) adds new material to the debate on the origin of mutualism between phototrophic and heterotrophic protists.

In this chapter, we first give an overview about the known benefits in algae-ciliate symbiosis, then we explore adaptations and responses of *P. bursaria* and its algal symbionts when they are exposed to photosynthetically active radiation (PAR; 400–700 nm) and UV radiation (UVR; 280–400 nm), and, finally, suggest future research topics to advance the understanding of the evolutionary and ecological adaptation of this mutualistic relationship. We think it is worth scrutinizing what other benefits are thought to form the basis of the mutualism, but we stress the idea of Lobban et al. (2007) that all benefits are important to the well-being of the symbiotic association, making it difficult to assert a "primary" one.

## 1.1 The "Classical" View of Mutual Benefits

Certainly, the "classical" understanding of the physiological basis for the existence of a mutualistic relationship between ciliates and *Chlorella* is the efficient transfer of inorganic elements from the ciliate to the algae and of photosynthate leaking from the endosymbionts to the host. Since the pioneering work of Muscatine et al. (1967), we know that endosymbiotic *Chlorella* excrete large concentrations of carbohydrates used by *P. bursaria* to maintain its metabolism. In this way, the ciliate becomes partially or totally independent of external food supply (Reisser et al. 1984). In return, the ciliate provides respiratory $CO_2$ that can be photosynthetically fixed by *Chlorella* (Reisser 1980). When light is sufficient to saturate the photosynthetic apparatus, the oxygen demand of the ciliates is covered entirely by their symbiotic algae and oxygen is also released into the medium.

The ultimate factor and the most accepted benefit of this type of mutualistic association as well as in plastid enslavement is undoubtedly the higher chance of surviving starvation and nutrient limitation. In oligotrophic marine and freshwater ecosystems, inorganic nutrients needed by photoautotrophs are scarce, as are prey for heterotrophs (Dolan 1992l; Dolan and Pérez 2000). Mixotrophy (i.e., the nutrition mode combining both phagotrophy and phototrophy) is generally considered as an adaptation allowing exploitation of oligotrophic environments (Norris 1996; Dolan and Pérez 2000). In fact, there are many other examples of mutualistic associations that illustrate the widespread occurrence of phototrophic–heterotrophic symbioses such as those of reef-building corals with the so-called zooxanthellae, sea anemones and jellyfish with zooxanthellae or zoochlorellae, and the freshwater solitary coelenterate *Hydra viridis* and sponges from shallow freshwaters with *Chlorella*.

Pringsheim (1928) was among the first to recognize that starved green symbiotic *P. bursaria* perform better than aposymbiotic populations of the same species. Detailed quantitative experiments with symbiotic and aposymbiotic *P. bursaria* and other ciliates such as *Coleps* sp. have confirmed this observation (Görtz 1988; Stabell et al. 2002). For example, when food is abundant, the growth rates of symbiotic and aposymbiotic *Coleps* sp. are not significantly different. Stabell et al. (2002) found that the gross growth rate of endosymbiotic *Chlorella* is always close to maximum, making it possible for the host to receive an increasing fraction of the total carbon supply from the algae with increasing food limitation. Direct measurements of the relative importance of photosynthate production by algal endosymbionts versus heterotrophic food uptake in *Stentor* spp. also suggest a relationship with food supply levels (Woelfl and Geller 2002). For example, photosynthate production by *Stentor araucanus* and *Stentor amethystinus* exceeds food uptake in autumn and winter, but not in summer at the time of higher food supply (Woelfl and Geller 2002). Berk et al. (1991) reported that ingestion rates on bacteria by “green” *P. bursaria* increase in relation to PAR levels (range 1–90 $\mu$mol $m^{-2}s^{-1}$). As this was not observed in the aposymbiotic counterparts, the authors suggested that the endosymbiotic *Chlorella* control the feeding rates of the host.

Though a plethora of laboratory experiments support the idea of the *Chlorella* -ciliate symbiosis as an adaptation to survive starvation at least during certain periods of time, results from natural aquatic habitats show that this is not always the case and that the factors controlling the relative contribution of mixotrophic species to the ciliate assemblage remain unclear (Dolan 1992). As argued by Stoecker (1998), “the cost and benefits of mixotrophy in different taxa and environments are still largely a subject of speculation” and this still applies nowadays.

Yet, though symbiosis in algae-bearing ciliates provides an advantage in the struggle against starvation, the occurrence of these organisms also in sunlit waters of eutrophic environments suggests the existence of other benefits in this mutualistic association (Dolan and Pérez 2000). For example, the common finding of high numbers of algae-bearing ciliates at the oxic–anoxic interface in eutrophic lakes has been related with the supply of oxygen by algal endosymbionts to the ciliate’s metabolic demand (Berninger et al. 1986; Finlay et al. 1996). This hypothesis has been challenged by Stabell et al. (2002), however, who argued that at oxic–anoxic boundaries, light is usually limiting and net diel oxygen production is close to zero. Occupation of deep water layers at the limit of the euphotic zone has also been observed in the colonial *Chlorella*-bearing ciliate *Ophrydium naumanii* living in well-oxygenated oligotrophic waters (Queimaliños et al. 1999). High food levels in eutrophic systems do not necessarily exclude the possibility of food limitation because the dominant prey cell size available may lie outside the range ingested by mixotrophic ciliate species, particularly if they are size-selective (Dolan 1992). Thus, it seems that surface avoidance by *Chlorella*-bearing ciliates may have different explanations, including higher availability of inorganic nutrients and reduction of metazoan predation pressure (e.g., in poorly oxygenated waters) as argued by Finlay et al. (1996).

Indeed, after a symbiosis has been established, mortality factors for the respective partners or for the whole symbiosis can be substantially minimized. For example, the presence of symbiotic *Chlorella* prevents the infection of *P. bursaria* by bacteria and yeasts which are not able to entirely support the ciliate's growth (Görtz 1982). Moreover, though *Chlorella* from other algae-bearing planktonic ciliates are ingested by *P. bursaria* and, further, stable symbioses are established (Niess et al. 1982), *P. bursaria* prefer their own *Chlorella* from a mixture of three strains (Summerer et al. 2007). In addition, *Chlorella*-bearing *P. bursaria* seems to be less predated than its aposymbiotic counterpart by other ciliates such as *Didinium nasutum* (Berger 1980). Similar results have been obtained with the *Chlorella*-bearing *Climacostomum virens* in experiments with the predatory ciliate *Dileptus margaritifer* (Miyake et al. 2003). The exact mechanism for the lower vulnerability of mixotrophic ciliates is unknown, but higher swimming speed and strong escape reactions have been argued as possible mechanisms (Pérez et al. 1997).

On the *Chlorella* side, the mutualistic association with a host represents a physical refuge and a way to avoid mortality, for example, by lytic chloroviruses that are widespread in freshwater systems (Van Etten 2003). Though generally less discussed in the scientific literature, the establishment of a symbiotic relationship between *Chlorella* and *P. bursaria* and other ciliates could have been driven by the additional selective pressure of mortality in an environment hostile at least to the symbiont. The information available on *Chlorella* viruses suggests they have a long evolutionary history of more than 1.2 billion years (Kang et al. 2003).

## 1.2 Photoprotection in *Chlorella*-Ciliate Symbiosis

To obtain the benefits of living in symbiosis with *Chlorella*, ciliate hosts need to expose themselves to solar radiation, which includes not only PAR but also damaging UVR. UVR is more rapidly attenuated in the water column than PAR; thus, exposure of organisms will be significant in shallow and transparent aquatic systems or in populations thriving in the upper water column. As evidenced by the several strategies developed among different forms of life to obtain protection and to repair damage, solar UVR has been an important selective factor during evolution of life on Earth, particularly for photosynthetic organisms (Cockell 1998). However, physiological and ecological work on *Chlorella*-bearing ciliates has largely ignored the effects of these potentially harmful wavelengths of the natural solar spectrum. In fact, UVR can exert negative effects on several cell targets such as DNA, proteins, pigments, and membranes that may be caused by direct absorption of UV-B radiation (280–315 nm) or indirectly through the generation of reactive oxygen species (photo-oxidative stress) induced by both UV-B and UV-A radiation (315–400 nm). Whether UV-B or UV-A is more effective in causing a certain type of damage will finally depend on the presence of endogenous photosensitizers and the absorption maximum properties of the chromophore(s) involved. Further, other characteristics, such as the presence of DNA rich in A + T, can make organisms particularly susceptible to UV damage owing to the higher probability of production of cyclobutane thymine dimers. We are not aware of A + T values for *P. bursaria,* but in *Paramecium tetraurelia* and *Paramecium primaurelia* the macronuclear DNA has a very high content of 72–75% A + T (McTavish and Sommerville 1980; Aury et al. 2006), suggesting a high susceptibility to UVR of this genus. By contrast, values for other ciliate species such as the marine representatives *Dysteria procera* (56%) and *Hartmannula derouxi* (55%) are lower (Li and Song 2006) and in the range of other species (Schlegel et al. 1991).

The occurrence of negative effects of UVR on ciliates has been known since the pioneering work of Giese (1945) with species of *Paramecium*. Interestingly, the few studies available on UV sensitivity of mixotrophic ciliates suggest that they are resistant. For example, Modenutti et al. (1998) reported that *S. araucanus*, a symbiotic planktonic freshwater ciliate widely distributed in the southern hemisphere, is resistant to exposure to surface solar UVR. The authors, however, attributed the resistance to the protection given by the dark ciliate pigment stentorin. By contrast, Lobban et al. (2007) cast doubt on hypericin-like compounds such as stentorin having any direct role in protecting cells against UVR. UV-exclusion experiments performed with natural ciliate assemblages including the photosynthetic species *Mesodinium rubrum* and the kleptoplast-bearing *Laboea strobila*, collected from a shallow transparent area (George Bank) on the continental shelf in the Northwest Atlantic, indicate a lower UV-B sensitivity for these species (Martin-Webb 1999). Summerer et al. (2009) experimentally tested the sensitivity against UVR of symbiotic *P. bursaria* compared with *Chlorella*-reduced (about half symbiont density) and aposymbiotic *P. bursaria*. Their results indicate that under UVR exposure, aposymbiotic *P. bursaria* had significantly higher mortalities than *Chlorella*-bearing individuals.

Most of our knowledge on the interaction between UVR and mutualistic associations is based on studies on algae–invertebrate symbiosis, particularly on scleractinian corals and their zooxanthellae of the genus *Symbiodinium* (Shick et al. 1996; Shick and Dunlap 2002). Several studies have shown that UVR severely depresses photosynthetic rates in freshly isolated zooxanthellae from corals or other reef organisms, but this effect is small or absent *in hospite* (Dionisio-Sese et al. 2001). The different sensitivity between free and *in hospite* forms appears to be related to protection given by the host through the accumulation in its tissues of sunscreen compounds known as mycosporine-like amino acids (MAAs) produced by the symbionts. MAAs are intracellular water-soluble compounds with high molar extinction coefficients having absorption maxima between 309 and 360 nm.

Until now, approximately 22 different MAAs have been identified in marine and freshwater organisms (Karentz et al. 1991; Dunlap and Shick 1998; Sommaruga and Garcia-Pichel 1999). The biogenesis of MAAs presumably originates in the shikimate pathway that is known to be only present in bacteria, fungi, and algae (Dunlap and Shick 1998; Shick et al. 1999). Independently of whether MAAs are synthesized or obtained from other sources, they will protect to a certain extent (depending on concentration and pathlength) important cell targets from deleterious effects of UVR and, thus, increase the overall resistance of phototrophic and heterotrophic organisms (Shick and Dunlap 2002).

Recently, the existence of MAAs in symbiotic ciliates has been reported for marine and freshwater species (Tartarotti et al. 2004; Sommaruga et al. 2006; Sonntag et al. 2007; Summerer et al. 2008). One of those species is the giant marine heterotrich ciliate *Maristentor dinoferus* inhabiting coral reefs of Guam (Lobban et al. 2002). *M. dinoferus* is a sessile species that hosts 500–800 *Symbiodinium* cells (Lobban et al. 2002). Methanolic extracts of this ciliate analyzed by high-performance liquid chromatography showed the presence of three different MAAs (Sommaruga et al. 2006). The symbiotic origin of MAAs in several freshwater *Chlorella*-bearing ciliates has been demonstrated (Sonntag et al. 2007). Moreover, Summerer et al. (2008) have shown that the *Chlorella* strains isolated from two *Askenasia chlorelligera* populations inhabiting lakes having a tenfold difference in underwater UV transparency presented a distinct physiological trait, such as the ability to synthesize MAAs.

The occurrence of MAAs in several *Chlorella*-bearing ciliate species suggests the existence of multiple alternative benefits in this type of mutualistic association. However, the symbiotic *Chlorella* from *P. bursaria* have tested negative for the presence of MAAs (Summerer et al. 2009), indicating that this physiological trait is not ubiquitous in *Chlorella*. However, this is not the only mechanism possible to obtain photoprotection, as we discuss in the next section.

## 2 UV-Induced Accumulation Behavior of Symbiotic *P. bursaria*

When symbiotic *P. bursaria* are exposed to PAR, the phenomenon of "photoaccumulation" can be observed (Engelmann 1882; Pado 1972; Saji and Oosawa 1974). During photoaccumulation, the light stimulus induces a depolarization of the ciliates' membrane potential and the individuals slow down or even come to a standstill (Machemer 1974; Nakaoka et al. 1987; Matsuoka and Nakaoka 1988). Generally, this behavior is a step-down photophobic response where the ciliates move from shaded regions into lighted areas where they are "trapped."

Summerer et al. (2009) studied the aggregation behavior of symbiotic and aposymbiotic *P. bursaria* in the presence of UVR and PAR. Ciliate cultures exposed to artificial UVR + PAR and "high" PAR (160 μmol $m^{-1}s^{-1}$) showed an immediate aggregation of symbiotic *P. bursaria* into several dense "spots" of about 3 mm in diameter. However, this aggregation occurred only when the culture density was 500 cells $ml^{-1}$ or higher (Summerer et al. 2009). In contrast to the "classical" photoaccumulation,

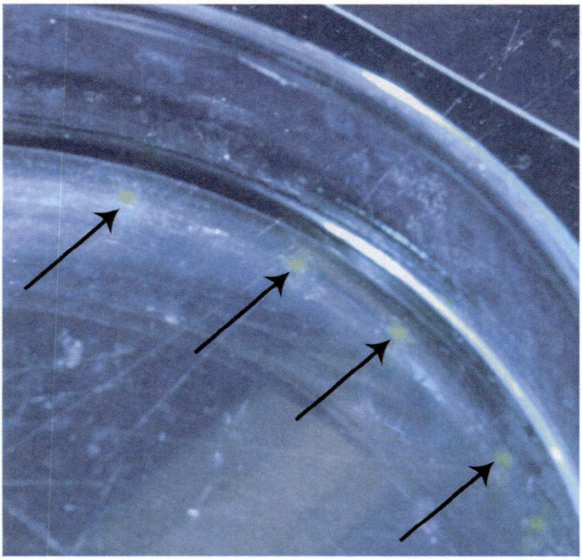

**Fig. 1** Photograph of symbiotic *Paramecium bursaria* aggregated into dense spots of 1–3 mm diameter (*arrows*) during exposure to artificial UV radiation (UVR)

this aggregation behavior of *P. bursaria* occurs within the completely illuminated experimental vessel (Fig. 1). Summerer et al. (2009) suggested that aggregation under UVR exposure is induced by chemical signaling and the encounter rate of individual *P. bursaria* cells.

In aposymbiotic *P. bursaria*, no spot-aggregation is observed under UVR and PAR exposure; thus, Summerer et al. (2009) argued that the *Chlorella* symbionts are responsible for the induction of the aggregation behavior in the symbiotic individuals. Because the pattern of photoaccumulation under PAR exposure is not observed in aposymbiotic or in *Chlorella*-reduced (fewer than 50 algae per ciliate), as well as in symbiotic *P. bursaria* treated with a photosynthesis inhibitor (dichlorophenyl dimethylurea), a light receptor is thought to be located in the algal symbionts (Iwatsuki and Naitoh 1981; Niess et al. 1981, 1982; Reisser and Häder 1984). Further, other *Chlorella*-bearing ciliates such as *Euplotes daidaleos* "photoaccumulate", whereas *Climacostomum virens* do not, though the aposymbiotic forms of both species show a step-up photophobic response (photodispersal), suggesting that the photoreceptor is located in the ciliates (Reisser and Häder 1984). Direct evidence for the presence of photoreceptors in the ciliate has been given by Tokioka et al. (1991), who extracted retinal, a chromophore of the visual pigment, from aposymbiotic *P. bursaria* and by Nakaoka et al. (1991), who detected a rhodopsin-like protein in the ciliary and somatic membranes of *P. bursaria*.

At present, the mechanisms that lead to the accumulation (behavior) under different wavelengths and intensities of the solar spectrum are not clear and further research is needed.

## 3 Photoprotection of *P. bursaria* by *Chlorella* Self-Shading

Another phenomenon observed when symbiotic *P. bursaria* are exposed to UVR + PAR is that their algal symbionts are dislocated and move to the posterior cell region (Summerer et al. 2009; Fig. 2). When the ciliates are returned to culture conditions without UVR, they disperse again in the medium and the symbionts are relocated to a "normal" distribution. However, it is unclear whether the displacement of the symbionts is controlled by the algae or the host (Summerer et al. 2009). Nishihara et al. (1999) detected cytoplasmatic movements of the symbionts in *P. bursaria* in connection with microtubule filaments. Usually, *P. bursaria* cells bear several hundred unicellular *Chlorella* of approximately 3.5 µm in diameter (Karakashian et al. 1968). Typically, within the host, the symbionts are arranged directly under the periphery of the cytoplasm, close to the macronucleus and micronucleus and to the buccal field (Karakashian et al. 1968). According to these distributions, especially underneath the cell surface and close to DNA-containing cell organelles, it seems likely that several cell layers of *Chlorella* and many food vacuoles, i.e., absorbing cell matter, increase the photoprotection of the host, as well as of the symbionts themselves (Giese 1968; Garcia-Pichel 1994; Summerer et al. 2009). On the basis of a bio-optical model to estimate the internal photoprotective self-shading potential of cell matter (Garcia-Pichel 1994), Summerer et al. (2009) calculated the transmittance and self-shading under UV exposure at 320 nm for one to several *Chlorella* cell layers within *P. bursaria*. When the *Chlorella* are distributed regularly, the screening efficiency achieved is 59% for one layer, and 83 and 93% for two and three layers, respectively. In *P. bursaria* with dislocated symbionts, 100% protection from UVR at 320 nm can only be achieved when at least eight layers are present (Fig. 3). The results of Summerer et al. (2009) indicate that *P. bursaria* can obtain protection against UV damage by accumulation, as well as by symbiont dislocation (Fig. 3).

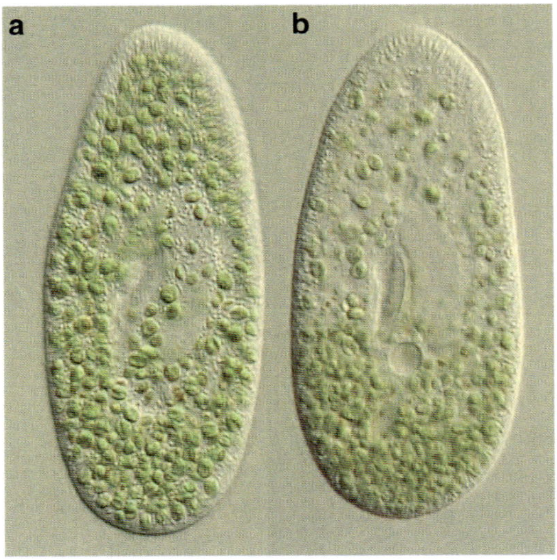


**Fig. 2** *P. bursaria* showing the distribution of *Chlorella* before (**a**) and after (**b**) exposure to UVR

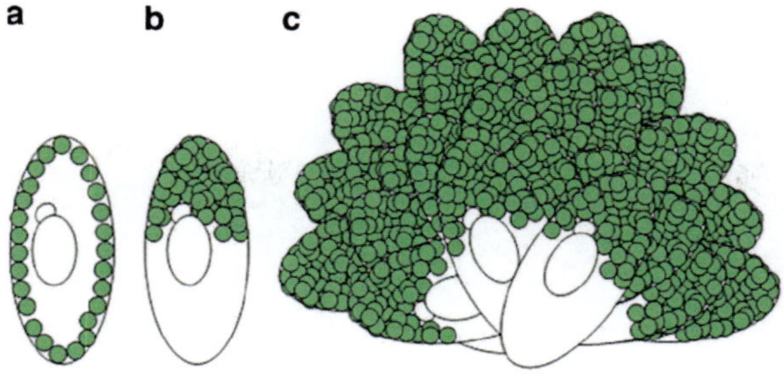


**Fig. 3** The *Chlorella* distribution within *P. bursaria* under PAR (**a**) and UVR (**b**) exposure. Hypothetical assemblage of *P. bursaria* with dislocated symbionts within dense spots ("collective shield") formed under UVR exposure (**c**)

## 4 Role of Endosymbiotic *Chlorella* in the (Photo-)oxidative Stress Balance of *P. bursaria*

Detailed consideration of oxidative stress is found in a number of reviews (Apel and Hirt 2004; Lesser 2006). It is therefore necessary to consider here only some basic aspects. Oxidative stress is defined as the production and accumulation of reactive oxygen species (ROS) beyond the capacity of an organism to quench these strong oxidants to a safe level (Halliwell and Gutteridge 1999; Lesser 2006). Oxidation of membrane lipids, DNA, proteins, and pigments and cellular death are the consequences of oxidative stress (Fridovich 1998). The formation of ROS is basically a consequence of redox reactions with $O_2$ in all photosynthetic and respiring cells. However, the basal level of oxidative stress can be increased by exposure of organisms to several types of photosensitizers (e.g., certain pollutants), as well as to excessive PAR and UVR. In aquatic ecosystems, extracellular and intracellular photo-oxidative production of ROS is known to impose a physiological burden on many organisms (Kieber et al. 2003; Lesser 2006).

The photosynthetic activity of endosymbiotic *Chlorella* leads not only to the fixation of carbon dioxide but also to the release of molecular oxygen, which because of its two unpaired electrons has a limited potential to react with organic molecules. However, the univalent reduction of molecular oxygen produces different partially reduced short-lived ROS, such as superoxide radical ($O_2^{\bullet -}$), which is the first free radical normally produced in aerobic cells, singlet oxygen ($^1O_2$), and hydroxyl radical ($HO\bullet$). Hydrogen peroxide ($H_2O_2$), which is not a radical, is produced by the dismutation of superoxide and its half-life is substantially longer than that for other ROS (Lesser 2006). In the chloroplast, the photosynthetic production of hydrogen peroxide, superoxide, and singlet oxygen proceeds through photorespiration, Mehler reaction (i.e., reduction of $O_2$ to $O_2^-$), and photodynamic actions, respectively (Asada 1999, Fig. 4a)

Similarly to other phototrophic symbioses, respiration of the ciliate and endosymbiotic algae at night and photosynthesis during daytime probably leads to dramatic changes in oxygen concentration within the host cell. Whereas photosynthetic organisms have efficient mechanisms to reduce oxidative stress, heterotrophic single cells

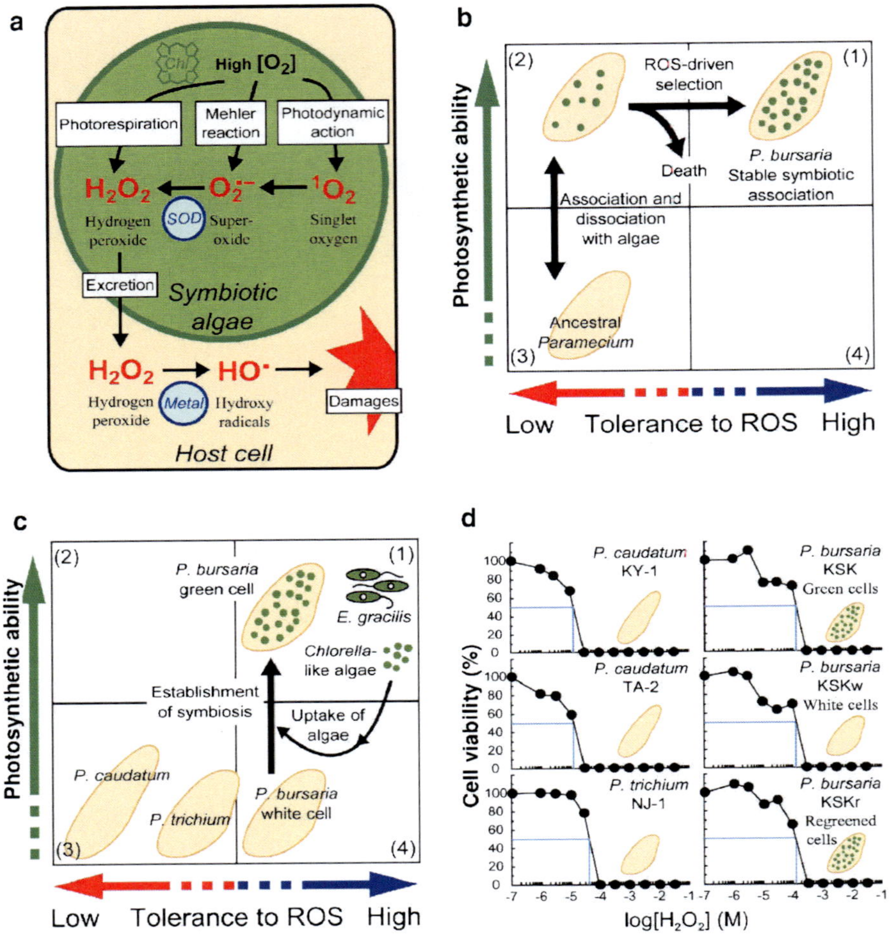


**Fig. 4** Oxidative stress in symbiotic *P. bursaria*. **a** Potential sources of reactive oxygen species (ROS) in symbiotic *P. bursaria*. In this scheme, singlet oxygen, superoxide, and hydrogen peroxide ($H_2O_2$) are produced mainly by symbiotic *Chlorella*'s photosynthetic activity. $H_2O_2$ may be excreted by *Chlorella* to the host and highly reactive hydroxyl radicals may be formed via $H_2O_2$ dissociation, which can damage the host. **b** Hypothetical model illustrating how cohabitation of the ancestral paramecia with algae was followed by the selection of ROS-tolerant species. **c** In this alternative model, emergence of ROS-tolerant paramecia enabled the symbiosis with algae. **d** Viability of different *Paramecium* species to the oxidative burst of $H_2O_2$. Data for *P. caudatum* (KY-1 and TA-2 strains), *P. trichium* (NJ-1 strain), and three cell lines derived from *P. bursaria* KSK-103 strain are shown. (Modified from Kawano et al. 2004)

may have had limited capacity to manage the excess of ROS caused by endosymbiotic algae. Kawano et al. (2004) presented two models to explain how the symbiosis between *P. bursaria* and *Chlorella* acquired the ability to cope with oxidative stress (Fig. 4b, c). In the first one, the symbiosis started by infection of *P. bursaria* cells having high tolerance to ROS that successfully acquired endosymbiotic *Chlorella*.

In the second model, the additional oxidative stress caused by the infection of the host by *Chlorella* selected for ROS-tolerant species during evolution. As Kawano et al. (2004) argued, in the first model the relation between the (re)greening ability of the paramecia and the tolerance to ROS is stressed, whereas in the second one evolutionary selection leading to emergence of ROS-tolerant *Paramecium* species is emphasized. The authors also discussed the possibility that ciliates could select for endosymbiotic *Chlorella* that are not so photosynthetically active and thus have lower oxidative stress potential. However, there is some circumstantial evidence suggesting that *P. bursaria* favor the symbiosis with active (i.e., high chlorophyll *a* and photosynthetic rates) *Chlorella* (Geraschenko et al. 2000).

Kawano et al. (2004) tested the tolerance to exogenous oxidative stress of diverse photosynthetic organisms including exsymbiotic and free-living *Chlorella* and different strains of green *P. bursaria*, aposymbiotic counterparts, as well as three other nonsymbiotic paramecia species. Cell viability of those organisms was measured in stationary cultures grown with PAR levels of approximately 33 μmol photons $m^{-2}s^{-1}$ after 12-h incubation in a range of hydrogen peroxide concentrations to establish the threshold lethal to 50% of the population ($LD_{50}$).

The results of those experiments (Fig. 4d) clearly demonstrated that free-living and exsymbiotic *Chlorella* have high tolerance ($LD_{50}$ 300–600 μM $H_2O_2$) to this extreme oxidative burst. Similarly, among the paramecia tested, the "green" *P. bursaria* was the most tolerant ($LD_{50}$ 120 μM), whereas the nonsymbiotic species *Paramecium trichium* ($LD_{50}$ 43 μM) and *Paramecium caudatum* ($LD_{50}$ 12 μM) were the most sensitive and cells burst out shortly after low $H_2O_2$ concentrations were added. Remarkably, the aposymbiotic *P. bursaria* strains and those reinfected with *Chlorella* had tolerance similar to the symbiotic ones. This result led Kawano et al. (2004) to conclude that endosymbiotic *Chlorella* do not contribute to the higher resistance observed in "green" *P. bursaria*. However, it is intriguing that the $LD_{50}$ these authors found for free-living *Chlorella* and exsymbiotic ones is 2.5–5 times higher than for the symbiotic *Chlorella-P. bursaria* association.

Hörtnagl and Sommaruga (2007) studied whether oxidative stress and UV-induced photo-oxidative stress are greater in *Chlorella*-bearing *P. bursaria* than in its aposymbiotic counterpart. In this study, the level of oxidative stress was directly determined by assessing ROS with two fluorescent probes (hydroethidine and dihydrorhodamine 123) by flow cytometry.

Dihydrorhodamine 123 is an indicator of hydrogen peroxide levels, whereas hydroethidine is a more general proxy for ROS because it reacts with intracellular superoxide and also with hydrogen peroxide. This approach makes possible the rapid detection of oxidative levels in a population of cells and also the visualization by epifluorescence microscopy of oxidative burst within cells (Fig. 5). These experiments indicated that oxidative stress (PAR 160 μmol photons $m^{-2}s^{-1}$) is higher in aposymbiotic ciliates than in "green" *P. bursaria* (Fig. 6). This oxidative stress was higher during the exponential growth of both populations, but particularly in the aposymbiotic *P. bursaria*. During exponential growth, high cell division rates are accompanied by an increase in metabolic activity that usually results in higher oxidative stress levels (Hörtnagl and Sommaruga 2007). Interestingly, no significant

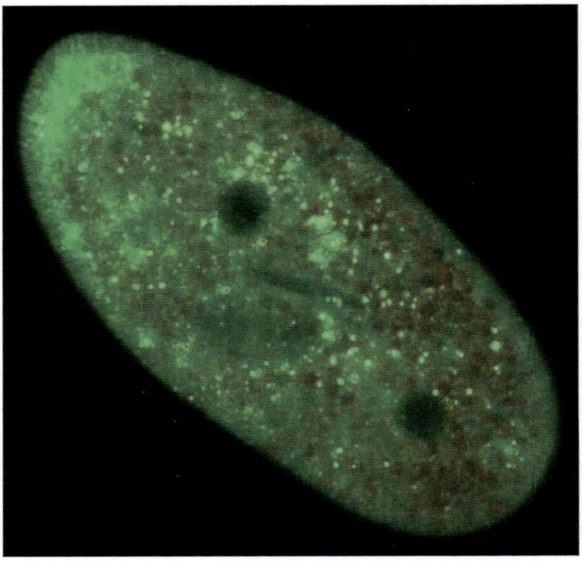

**Fig. 5** *P. bursaria* showing spots of oxidative stress (*brilliant green*) as detected by the fluorescent probe 2′, 7′-dichlorodihydrofluorecein diacetate under the epifluorescence microscope. The autofluorescence of the chlorophyll pigments of *Chlorella* appears in *red*. The two *dark spots* are the contractile vacuoles. Note that oxidative spots are found in the ciliate's cell but not inside *Chlorella*

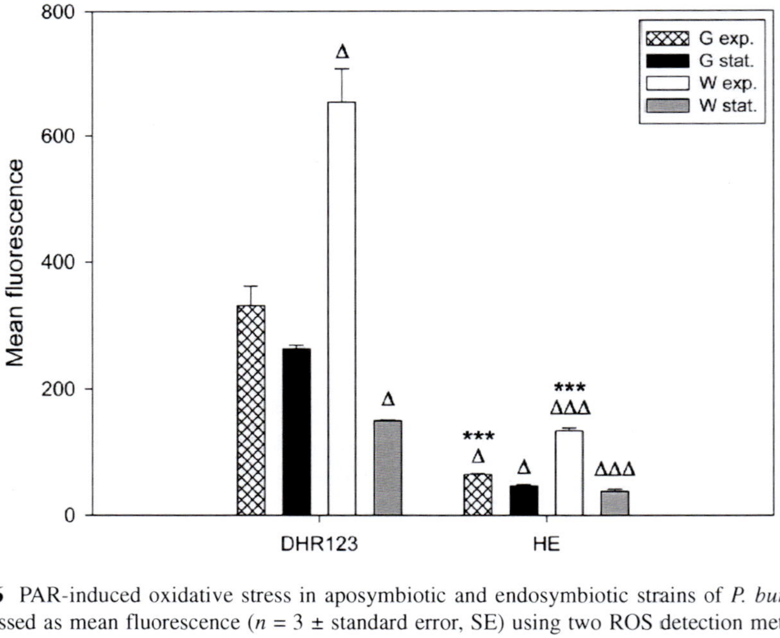


**Fig. 6** PAR-induced oxidative stress in aposymbiotic and endosymbiotic strains of *P. bursaria* expressed as mean fluorescence ($n = 3 \pm$ standard error, SE) using two ROS detection methods. *Stars* denote significant differences between endosymbiotic (*G*) and aposymbiotic (*W*) strains, whereas *triangles* indicate significant differences between exponential (*exp*) and stationary (*stat*) growth phases. $^{*/\Delta}p < 0.05$, $^{***/\Delta\Delta\Delta}p < 0.001$. (From Hörtnagl and Sommaruga 2007)

difference in oxidative stress is found within growth phases of free-living *Chlorella vulgaris* (Malanga and Puntarulo 1995). However, it is questionable whether this result applies to endosymbiotic *Chlorella* too.

The results obtained by Hörtnagl and Sommaruga (2007) contradict somewhat the conclusion drawn by Kawano et al. (2004) that "green" and "white" *P. bursaria* are not different regarding their response to oxidative stress. However, it is important to note that the response measured by Kawano et al. (2004) relates mainly to the enzymatic defenses of the "green" and "white" cell lines of *P. bursaria*, particularly of the levels of catalase, which is one of the enzymes responsible for deactivating hydrogen peroxide. Hörtnagl and Sommaruga (2007), however, measuring the endogenous levels of oxidative stress, concluded that endosymbiotic *Chlorella* do not burden the host and there are clear differences between these two types of cell lines. As discussed below, the reason for the higher oxidative stress in aposymbiotic *P. bursaria* partially relates to differences in their antioxidative defenses. However, Gu et al. (2002) studying the physiological changes that occur after 1 month of dark cultivation of *P. bursaria* found an increase in the activity of several enzymes (notably of acid phosphatase and ATPase, but also of glucose 6-phosphatase and succinate dehydrogenase) compared with light-cultured cells. These results suggest that the aposysmbiotic strain increases its metabolic activity (and oxidative stress) to compensate for the lack of endosymbiotic *Chlorella*.

Hörtnagl and Sommaruga (2007) also measured the response of "green" and "white" cell lines of *P. bursaria* to artificial UVR exposure (i.e., photo-oxidative stress). As mentioned in the introduction of this section, UVR stimulates the generation of ROS in both heterotrophic and photoautotrophic organisms. Again, photo-oxidative stress caused by UVR exposure was higher in "white" than in "green" *P. bursaria* (Fig. 7).

Hörtnagl and Sommaruga (2007) argued that the lower UV-induced photo-oxidative stress in the "green" strain of *P. bursaria* is related to the screening of UV wavelengths by several layers of *Chorella* (see Sect. 3).

## 5 Antioxidative Capacity of the *P. bursaria–Chlorella* Symbiosis

Aquatic organisms have different defense mechanisms to minimize the harmful effects of ROS that include various antioxidative enzymes and nonenzymatic antioxidants. In fact, ROS trigger a series of responses that provide increased protection against the damaging effect of these strong oxidants. Among the antioxidative enzymes, superoxide dismutase (SOD; EC 1.15.1.1) is the first line of defense removing superoxide but producing hydrogen peroxide, which is eliminated by catalase (EC 1.11.1.6) and several peroxidases. Another antioxidative enzyme, glutathione reductase (EC 1.6.4.2), is involved in the conversion of oxidized glutathione to its reduced form with consumption of NADPH (Dunlap et al. 2000). The activity of catalase is crucial in keeping low oxidative damage within cells because hydrogen peroxide is uncharged, making possible its easy diffusion across biological membranes (Asada 1999; Lesser 2006). Consequently, significant damage can occur in different cell

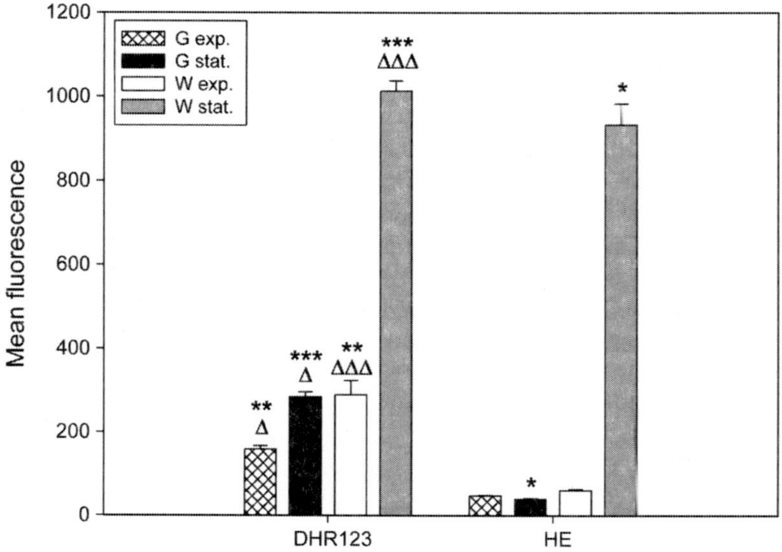


**Fig. 7** UV-induced oxidative stress in endosymbiotic and aposymbiotic *P. bursaria* strains expressed as mean fluorescence ($n$ = 3 ± SE) for two different ROS detection methods. *Stars* denote significant differences between endosymbiotic (*G*) and aposymbiotic (*W*) strains, whereas *triangles* indicate significant differences between exponential (*exp*) and stationary (*stat*) growth phases of the ciliates from the endosymbiotic or the aposymbiotic strain. $^{*/\Delta}p < 0.05$, $^{**/\Delta\Delta}p < 0.01$, $^{***/\Delta\Delta\Delta}p < 0.001$. (From Hörtnagl and Sommaruga 2007)

compartments because the reaction of hydrogen peroxide with different cell targets is not restricted to its place of synthesis (Lesser 2006). In fact, Kawano et al. (2004) argued that hydrogen peroxide levels within symbiotic *Chlorella* are probably reduced to safe levels by diffusion, but the host ciliate will have to cope with oxidative stress. In addition to antioxidative enzymes, glutathione, ascorbic acid, α-tocopherol, β-carotene, and uric acid can protect many cellular constituents from ROS because these substances interrupt the spreading of autocatalytic radical reactions (Lesser 2006).

Hörtnagl and Sommaruga (2007) screened the antioxidant defenses of "white" and "green" *P. bursaria* by measuring the activity of catalase, SOD, and glutathione reductase in cells exposed to either UVR + PAR or just PAR. Whereas glutathione reductase activity was undetectable in all cases, UVR strongly decreased catalase activity during the exponential and stationary growth phases of both strains (Fig. 8). Catalase structure and function appears to be very sensitive to UVR (Zigman et al. 1996) and similar negative effects have been observed in other aquatic organisms, but this is not a general pattern. For example, catalase activity is not reduced by UVR in free-living *C. vulgaris* (Malanga and Puntarulo 1995). However, the enzymatic activity measured by Hörtnagl and Sommaruga (2007) should be regarded as that from the symbiosis, since the authors did not differentiate between symbiont and host sources. However, the catalase gene has been found in *P. bursaria* (Tonooka et al. 2008) and gene expression analysis in host paramecia in response to oxidative stress has been proved positive (Y. Tonooka, personal communication).

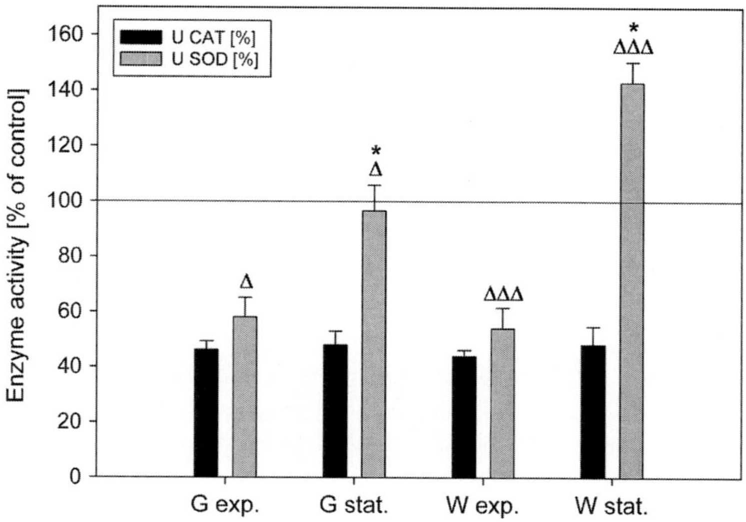


**Fig. 8** Relative activity of the antioxidant scavenging enzymes catalase (*CAT*) and superoxide dismutase (*SOD*) in endosymbiotic (*G*) and aposymbiotic (*W*) strains of *P. bursaria* exposed for 2 h to PAR + UVR. Enzyme activity ($n = 3 \pm$ SE) is expressed as the percentage of the PAR control (*horizontal line* reference) for the respective strains and growth phases. *Stars* denote significant differences between endosymbiotic and aposymbiotic strains, whereas *triangles* indicate significant differences between exponential (*exp*) and stationary (*stat*) growth phases of the ciliates from the endosymbiotic or the aposymbiotic strain. $^{*/\Delta}p < 0.05$, $^{***/\Delta\Delta\Delta}p < 0.001$. (From Hörtnagl and Sommaruga 2007)

A negative effect of UVR was also detected on SOD activity but only during the exponential growth phase of both strains (Hörtnagl and Sommaruga 2007). By contrast, SOD activity significantly increased in the aposymbiotic strain during the stationary growth phase (Fig. 8). As discussed by Hörtnagl and Sommaruga (2007), the increase in SOD activity did not counteract the direct and indirect (i.e., ROS) damaging effects of UVR on aposymbiotic ciliates as suggested by the authors' observation of an erratic swimming behavior after 2-h exposure. Further, they concluded that the higher tolerance of the symbiotic strain of *P. bursaria* can also be attributed to the antioxidant defenses from the algae.

## 6 Conclusions and Perspectives

In this chapter, we have shown new aspects on the *P. bursaria–Chlorella* association in relation to the photoprotective potential of the symbiosis against direct and indirect (photo-oxidative stress) damaging effects of solar irradiation. The distinct responses of symbiotic *P. bursaria* when they are exposed to artificial UVR and PAR include physical protection by spot-accumulation together with the phenomenon of symbiont dislocation, as well as physiological adaptations in the oxidative stress balance between the algae and the host. At present, the exact mechanisms that trigger these reactions are largely unknown. Despite *P. bursaria* certainly being the best studied *Chlorella*-bearing ciliate species for over more than one century, possible effects of solar radiation and associated photoprotective strategies have been considered only marginally. However, their exploration would significantly contribute to the mostly unknown ciliate's photobiology and to the understanding of the environmental adaptation and (co-)evolution of symbiotic species, as well as of the establishment of mutualistic associations.

Many gaps in the photobiological knowledge of these protists can hopefully be filled in future studies when answering questions such as the following: What is the (chemical?) signal that induces the accumulation behavior of symbiotic *P. bursaria* when exposed to UVR and PAR? Do the algae or does the host trigger the symbiont displacement? Would *Chlorella*-bearing ciliate species from highly transparent habitats have different numbers of symbiotic algae than those from less transparent ones?

We hope this short review will stimulate the study of at least some of these and other photobiological aspects in the near future to better understand the specific relationship between the algae and the host in *P. bursaria* but also in other symbiont-bearing ciliate species.

**Acknowledgements** We thank M.T. Pérez and M. Summerer for critical reading of the original manuscript. We are also indebted to M. Summerer for providing Figs. 1–3, P. Hörtnagl for providing Fig. 5, and T. Kawano for providing Fig. 4 in color. This work was supported by the Austrian Science Fund (project FWF 16559) granted to R.S.

## End Note

The original light intensity in the paper by Kawano et al. (2004) was given in lux, which can only be converted to micromole photons per square meter per second when the wavelength dependency is considered. Here a factor of 0.0165 was applied to facilitate comparison with other studies.

## References

Apel K, Hirt H (2004) Reactive oxygen species: metabolism, oxidative stress, and signal transduction. Annu Rev Plant Biol 55:373–399

Asada K (1999) The water-water cycle in chloroplasts: scavenging of active oxygen and dissipation of excess photons. Annu Rev Plant Physiol Plant Mol Biol 50:601–639

Aury J-M, Jaillion O, Duret L et al (2006) Global trends of whole-genome duplications revealed by the ciliate *Paramecium tetraurelia*. Nature 444:171–178

Berger J (1980) Feeding behavior of *Didinium nasutum* on *Paramecium bursaria* with normal or apochlorotic zoochlorellae. J Gen Microbiol 118:397–404

Berk SG, Parks LH, Ting RS (1991) Photoadaptation alters the ingestion rate of *Paramecium bursaria*, a mixotrophic ciliate. Appl Environ Microbiol 57:2312–2316

Berninger U-G, Finlay BJ, Canter HM (1986) The spatial distribution and ecology of zoochlorellae-bearing ciliates in a productive pond. J Protozool 33:557–563

Cockell CS (1998) Ultraviolet radiation, evolution and the π-electron system. Biol J Linn Soc 62:449–457

Dionisio-Sese L, Maruyama T, Miyachi S (2001) Photosynthesis of *Prochloron* as affected by environmental factors. Mar Biotechnol 3:74–79

Doebeli M, Knowlton N (1998) The evolution of interspecific mutualisms. Proc Natl Acad Sci USA 95:8676–8680

Dolan J (1992) Mixotrophy in ciliates: a review of *Chlorella* symbiosis and chloroplast retention. Mar Microb Food Webs 6:115–132

Dolan J, Pérez MT (2000) Costs, benefits and characteristics of mixotrophy in marine oligotrichs. Freshwat Biol 45:227–238

Dunlap WC, Shick JM (1998) Ultraviolet radiation-absorbing mycosporine-like amino acids in coral reef organisms: a biochemical and environmental perspective. J Phycol 34:418–430

Dunlap WC, Shick JM, Yamamoto Y (2000) UV protection in marine organisms. I. Sunscreens, oxidative stress and antioxidants. In: Yoshikawa T, Toyokuni S, Yamamoto M, Naito Y (eds) Free radicals in chemistry, biology and medicine. OICA International, London, pp 200–214

Engelmann TW (1882) Ueber Licht- und Farbenperception niederster Organismen. Arch Ges Physiol 29:387–400

Finlay BJ, Maberly SC, Esteban GF (1996) Spectacular abundance of ciliates in anoxic pond water: contribution of symbiont photosynthesis to host respiratory oxygen requirements. FEMS Microbiol Ecol 20:229–235

Fridovich I (1998) Oxygen toxicity: a radical explanation. J Exp Biol 201:1203–1209

Garcia-Pichel F (1994) A model for internal self-shading in planktonic organisms and its implications for the usefulness of ultraviolet sunscreens. Limnol Oceanogr 39:1704–1717

Geraschenko BI, Nishihara N, Ohara T, Tosuji H, Kosaka T, Hosoya H (2000) Flow cytometry as a strategy to study the endosymbiosis of algae in *Paramecium bursaria*. Cytometry 41:209–215

Giese AC (1945) The ultraviolet action spectrum for retardation of *Paramecium*. J Cell Comp Physiol 26:47–55

Giese AC (1968) Ultraviolet action spectra in perspective: with special reference to mutation. Photochem Photobiol 8:527–546

Görtz H-D (1982) Infection of *Paramecium bursaria* with bacteria and yeasts. J Cell Sci 58:445–453

Görtz H-D (1988) Endocytobiosis. In: Görtz H-D (ed) *Paramecium*. Springer, Berlin, pp 393–405

Gu F, Chen L, Ni B, Zhang, X (2002) A comparative study on the electron microscopic enzymo-cytochemistry of *Paramecium bursaria* from light and dark cultures. Eur J Protistol 38:267–278

Halliwell B, Gutteridge JMC (1999) Free radicals in biology and medicine. Oxford University Press, New York

Herre EA, Knowlton N, Mueller UG, Rehner SA (1999) The evolution of mutualisms: exploring the paths between conflict and cooperation. Trends Ecol Evol 14:49–53

Hörtnagl P, Sommaruga R (2007) Photo-oxidative stress in symbiotic and aposymbiotic strains of the ciliate *Paramecium bursaria*. Photochem Photobiol Sci 6:842–847

Iwatsuki K, Naitoh Y (1981) The role of symbiotic *Chlorella* in photoresponses of *Paramecium bursaria*. Proc Jpn Acad Ser B 57:318–323

Jones RI (1994) Mixotrophy in planktonic protists as a spectrum of nutritional strategies. Mar Microb Food Webs 8:87–96

Kang M, Moroni A, Gazzarrini S, Van Etten JL (2003) Are chlorella viruses a rich source of ion channel genes? FEBS Lett 552:2–6

Karakashian SJ, Karakashian MW, Rudzinska MA (1968) Electron microscopic observations on the symbiosis of *Paramecium bursaria* and its intracellular algae. J Protozool 15:113–128

Karentz D, McEuen FS, Land MC, Dunlap WC (1991) Survey of mycosporine-like amino acid compounds in Antarctic marine organisms: potential protection from ultraviolet exposure. Mar Biol 108:157–166

Kawano T, Kadono T, Kosaka T, Hosoya H (2004) Green paramecia as an evolutionary winner of oxidative symbiosis: a hypothesis and supportive data. Z Naturforsch 59:538–542

Kieber DJ, Peake BM, Scully NS (2003) Reactive oxygen species in aquatic ecosystems. In: Helbling EW, Zagarese H (eds) UV effects in aquatic organisms and ecosystems. Royal Society of Chemistry, Cambridge, pp 251–288

Landis WG (1988) Ecology. In: Görtz H-D (ed) *Paramecium*. Springer, Berlin, pp 419–436

Lesser MP (2006) Oxidative stress in marine environment: biochemistry and physiological ecology. Annu Rev Physiol 68:253–278

Li L, Song W (2006) Phylogenetic positions of two cyrtophorid ciliates, *Dysteria procera* and *Hartmannula derouxi* (Ciliophora: Phyllopharyngea: Dysteriida) inferred from the complete small subunit ribosomal RNA gene sequences. Acta Protozool 45:265–270

Lobban CS, Schefter M, Simpson AGB, Pochon X, Pawlowski J, Foissner W (2002) *Maristentor dinoferus* n. gen., n. sp., a giant heterotrich ciliate (Spirotrichea: Heterotrichida) with zooxanthellae, from coral reefs on Guam, Mariana Islands. Mar Biol 140:411–423

Lobban CS, Hallam SJ, Mukherjee P, Petrich JW (2007) Photophysics and multifunctionality of hypericin-like pigments in heterotrich ciliates: a phylogenetic perspective. Photochem Photobiol 83:1074–1094

Machemer H (1974) Frequency and directional responses of cilia to membrane potential changes in *Paramecium*. J Comp Physiol 92:293–316

Malanga G, Puntarulo S (1995) Oxidative stress and antioxidant content in *Chlorella vulgaris* after exposure to ultraviolet-B radiation. Physiol Plant 94:672–679

Martin-Webb EB (1999) The effect of solar ultraviolet radiation on marine planktonic ciliates survival and the penetration of ultraviolet irradiance on Georges Bank, Ph. D. thesis, University of Rhode Island, Kingston

Matsuoka K, Nakaoka Y (1988) Photoreceptor potential causing phototaxis of *Paramecium bursaria*. J Exp Biol 137:477–485

McTavish C, Sommerville J (1980) Macronuclear DNA organization and transcription in *Paramecium primaurelia*. Chromosoma 78:147–164

Meier R, Wiessner W (1988) Infection of algae-free *Paramecium bursaria* with symbiotic *Chlorella* sp., isolated from green paramecia, I: effect of incubation period. Eur J Protistol 24:69–74

Meier R, Wiessner W (1989) Infection of algae-free *Paramecium bursaria* with symbiotic *Chlorella* sp., isolated from green paramecia, II: a timed study. J Cell Sci 93:571–579

Miyake AF, Buonanno P, Saltalamacchia P, Masaki ME, Iio H (2003) Chemical defense by means of extrusive cortical granules in the heterotrich ciliate *Climacostomum virens*. Eur J Protistol 39:25–36

Modenutti BE, Balseiro EG, Moeller R (1998) Vertical distribution and resistance to ultraviolet radiation of a planktonic ciliate, *Stentor araucanus*. Verh Int Ver Limnol 26:1636–1640

Muscatine L, Karakashian SJ, Karakashian MW (1967) Soluble extracellular products of algae symbiotic with a ciliate, a sponge and a mutant *Hydra*. Comp Biochem Physiol 20:1–12

Nakaoka Y, Kinugawa K, Kurotani T (1987) $Ca^{2+}$-dependent photoreceptor potential in *Paramecium bursaria*. J Exp Biol 131:107–115

Nakaoka Y, Tokioka R, Shinozawa T, Fujita J, Usukura J (1991) Photoreception of *Paramecium* cilia: localization of photosensitivity and binding with anti-frog-rhodopsin IgG. J Cell Sci 99:67–72

Niess D, Reisser W, Wiessner W (1981) The role of endosymbiotic algae in photoaccumulation of green *Paramecium bursaria*. Planta 152:268–271

Niess D, Reisser W, Wiessner W (1982) Photobehaviour of *Paramecium bursaria* infected with different symbiotic and aposymbiotic species of *Chlorella*. Planta 156:475–480

Nishihara N, Horjike S, Oka Y, Takahashi T, Kosaka T, Hosoya H (1999) Microtubule-dependent movement of symbiotic algae and granules in *Paramecium bursaria*. Cell Motil Cytoskeleton 43:85–98

Norris, RD (1996) Symbiosis as an evolutionary innovation in the radiation of Paleocene planktic foraminifera. Paleobiology 22:461–480

Pado R (1972) Spectral activity of light and phototaxis in *Paramecium bursaria*. Acta Protozool 11:387–393

Pérez MT, Dolan JR, Fukai E (1997) Planktonic oligotrich ciliates in the NW Mediterranean growth rates and consumption by copepods. Mar Ecol Prog Ser 155:89–101

Pringsheim EG (1928) Physiologische Untersuchungen an *Paramecium bursaria*. Ein Beitrag zur Symbioseforschung. Arch Protistenkd 64:289–418

Queimaliños CP, Modenutti BE, Balseiro EG (1999) Symbiotic association of the ciliate *Ophrydium naumanni* with *Chlorella* causing a deep chlorophyll a maximum in an oligotrophic South Andes lake. J Plankton Res 21:67–178

Reisser W (1980) The regulation of the algal population size in *Paramecium bursaria*. In: Schenk HEA, Schwemmler W (eds) Endocytobiology, vol 2. de Gruyter, Berlin, pp 97–104

Reisser W, Häder D-P (1984) Role of endosymbiotic algae in photokinesis and photophobic responses of ciliates. Photochem Photobiol 39:673–678

Reisser W, Fisher-Defoy D, Staudinger J, Schelling N, Hausmann K (1984) The endosymbiotic unit of *Climacostomum virens* and *Chlorella* sp. 1: morphological and physiological studies of the algal partner and its localization in the host cell. Protoplasma 119:93–99

Reyes-Prieto A, Moustafa A, Bhattacharya D (2008) Multiple genes of apparent algal origin suggest ciliates may once have been photosynthetic. Curr Biol 18:956–962

Saji M, Oosawa F (1974) Mechanism of photoaccumulation in *Paramecium bursaria*. J Protozool 21:556–561

Schlegel M, Elwood HJ, Sogin ML (1991) Molecular evolution in hypotrichous ciliates: sequence of the small subunit RNA genes from *Onychodromus quadricornutus* and *Oxytricha granulifera* (Oxytrichidae, Hypotrichida, Ciliophora). J Mol Evol 32:64–69

Shick JM, Dunlap WC (2002) Mycosporine-like amino acids and related gadusols: biosynthesis, accumulation, and UV-protective functions in aquatic organisms. Annu Rev Physiol 64:223–262

Shick JM, Lesser MP, Jokiel PJ (1996) Effects of ultraviolet radiation on corals and other coral reef organisms. Global Change Biol 2:527–545

Shick JM, Romaine-Lioud S, Ferrier-Pages C, Gattuso J-P (1999) Ultraviolet-B radiation stimulates shikimate pathway-dependent accumulation of mycosporine-like amino acids in the coral *Stylophora pistillata* despite decreases in its population of symbiotic dinoflagellates. Limnol Oceanogr 44:1667–1682

Sommaruga R, Garcia-Pichel F (1999) UV-absorbing mycosporine-like compounds in planktonic and benthic organisms from a high-mountain lake. Arch Hydrobiol 144:255–269

Sommaruga R, Whitehead K, Shick JM, Lobban CS (2006) Mycosporine-like amino acids in the zooxanthella-ciliate symbiosis *Maristentor dinoferus*. Protist 157:185–191

Sonntag B, Summerer M, Sommaruga R (2007) Sources of mycosporine-like amino acids in planktonic *Chlorella*-bearing ciliates (Ciliophora). Freshwat Biol 52:1476–1485

Stabell T, Andersen T, Klaveness D (2002) Ecological significance of endosymbionts in a mixotrophic ciliate – an experimental test of a simple model of growth coordination between host and symbiont. J Plankton Res 24:889–899

Stoecker DK (1998) Conceptual models of mixotrophy in planktonic protists and some ecological and evolutionary implications. Eur J Protistol 34:281–290

Summerer M, Sonntag B, Sommaruga R (2007) An experimental test of the symbiosis specificity between the ciliate *Paramecium bursaria* and strains of the unicellular green alga *Chlorella*. Environ Microbiol 9:2117–2122

Summerer M, Sonntag B, Sommaruga R (2008) Ciliate-symbiont specificity of freshwater endosymbiotic *Chlorella* (Trebouxiophyceae, Chlorophyta). J Phycol 44:77–84

Summerer M, Sonntag B, Hörtnagl P, Sommaruga R (2009) Symbiotic ciliates receive protection against UV damage from their algae: A test with *Paramecium bursaria* and *Chlorella*. Protist doi:10.1016/j.protis.2008.11.005

Tartarotti B, Baffico G, Temporetti P, Zagarese HE (2004) Mycosporine-like amino acids in planktonic organisms living under different UV exposure conditions in Patagonian lakes. J Plankton Res 26:753–762

Tokioka R, Matsuoka K, Nakaoka Y, Kito Y (1991) Extraction of retinal from *Paramecium bursaria*. Photochem Photobiol 53:149–151

Tonooka Y, Watanabe T (2002) A natural strain of *Paramecium bursaria* lacking symbiotic algae. Eur J Protistol 38:55–58

Tonooka Y, Mizukami Y, Fujishima M (2008) One-base excess adaptor ligation method for walking uncloned genomic DNA. Appl Microbiol Biotechnol 78:173–180

Van Etten JL (2003) Unusual life style of giant chlorella viruses. Annu Rev Genet 37:153–195

Woelfl S, Geller W (2002) *Chlorella*-bearing ciliates dominate in an oligotrophic North Patagonian Lake (Lake Pirehueico, Chile): abundance, biomass and symbiotic photosynthesis. Freshwat Biol 47:231–242

Zigman S, Reddan J, Schultz JB, McDaniel T (1996) Structural and functional changes in catalase induced by near-UV radiation. Photochem Photobiol 63:818–824

# Diversity of Endosymbiotic Bacteria in *Paramecium*

**Hans-Dieter Görtz and Sergei I. Fokin**

**Contents**



H.D. Görtz (✉)
Department of Zoology, Biologisches Institut, Universität Stuttgart, Pfaffenwaldring 57, 70569 Stuttgart, Germany
e-mail: goertz@bio.uni-stuttgart.de

S.I. Fokin
Department of Zoology Invertebrate, Biological and Soil Faculty, St. Petersburg State University, Universitetskaya emb. 7/9, 199034 St. Petersburg, Russia
Department of Biology, Pisa University, Via A. Volta 4, 56126 Pisa, Italy

M. Fujishima (ed.), *Endosymbionts in Paramecium*, Microbiology Monographs 12,
DOI: 10.1007/978-3-540-92677-1_6, 

**Abstract** For about one and a half centuries now microbial symbioses in *Paramecium* have been observed and investigated. Meanwhile a great diversity of endosymbiotic bacteria is known in many of the different species of *Paramecium*. *Paramecium* is a unicellular but complete eukaryotic organism. For microbial symbionts, the large cells offer plenty of space and a variety of niches of different metabolic conditions. Bacteria living in the cytoplasm, in micronuclei or macronuclei, or even in the perinuclear space show quite different life strategies. Peculiar adaptations to the symbiotic mode of life have stimulated research in different fields, such as microbiology, cell biology, physiology, ecology, and phylogenetics. The symbionts of *Paramecium* not only turn out to belong to different taxa of *Eubacteria*, but some appear to be related to human pathogens. As a special highlight, an endonuclear symbiont of *Paramecium*, *Holospora obtusa*, may be the closest relative of mitochondria known to date.

## 1 Introduction

*Paramecium* like many other ciliates is a bacterial feeder. This bears the risk of being infected by bacteria that may withstand acidification and attacks by hydrolytic enzymes in the phagosome (Fok and Allen 1988). *Paramecium* may therefore be infected and colonized by bacteria more often than other protists, namely, autotrophic ones. Compared with other eukaryotic cells *Paramecium* is large, 100–250 μm long, and with its different compartments offers good conditions for a variety of intracellular bacteria (Wichterman 1986; Allen 1988). Traditionally, intracellular bacteria in ciliates have been termed "symbionts," "endosymbionts" (Preer et al. 1974), or "endocytobionts" (Sitte 1993). These terms are used in a neutral way, not necessarily in the sense of mutualism. In fact, some of these bacteria are rather parasitic. It makes sense to use such neutral terms as it is not easy in many of the symbioses to clarify whether their nature is mutualistic or rather parasitic. Killer symbionts, e.g., produce toxins that may kill competitors for the host cell, but under unfavorable conditions certain killer bacteria may overgrow the host cell (Schmidt et al. 1987).

We should be aware that in well-established, intimate symbioses, symbionts and the host represent a new entity that is exposed to selection and evolution as a unit. The question of whether or not each of the partners has an advantage from symbiosis may be put wrongly in such cases. Neither the host nor the symbiont can usually get rid of the partner. The unit of host and symbionts may occupy a new ecological niche or simply has to compete with uninfected individuals (Görtz et al. 2008).

On the basis of the diversity of free-living bacteria one would expect a great variety of bacterial symbionts in bacteria feeders. It is the aim of this chapter to review the diversity of bacterial symbionts in *Paramecium* and the wide spectrum of adaptations found in these symbioses.

## 2 Symbiont Diversity in *Paramecium* and its Research

*Paramecium* has been studied since the beginning of microscopy. Certainly it is one of the best investigated free-living protists. Not surprisingly, about 60 microbial symbionts of *Paramecium* have been detected and described in varying detail. Earlier work on microbial symbionts has been reviewed repeatedly (Preer et al. 1974; Gibson 1974, Ossipov 1981, Quackenbush 1988, Heckmann and Görtz 1991; Fokin 2004, Görtz and Schmidt 2005; Görtz 2007, 2008). Up to the 1970s biological and ultrastructural data had been used for the characterization of symbiont species almost exclusively. Methods of classical microbiology could not be applied, as the symbionts could, and still can, not usually be cultured outside host organisms. In recent years with the rise of molecular phylogenetics the diversity of intracellular bacteria in *Paramecium* is again being investigated and new techniques in molecular biology and biochemistry have brought about new insights into the biology of obligate endocytobionts and the symbiotic way of life.

### 2.1 Symbiont Diversity

Bacterial endocytobionts of ciliates comprise a variety of bacteria in different subgroups of *Proteobacteria* and other phyla. The common feature of these symbionts is that their habitat is the cytoplasm or a nucleus of a ciliate cell. Many appear to be well adapted to their environment; they are no longer found free-living and their genomes prove to be reduced in size, possibly indicating a lengthy period of symbiont–host association. Most of the symbionts are not infectious, but some are.

From the view of a cell biologist, it may not be so surprising that these infectious symbionts have developed specific features that guarantee uptake and transport to the intracellular sites where they can multiply. It is, however, quite unexpected that some bacteria have developed mechanisms to leave the host cells and reproducibly with each infection find their way into the cytoplasm, nuclei, or even the perinuclear space (the space within the membranes of the nuclear envelope) (Fokin 1989a; Fokin and Karpov 1995; Fokin 1988).

There are many open questions concerning invasion. How can a bacterium of 0.5-μm width pass the nuclear envelope? Nuclear pores have an inner width of about 9 nm. *Holospora* bacteria seem to make use of membrane fusion. Membranes of a transport vesicle, which takes the invading bacterium towards the nucleus, fuse with the membranes of the nuclear envelope (Görtz and Wiemann 1989). For other intranuclear bacteria, invasion into the nuclei is not understood. The mode of entry into the nucleus is unknown in case of rickettsias too. Rickettsias such as *Rickettsia rickettsii* may sometimes be found in host nuclei in early stages of infection (Yu and Walker 2005).

There are many reports of symbionts in ciliates. Some bacteria have only been mentioned briefly in ultrastructure reports; other symbionts have been investigated extensively. More than 60 intracellular bacteria have been found in *Paramecium* and many more will be detected in the next few years. When paramecia freshly isolated from nature are investigated, they frequently bear bacteria in their cytoplasm or in the nuclei. Many of the symbionts in isolates from nature are known species, but others are new. In Germany *Holosporacaryophila, H. obtusa*, and *Caedibacter caryophilus* are the symbiont species found most often in *Paramecium*. However, when checked carefully, unknown bacteria are found at considerable frequencies. Some of the bacteria in *Paramecium* have been deposited in stock cultures and/or have been investigated and described in such a way that they can be identified when found again. However, owing to the availability of only poor characters, many of the symbionts from *Paramecium* described earlier may not be identified again. This is regrettable, the more so as the old descriptions include valuable information about the biological nature of the symbioses, e.g., the production of killer toxins and the reaction of sensitive paramecia on those toxins, and the site of location in the host cell. Some symbionts are encircled by the host membrane, others are not (Beale et al. 1969; Preer et al. 1974). Only a few symbionts have been described with binomial names in keeping with the international rules of nomenclature.

The lag phase in the description of new symbionts in ciliates for several years after the publication of the review by Preer et al. (1974) may be partly due to the lack of definite recognition of the symbionts as bacteria. Intracellular bacteria could not be investigated with classical microbiological methods, since they cannot not be grown outside their host cells. It became obvious that mere observations on morphology and biology (e.g., host specificity, life cycle) complemented by information on G + C content were not sufficient for species descriptions. Now, with the new powerful techniques for detection and phylogenetic classification at hand, such as PCR and fluorescence in situ hybridization (FISH), intracellular bacteria may be classified definitely and unequivocally even without growing them in vitro (Amann et al. 1991; Fritsche et al. 1993; Springer et al. 1996).

## 2.2 Research Fields

The reasons for research on bacterial symbionts in *Paramecium* have been changing over the decades. In the beginning, it was the fascination of microorganisms, of diving with the microscope into a new world of tiny creatures. With the discovery of the killer trait by Sonneborn (1938, 1943) in *Paramecium*, a new field of research was opened. When paramecia of a so-called killer strain are kept together with paramecia of a nonkiller strain, the latter cells die. Sonneborn found that the killer activity was an example of cytoplasmic inheritance; at that time it was not known that bacterial symbionts were responsible for the killing activity. It was found that a killer phenotype was frequently associated with intracellular bacteria (see reviews by Preer et al. 1974; Quackenbush 1988). Discovery of the killer trait resulted in an enormous interest in symbionts of *Paramecium* for some decades and as a side product of this interest other topics concerned with intracellular bacteria in ciliates were investigated too. The killer trait is still being investigated. Objectives are the search for the toxins, the mechanisms of resistance of killer paramecia against their toxins, and ecological and other questions (see Schrallhammer and Schweikert, this volume).

Another field of research on bacterial symbionts in *Paramecium* is the field of host-symbiont interactions. Several symbionts of *Paramecium* have been shown to require the presence of specific *Paramecium* genes for their maintenance (Sonneborn 1943; Schneller et al. 1959; Gibson and Beale 1961; Fujishima and Fujita 1985). The function of genes crucial for maintenance of symbionts are not known yet. Maintenance genes may be part of a preadaptation of certain strains and species of *Paramecium* that frequently bear symbionts. Preer and Preer (1984) mentioned that in a number of species of the *Paramecium aurelia* complex symbionts have never been observed. Among those they mentioned, *Paramecium novaurelia* has meanwhile been found infected with *C. caryophilus* and with *H. caryophila* in Germany (Görtz 1987; Kusch et al. 2000; Fokin et al. 2004).

It has been argued that symbionts profit from living inside a paramecium by being better protected from digestion, as compared with free-living species of bacteria, and that symbionts are provided with a convenient supply of abundant nutrients (Beale et al. 1969; Görtz 1983). This view may be misleading, however. Intracellular bacteria are not in competition with bacteria living outside organisms. By invading a host cell, a bacterium is introduced or has migrated into a new habitat and faces new selective pressures there. In most cases bacteria do not have a choice to leave the host cell again, they are trapped. Those that have the means of taking up nutrients from the host cell may survive, others simply starve. Which metabolites of the host are used is not known for any symbiont of *Paramecium* with the exception of nucleoside triphosphates (see below).

Many symbionts of ciliates have a smaller genome size than free-living bacteria (Soldo and Godoy 1973b) and this is one reason why certain associations were suggested to be very ancient (Preer 1977). However, as was observed for *Polynucleobacter* symbionts of the ciliate *Euplotes*, there are good indications for a more recent origin at least of that symbiosis, although the genome of the bacterial symbiont is small (Schmidt 1982; Heckmann and Schmidt 1987; Vannini et al. 2005). Indications for a transfer of genes from symbiont to host nuclei have not been found in *Paramecium*. Schmidt (1984) studied the symbiosis of *Caedibacter varicaedens* with *Paramecium biaurelia*. He did not obtain evidence for a sharing of the translational systems of host and symbiont. Though his observations provide evidence that all major proteins found in *Caedibacter* are synthesized in the symbiont itself, it seems obvious that the bacteria profit from host metabolites. Linka et al. (2003) showed the presence of a nucleoside triphosphate transporter (NTT) in *Caedibacter* and *Holospora*. The authors also presented good reasons for a lateral gene transfer of NTTs in intracellular bacteria (Linka et al. 2003). For the host, energy parasitism may be a problem under unfavorable conditions. In fact, *C. caryophilus* may overgrow and even kill its host (Schmidt et al. 1987).

In only one case of symbiosis in *Paramecium* has it been demonstrated that host cells may profit in their metabolism from a symbiont. Soldo (1963) and Soldo and Godoy (1973a) found that the bacterial symbiont *Lyticum flagellatum* could provide *Paramecium* with folic acid, a vitamin that cannot be produced by the ciliate itself. For all the other bacterial symbionts of *Paramecium* we do not know whether they contribute to the metabolism of the hosts. However, the observation of Fujishima and his group of enhanced temperature resistance of *Paramecium caudatum* bearing *H. obtusa* indicates that there may be other than "simple" metabolic interactions that are favorable for the host (Dohra et al 1998). Strains of *P. caudatum* may react differently on bacterial infections. While some may take advantage of certain intracellular bacteria, others may suffer (Dohra et al 1998; Fels and Kaltz 2006; Kusch and Görtz 2006). Differences in host tolerance and sensitivity towards bacterial infections in connection with intraspecific and interspecific competition point to the field of microbial ecology. While the significance of killer symbioses for *Paramecium* populations was acknowledged long ago (Landis 1981, 1988), recent investigations have shown that bacterial symbionts may be of wider ecological importance (Lohse et al. 2006; Nidelet and Kaltz 2007; Görtz 2008), e.g., in defense of the host against competitors or (additional) microbial infections (Görtz 1982, Görtz et al. 2008). Bacterial symbioses in protozoa even have a potential risk for human health (Görtz and Michel 2003; Ferrantini et al. 2007). These observations were additional reasons for phylogenetic investigations and it appears that bacterial symbionts of *Paramecium* as of other protists belong to various higher taxa of *Eubacteria* (see later).

## 3 Compartments of the *Paramecium* Cell Colonized by Bacterial Symbionts

*Paramecium* is a ciliate. Ciliates belong to the phylum Alveolata. Flat membrane-bounded cisternae, termed "alveoles," underlying of the cell membrane are characteristic of this phylum (including dinoflagellates and apicomplexans too). Paramecia are slender cells, about 100–200 μm long, and ciliated all over. With their basal bodies cilia are part of the cell cortex. The cell cortex is bounded by three membranes, the cell membrane and underneath it flat cisterns called alveoles, and multiple cytoskeletal systems with various organelles.

Depending on the species, paramecia contain one, two, or several micronuclei and a single macronucleus. The oral apparatus with its typical structure of ciliary membranelles (rows of tightly arranged ciliary batteries) is positioned ventral at roughly the middle of the cell. Further organelles are mitochondria, peroxisomes, Golgi apparatus, contractile vacuoles, lysosomes, and trichocysts (Wichterman 1986; Allen 1988; Hausmann et al. 2003).

## 3.1 Cytoplasm

The cytoplasm of the *Paramecium* provides space for the aforementioned organelles and sometimes hundreds, even thousands, of endocytobiotic bacteria. It is a continuous space that is highly ordered. The larger organelles are positioned at certain sites in the cell. Positioning is achieved and maintained by the cytoskeleton, which is complex and highly developed in *Paramecium* (Allen 1988; Cohen and Beisson 1988). A rigid, nonetheless flexible network of microtubules, actin microfilaments, and other components shapes the cell and holds the organelles. The main structures and organelles are the oral apparatus, the cytoproct, contractile vacuoles, the macronucleus and the micronuclei, basal bodies, trichocysts, and mitochondria (Allen 1988). Microtubules extending from basal bodies may reach far into the cell. There is good evidence that microtubules and microfilaments may guide the traffic of cell organelles and endocytobionts (Cohen and Beisson 1988; Ossipov 1981; Görtz and Wiemann 1989; Dohra and Fujishima 1999).

Ciliates are heterotrophic, most feed on bacteria or other microorganisms. Phagocytosis may only occur at the oral apparatus, while feces and sometimes symbionts may be released at the cytoproct (Fig. 1). The oral apparatus is a complex metaorganelle consisting of a funnel-like pouch with batteries of cilia and a highly ordered cytoskeletal basket at the bottom of which new phagosomes are formed. Newly formed phagosomes are acidified and fuse with primary lysosomes (Fok and Allen 1988). As known from other professional phagocytes such as macrophages, ingested bacteria are attacked by acidification, oxidative burst, and lysosomal enzymes. In spite of these attacks, bacteria with appropriate features may survive and even escape from the phagosomes. Bacteria may stay in symbiontophorous

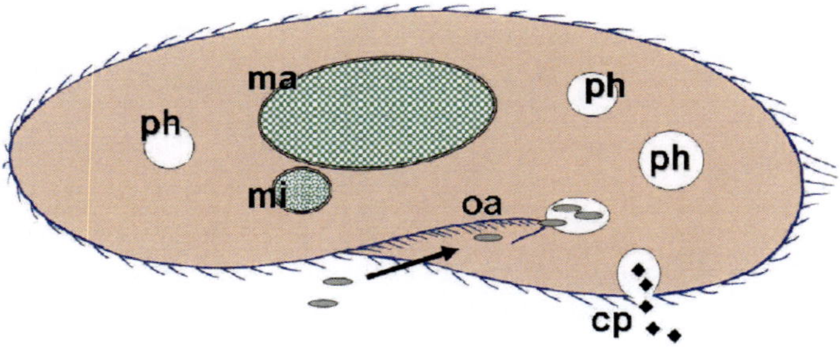


**Fig. 1** The *Paramecium* cell. The *Paramecium* cell contains one macronucleus and one to several micronuclei. The number of micronuclei is a species–specific character. Bacteria – prey as well as potential symbionts - are ingested via the oral apparatus into newly formed phagosomes. Bacteria may leave the phagosomes to get into the cytoplasm or to be taken into the nuclei. At the cytoproct, feces or even endocytobiotic bacteria may be released. *cp* cytoproct (only at the cytoproct exocytosis of feces as well as release of endocytobionts may occur), *ma* macronucleus, *mi* micronucleus, *oa* oral apparatus (only here phagocytosis may occur), *ph* phagosome, *arrow* bacteria are ingested into a new phagosome. Other organelles have been omitted

vesicles or naked in the cytoplasm – not encircled by host membranes (Sitte 1993). Other symbionts may be taken up into the nuclei. Endonucleobionts are found to be nuclear-specific, invading either micronuclei or macronuclei. Even the perinuclear space and the endoplasmic reticulum may be colonized by bacteria (Fokin and Karpov 1995).

Generally, internalization of bacteria into ciliates is known to occur by phagocytic uptake. There is one exception in the symbiosis of the ciliate *Parauronema acutum* and its bacterial endosymbionts (so-called xenosomes), where (Soldo (1987) found evidences of invasion through the host cell membrane. For infectious bacteria, uptake by phagocytosis is frequently observed; for some permanent endocytobionts, uptake is considered to have taken place long ago. The possibility for introduction of noninfectious endocytobionts into a host cell could be an event of coinfection together with infectious bacteria. This has been shown by (Fokin et al. (2004).

While endonucleobionts are not usually surrounded by host membranes in the nuclei, for symbionts in the cytoplasm both possibilities were observed. Most symbionts are encircled by a host membrane (Fig. 2), and are thus situated in a symbiontophorous vacuole, but others are "naked" (Fig. 3). Chiefly owing to the comparison with mitochondria and chloroplasts, the host-membrane-bounded state may be regarded as further developed concerning host–symbiont communication. On the other hand, the loss of a host membrane that has ultimately been derived from the membrane of the endosome (phagosome) may not be easy to understand. Symbionts in a host vacuole may be released from the host cell and in fact are released from time to time (Nobili 1961; Jurand et al. 1971). The surrounding host membrane may fuse with the cell membrane at the cytoproct – the site exocytosis occurs in ciliates, so liberating the endocytobiont. The release of "naked" symbionts by exocytosis seems difficult. How "naked" endocytobionts have escaped a phagosome

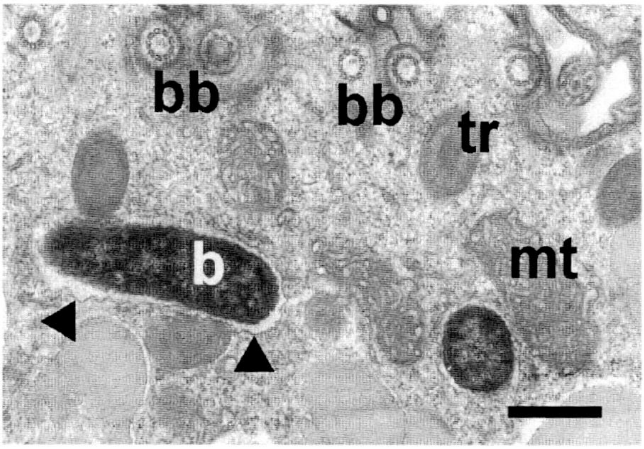


**Fig. 2** Bacterial symbionts in vacuoles. *Paramecium bursaria* with bacteria in the cytoplasm. The bacteria dwell in host vacuoles (symbiontophorous vacuoles). *Arrowheads* membranes of the host vacuole, *b* bacterial symbiont, *bb* ciliary basal bodies, *mt* mitochondria, *tr* trichocyst (extrusive organelle). *Bar* 0.5 μm

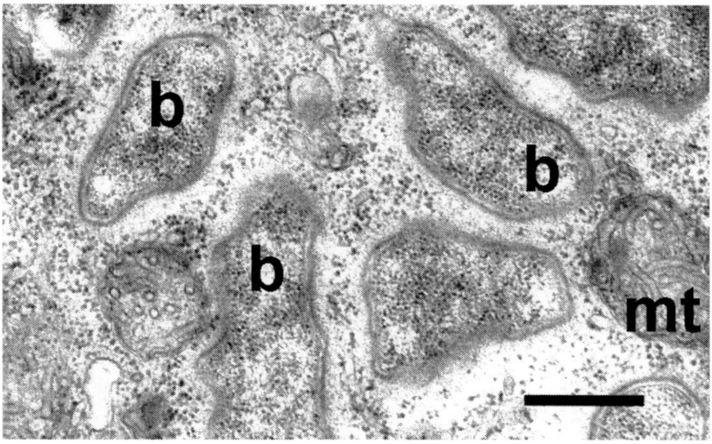


**Fig. 3** Bacterial symbiont not encircled by host membranes. *Paramecium sexaurelia* with bacteria in its cytoplasm. The bacteria are "naked" – not dwelling in host vacuoles. *b* bacterial symbionts, *mt* mitochondrion. *Bar* 0.5 μm

is not known. Escape from a phagosome or a symbiontophorous vacuole may be regarded as a step for advanced safety in the sometimes hostile environment with acidosomes and lysosomes threatening to fuse with a symbiont-bearing vacuole. Examples of "naked" symbionts as well as the occurrence of bacteria in a host vacuole are observed and apparently both are stabile.

## 3.2 Micronuclei

Ciliates have two types of nuclei, micronuclei and macronuclei (Beale 1954; Raikov 1996; Nanney 1980; Ossipov 1981; Paulin 1996; Katz 2001). In the micronucleus, the DNA is organized in typical long chromosomes that contain considerable amounts of nongenic sequences. Paramecia divide by binary division, the micronucleus dividing mitotically. It is a closed mitosis with the nuclear envelope being preserved, the mitotic spindle originating from the nuclear envelope (Cohen and Beisson 1988). Closed nuclear divisions are a precondition of permanent endonucleobioses, with symbionts being maintained over cell divisions. This may be the chief reason for the frequent occurrence of endonucleobioses in ciliates, namely, *Paramecium*. Whatever (e.g., RNA, ribosomal subunits) is inside the micronucleus, it is thought to leave the nucleus via nuclear pores only. Vice versa, for anything to get into the nucleus it has to pass through a nuclear pore too. Though the micronucleus is transcriptionally inactive in vegetative life, the number per area of nuclear pores is about the same for micronuclear as for macronuclear envelopes (Fig. 4).

The micronucleus is the generative nucleus thought to contain and preserve the complete diploid genome. Micronuclei take part in genetic exchange processes during

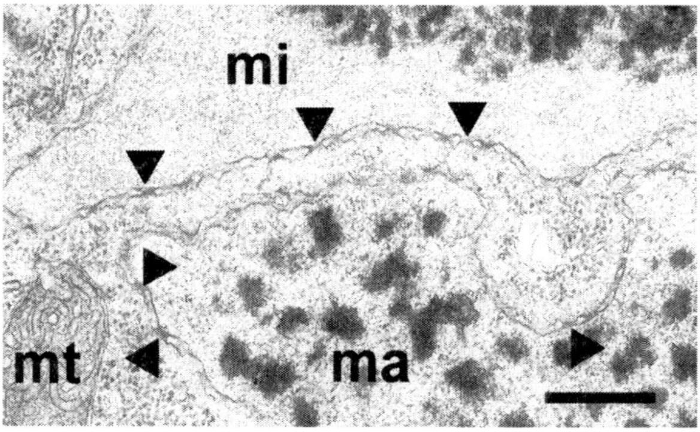


**Fig. 4** Nuclear envelopes, *Paramecium bursaria*. The distribution of nuclear pores (*arrowheads*) are about the same in the envelopes of micronuclei and macronuclei. *ma* macronucleus, *mi* micronucleus, *mt* mitochondrion. *Bar* 0.5 μm

sexual propagation, in ciliates called "conjugation" (Wichterman 1986; Hausmann et al. 2003). Micronuclei undergo meiosis; macronuclei degenerate during and after conjugation. All but one of the haploid nuclei resulting from meiosis degenerate too. One divides by mitosis, so producing the pronuclei – a stationary (female) and a migratory (male) pronucleus. After exchange of (male) pronuclei between the mating cells of a pair, in each cell the pronuclei fuse (fertilization). The resulting synkaryon divides mitotically. One division product gives rise to the new micronuclei; the other develops into the new macronucleus (Beale 1954; Wichterman 1986; Hausmann et al. 2003). The evolution and genetics of macronuclear development are being understood more and more (Prescott 1994; Katz 2001; Beale and Preer 2008).

Meiotic and mitotic processes have direct consequences for endonuclear symbionts. Bacteria in micronuclei may profit from the mechanisms of chromosome segregation in mitosis and meiosis. Namely, in the micronuclear endonucleobionts horizontal propagation of symbionts is highly probable compared with mitotic divisions. Infections may impede or even inhibit micronuclear divisions and so prevent successful sexual propagation (Ossipov 1981; Görtz and Fujishima 1983). Micronuclei are morphologically different in the species of *Paramecium* (Fokin 1997). The micronuclei of the *P. aurelia* species may be too tiny for bacterial infections and have never been found infected. The micronuclei of *P. caudatum* and the *Paramecium bursaria*-type micronuclei are frequently infected. Highly infectious micronucleus-specific *Holospora* species are found as well as bacteria that do not seem to be highly infectious.

## 3.3 Macronuclei

Macronuclei contain all the genes needed for vegetative life of a ciliate. Macronuclei are transcriptionally active. They contain only part of the sequences of micronuclei. Those genes maintained are amplified in ciliate macronuclei up to 1,000 copies

(Meyer and Lipps 1980; Steinbrück 1986). During development micronuclear chromosomes are cut to pieces, in some ciliates of only gene size (Prescott 1994). Thus, macronuclear chromosomes are typically shorter than micronuclear chromosomes, do not have centromers, and during macronuclear development are supplied with new telomeres (Blackburn 1991). It had been shown that transformation of *Paramecium* is possible by microinjection of a cloned gene (Godiska et al. 1987). This behavior should be a good precondition for exchange of genetic material from symbiont to host, but up to now evidence for a gene transfer from symbiont to host has not been found in ciliates (Schmidt 1982).

The classical view of the micronucleus being the genetic reservoir and the macronuclear genome being established anew from micronuclear genes after each conjugation has to be modified. Observations first made in *Oxytricha* (DuBois and Prescott 1995; Prescott and Greslin 2005) showed that at least some micronuclear genes are scrambled: the order of macronucleus-destined sequences of a micronuclear gene is different from the arrangement of the resulting macronuclear gene. Nowacki et al. (2008) found evidence that macronucleus-destined micronuclear genes are rearranged using macronuclear RNA being transcribed just during differentiation of macronuclear anlagen. A macronuclear RNA may serve as a matrix to ensure the correct order of the new macronuclear gene (Nowacki et al. 2008; Yao 2008). Thus, symbiont genes introduced into the micronucleus would not have a chance to be established in the macronucleus.

Like micronuclei, macronuclei too have a closed division with the nuclear envelope persisting. Macronuclear division has been called "amitosis" as it lacks the spindle typical for mitosis. We still have little knowledge of the division process. In *P. biaurelia* it has been shown that division may yield daughter macronuclei of different size and with different amounts of DNA. Extrareplications may correct for low levels of DNA/genes (Berger 1988). Namely, uneven distribution of chromatin to the daughter nuclei should result in the loss of genes after some divisions, but there are indications from other ciliates that the cells may "count" the copy numbers of genes and may regulated them (Steinbrück 1983). Macronuclei may also have special repair mechanisms that are kinds of preadaptations for endonuclear symbioses. Freiburg (1985) found that the activity of ribosomal RNA (rRNA) polymerase is enhanced fivefold in macronuclei bearing *H. obtusa*.

Within the macronucleus bacterial symbionts may be evenly distributed as is observed for macronuclear *Holospora* species, or may occur in clusters of sometimes several hundred bacteria (Fig. 5). Like in micronuclei, endonucleobionts may be trapped in the macronuclei unless they have developed mechanisms to leave the nuclei, as have *Holospora* bacteria. On the other hand, living within nuclei should ensure the optimal food supply. Also, inside a nucleus bacteria may be better protected from cellular defense mechanisms such as autophagy. As mentioned before, whereas cytoplasmic symbionts may be simply distributed to daughter cells during binary division, endonuclear symbionts may make use of the nuclear division machinery. However, resorption of the old macronucleus during conjugation of the host cell may be deleterious for endonuclear symbionts. New macronuclei being developed after conjugation are free of bacteria unless they are infected anew.

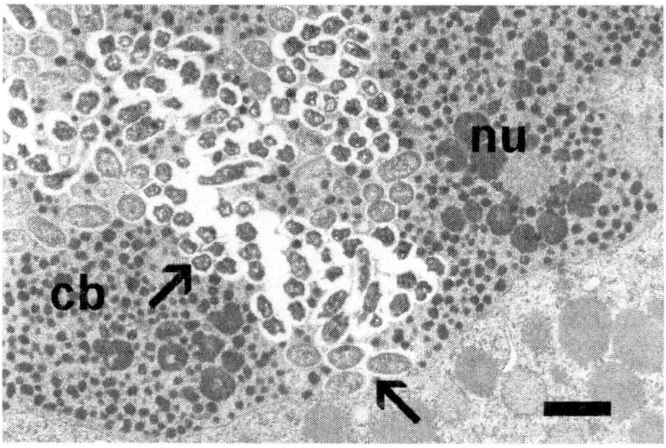


**Fig. 5** Macronucleus with clusters of bacteria. Clusters (*arrows*) of two different types of bacteria are found in the macronucleus of a *Paramecium caudatum* cell. *cb* chromatin bodies, *nu* nucleoli. *Bar* 1 μm

# 4 Bacteria in *Paramecium* Given Binomial Names (Except *Holospora*)

## 4.1 *Caedibacter*

The valid description of the genus *Caedibacter* was given by Preer and Preer (1982). According to that description the ability of forming refractile inclusion bodies (R bodies) and thereby conferring a killer toxicity upon their *Paramecium* hosts is a crucial feature of the genus. Preer and Preer state that usually less than 10%, but sometimes up to 50% of the symbionts contain R bodies, cells containing R bodies usually being longer than cells that do not contain R bodies. In addition to R bodies, many spherical phagelike structures or covalently closed circular DNA plasmids are present. The symbionts are straight rods or coccobacilli, 0.4–1.0 μm in diameter and 1.0–4.0 μm in length. They are Gram-negative and nonmotile. The GC content of the DNA is 40–44 mol%. Earlier work on the genus has been reviewed by Preer et al. (1974) and Preer and Preer (1984).

The genus *Caedibacter* turned out to be polyphyletic. Whereas *C. caryophilus* was revealed to belong to *Alphaproteobacteria* (Springer et al. 1993), *Caedibacter taeniospiralis*, the type species of the genus *Caedibacter*, belongs to *Gammaproteobacteria* (Beier et al. 2002). An endosymbiont from *Acanthamoeba polyphaga*, "*Candidatus* Caedibacter acanthamoebae," does not establish R bodies but showed 93% sequence similarity to *C. caryophilus* and had therefore been included in the genus *Caedibacter* (Horn et al. 1999). The phylogenetic position of other species has not been investigated. For more information, see the chapter by Schrallhammer and Schweikert in this volume.

The species of *Caedibacter* found in *Paramecium* are:

1. *C. taeniospiralis* Preer and Preer 1982. It is known from the cytoplasm of *Paramecium tetraurelia* only. *C. taeniospiralis* is a killer symbiont able to form R bodies that unroll from the inside and contain plasmids. For details on the mode of action of the killer toxin, see Preer et al. (1974). It belongs to *Gammaproteobacteria* (Beier et al. 2002). It occurs as rods 0.4–0.7 μm wide and 1.0–2.5 μm long. The GC content is 41 mol%. For further information, see the chapter by Schrallhammer and Schweikert in this volume.
2. *C. varicaedens* Quackenbush 1982. It is known from the cytoplasm of *P. biaurelia. C. varicaedens* is a killer symbiont able to form R bodies that unroll from the outside. R bodies are usually associated with bacteriophage capsids. For details on the mode of action of the killer toxin, see Preer et al. (1974). It occurs as rods 0.4–1.9 μm wide and 2.0–4.0 μm long. The GC content of the DNA is 40–41 mol%. For further information, see the chapter by Schrallhammer and Schweikert in this volume.
3. *Caedibacter pseudomutans* Quackenbush 1982. It is known from the cytoplasm of *P. tetraurelia. C. pseudomutans* is a killer symbiont able to form R bodies that unroll from the outside. For details on the mode of action of the killer toxin, see Preer et al. (1974). It occurs as cigar-shaped rods, approximately 0.5 μm in diameter and 1.5 μm long. The GC content is 44 mol%. For further information, see the chapter by Schrallhammer and Schweikert in this volume.
4. *Caedibacter paraconjugatus* Quackenbush 1982. It is known from the cytoplasm of *P. biaurelia. C. paraconjugatus* is a killer symbiont able to form R bodies of the *C. varicaedens* type. Cell-to-cell contact of during sexual propagation of the host *Paramecium* is required for toxic effects to be observed in the sensitive paramecia (mate killers). For details on the mode of action of the killer toxin, see Preer et al. (1974). For further information, see the chapter by Schrallhammer and Schweikert in this volume.
5. *C. caryophilus* Euzeby 1997 (Euzeby 1997; Schmidt et al. 1987). It is known from the macronucleus of *P. caudatum. C. caryophilus* belongs to *Alphaproteobacteria* (Springer et al. 1993). Later, *C. caryophilus* was found in *P. novaurelia* (Kusch et al. 2000) and in the dinoflagellate *Peridinium cinctum* (Schweikert and Meyer 2001) too. *C. caryophilus* is a killer symbiont able to form R bodies unrolling from the inside. R bodies are associated with phages. For details on the mode of action of the killer and on the R bodies, see Schmidt et al. (1987). It occurs as rods 0.4–0.7 μm wide and up to 2.5 μm long. For further information, see the chapter by Schrallhammer and Schweikert in this volume.
6. *Caedibacter macronucleorum* Fokin and Görtz 1993. *C. macronucleorum* belongs to *Alphaproteobacteria*. The 16S rRNA genes of *C. macronucleorum* and *C. caryophilus* show a similarity of 99%; this and the lack of a specific oligonucleotide probe for discriminating between the two species did not allow validation of "*C. macronucleorum*" as a provisional taxon (Schrallhammer et al. 2006). However, the authors pointed out that the two species can be discriminated on the basis of a highly variable stretch of nucleotides that interrupts their single-stranded rRNA genes. For further information, see the chapter by Schrallhammer and Schweikert in this volume.

## 4.2 *Pseudocaedibacter*

The valid description of the genus *Pseudocaedibacter* was given by Quackenbush (Quackenbush 1982, 1978) according to the earlier denomination as *Pseudocaedobacter* by Preer et al. (1974). According to that description, the chief difference from the genus *Caedibacter* was the lack of ability of *Pseudocaedibacter* to produce R bodies and to confer killer activity upon the host. All species of *Pseudocaedibacter* are intracellular bacteria in various species of *Paramecium.* The symbionts are Gram-negative rods, nonmotile, 0.25–0.7 μm wide, and 0.5–4 μm long. The G + C content of the DNA is 53–39 mol% buoyant density (BD). No gene sequences are known for any of the *Pseudocaedibacter* species and the phylogenetic positions must be regarded as uncertain, the more so as it does not appear justified to separate *Pseudocaedibacter* from the genus *Caedibacter* only because of the above-mentioned features – R body production and killer trait. R body proteins are found encoded either by plasmid or by phage genes and, thus, can neither be a positive nor a negative characteristic for phylogenetics or taxonomy.
The species of *Pseudocaedibacter* are:

1. *Pseudocaedibacter conjugatus* Quackenbush 1978 (Preer et al. 1974). It occurs as rods 0.3–0.5 μm in diameter and 1–4 μm long. *P. conjugatus* like *C. paraconjugatus* is a symbiont responsible for the mate-killer trait in the *P. aurelia* complex. Cell-to-cell contact during sexual propagation of the host *Paramecium primaurelia* or *Paramecium octaurelia* is required for toxic effects to be observed in the sensitive paramecia (mate killers). The G + C content of the DNA is 35–37 mol% (BD). For details on the mode of action of the killer toxin, see Preer et al. (1974). For further information, see the chapter by Schrallhammer and Schweikert in this volume.
2. *Pseudocaedibacter minutus* Quackenbush 1978 (Preer et al. 1974). It is known to occur in the cytoplasm of *P. octaurelia.* It does not produce R bodies but is a very strong killer. For details on the mode of action of the killer toxin, see Preer et al. (1974). It occurs as rods, often double, 0.25–0.35 μm in diameter and 0.5–1.0 μm long (singles). The symbiont is surrounded by an extra set of membranes, apparently continuous with the endoplasmic reticulum of the host. The G + C content of the DNA is 38 mol% (BD). For further information, see the chapter by Schrallhammer and Schweikert in this volume.
3. *Pseudocaedibacter falsus* Quackenbush 1978 (Preer et al. 1974). It is known to occur in the cytoplasm of *Paramecium pentaurelia* and from the cytoplasm of *P. biaurelia.* Though the symbiont itself appears to be nontoxic, it may increase the resistance of its host to the toxin produced by the killer symbiont *Lyticumflagellatum* (Holtzman 1959). It occurs as rods 0.4–0.7 μm in diameter and 1.0–1.5 μm long. The G + C content of the DNA is 36 mol% (BD).
4. "*Pseudocaedibacter glomeratus*" Fokin and Ossipov 1986. It is known from *P. pentaurelia.* No killer activity has been observed. It occurs as rods 1.0–1.2 μm long, about 0.3 μm wide. No flagella have been observed. It is Gram-negative, and has no central nucleoid. The bacteria are encircled by host

membranes in a symbiontophorous vacuole. These vacuoles are surrounded by lacunas of endoplasmic reticulum with numerous ribosomes. No killing or toxic actions are known.

## 4.3 *Lyticum*

The valid description of the genus *Lyticum* was given by Preer and Preer (1984). It occurs in *P. biaurelia*, *P. tetraurelia*, and *P. octaurelia*. The symbionts produce labile toxins which kill sensitive strains of paramecia very quickly by lysis (less than 30 min at room temperature). It occurs as large rods 0.6–0.8 μm in diameter, straight curved or spiral. The length of single forms is 3.0–5.0 μm. There are numerous peritrichous flagella. It is Gram-negative. Within the host and outside the host it is nonmotile or almost so, in spite of having numerous, well-developed flagella. The G + C content is 27 and 45–49 mol% (BD). The nature of the toxin is unknown.
The species of *Lyticum* are:

1. *Lyticum flagellatum* (ex Preer et al. 1974) Preer and Preer 1982. It is known from the cytoplasm of *P. tetraurelia* and *P. octaurelia*. It occurs as straight rods 0.6–0.8 μm in diameter and 2.0–4.0 μm long, with many peritrichous flagella. Stock 299 of *P. octaurelia* containing *L. flagellatum* (lambda particles) does not require folic acid, whereas symbiont-free lines of 299 do (Soldo and Godoy 1973a).
2. *Lyticum sinuosum* (ex Preer et al. 1974) Preer and Preer 1982. It is known from the cytoplasm of *P. biaurelia*. It occurs as curved or spiral rods 0.7–0.9 μm in diameter and 2.0–10.0 μm long, sometimes forming chains of two or three cells. The G + C content of the DNA is 45 mol% (BD).

## 4.4 *Pseudolyticum*

The description of "*Pseudolyticum*" was given by Boss et al. (1987). *Pseudolyticum* are nonkiller symbionts. They occurs in *P. caudatum* as straight rods 2.0 μm in diameter, 3.5–14.0 μm long. The bacteria are not infective, they are Gram-negative, and are nonmotile in spite of having numerous flagella.
The species of "*Pseudolyticum*" are:

1. "*Pseudolyticum multiflagellatum*" Bosset al. (1987). The bacteria are known from the cytoplasm of *P. caudatum*, they not infective, and do not confer a killer trait on their host. They occur as straight rods 2.0 μm wide, 3.5–14.0 μm long. They are Gram-negative and are nonmotile in spite of having numerous flagella. Similar endosymbiotic bacteria in *P. caudatum* have been described by Dieckmann (1977). The bacteria described by Dieckmann were, however, infectious and motile when outside their host cells. An endosymbiont possibly

identical or closely related to *P. multiflagellatum* was found by Görtz (unpublished work) in a small pond near Feldberg in Brandenburg, Germany. These bacteria, too, were nonmotile within or outside the host cell in spite of having many flagella. Most impressive was the close association of the bacteria with host mitochondria (Fig. 6). The culture was not sufficiently stable for further work, e.g., to search of an NTT-shuttle system.

2. "*Pseudolyticum minutus*" described by Fokin (1989b). It was represented by rod-like symbionts in the cytoplasm of *Paramecium nephridiatum* (at that time identified as *Paramecium woodruffi*). This bacterium is 1.5–4.0 μm × 0.6–0.7 μm in size with numerous immobile flagella and an R body enclosed in an individual symbiontophorous vacuole.

## 4.5 *Tectibacter*

The valid description of the genus *Tectibacter* was given by Preer and Preer (1982) (ex Preer et al. 1974). *Tectibacter* is known from *P. primaurelia*, *P. biaurelia*, *P. tetraurelia*, *Paramecium sexaurelia*, and *P. octaurelia*, often together with other symbionts. No killer activity has been observed. It occurs as straight rods 0.4–0.7 μm in diameter, 1–2 μm long. It is distinguished by an outer covering around its cell

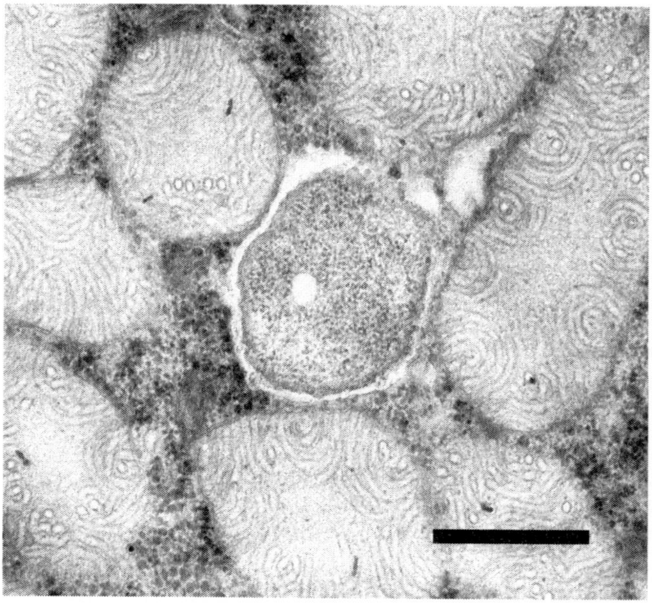

**Fig. 6** "Pseudolyticum flagellatum" in *Paramecium caudatum*. The bacterium inside its host vacuole is tightly surrounded by mitochondria. *Bar* 0.5 μm

wall visible in sections with the transmission electron microscope. It is sparsely peritrichous and is Gram-negative. It is often observed to be motile.

The species of *Tectibacter* is *Tectibacter vulgaris* (ex Preer et al. 1974) Preer and Preer 1982. The characteristics are described for the genus.

### 4.6 *Nonospora*

The description of the genus *Nonospora* was given by Fokin et al. (1987b). It occurs as rod-like symbionts living in the macronucleus. Flagella have not been observed. No toxic actions are known. The symbionts are retained in macronuclear fragments of exconjugants and enter the macronuclear anlagen by fusion of old fragments with the anlagen (Fokin et al. 1987b). Only one species has been described: *Nonospora macronucleata* Fokin, Ossipov, Skoblo, Rautian, and Sabaneyeva 1987. It is known from the macronucleus of *P. caudatum*, often clustered in the center of the nucleus. The symbionts are rod-like, 0.2–0.3 μm wide, and mostly about 1.0 μm long, sometimes forming chains up to 10 μm long. The surface of the symbionts appears irregularly wavy in the electron microscope. The symbiont can change the sequence of nuclear reorganization processes of the host paramecia.

## 5 Bacteria in *Paramecium* Not Given Binomial Names

Occurrence of bacteria has been mentioned for many species of *Paramecium*. From those species that are intensely investigated, such as *P. biaurelia*, *P. tetraurelia*, *P. caudatum*, and *P. bursaria*, more symbiotic bacteria are known than from little investigated species. Recent investigations by the authors of this chapter indicate that new symbionts may be found by doing more systematic searches – not only in *Paramecium*.

In two *Paramecium* spp. bacterial symbionts were just mentioned, but not yet even briefly investigated. These are *P. polycaryum* and *P. jenningsi*. In some other paramecia, namely, all African endemic species (*P. africanum, P. ugandae*, *P. pseudotrichium*, and *P. jankowskii*) and other rare members of the genus such as *P. sonnebornii*, *P. wichtermani*, and *P. schewiakoffi* bacterial symbionts have never been revealed, at least partly because these species have not been studied enough.

### 5.1 Bacteria in *Paramecium bursaria*

One cytoplasmic bacterium has been found in the species and was called "*Caedobacter chlorellopellens*" (Skoblo et al. 1985). It is a Gram-negative rod-like symbiont 1.4 μm × 0.35 μm in size with no flagella located within a symbiontophorous vacuole. It manifested antagonistic relationships with *Chlorella* sp. – the common eukaryotic symbiont of *P. bursaria*. As the bacterium has no R body, killer or mate-killer capacity, apparently, the proposed name does not fit the original description of the symbiont and it should be considered as unnamed with uncertain phylogenetic relationship. Thus, we include it in this section.

At least two unnamed bacterial symbionts were found in the cytoplasm of the species (Ossipov et al. 1994). The first one is a Gram-positive rod-shaped bacterium without flagella with a compact nucleoid. Its size was 3–30 μm × 0.5–0.7 μm. The symbiont was found in a strain infected with *Holospora acuminata* but had no *Chlorella* sp. The symbiont formed groups in the host endoplasm without a symbiontophorous vacuole and it seemed to have no infectivity (Ossipov et al. 1994). Another bacterium was found in the cytoplasm of *P. bursaria* and was identified as Gram-negative rod-like particles with two life forms: reproductive and sporelike ones. The last forms, 1.0–2.0 μm × 0.5 μm in size, had heterogenous cytoplasmic areas with polar periplasmic regions. These forms were located strictly in the cortex of the host and manifested pili on their surfaces (Ossipov et al. 1994).

A so far unnamed symbiont living in the micronucleus of *P. bursaria* was found by Görtz and Freiburg (1984). It is a small rod, 0.5 μm in diameter and up to 2 μm long. Its ultrastructure suggests that it is a Gram-negative bacterium. No flagella were found and no killing capacity of its host was observed. Chen (1955, 1956) found isolates of *P. bursaria* displaying a killer activity. Though it is likely that paramecia of both strains bore killer symbionts, no observations of endocytobionts were reported in these cases. It should be emphasized that *P. bursaria* was frequently observed infected by *Holospora*; see the chapter by Fokin and Görtz in this volume.

*P. bursaria* is the green *Paramecium*, and cells harbor up to 300 *Chlorella* symbionts. There is evidence that the algal symbionts prevent bacterial infections of green *P. bursaria*. Once the paramecia have become aposymbiotic – free of algal symbionts – infections by bacteria or yeasts may happen (Görtz 1982, 2008).

## 5.2 Bacteria in *Paramecium calkinsi*

In the brackish water species *Paramecium calkinsi* some bacteria were found in the cytoplasm and the macronucleus as well (Fokin and Sabaneyeva 1993; Fokin et al. 2000). One bacterium was revealed naked in the cytoplasm. These Gram-negative symbionts were 1–3 μm × 0.6–0.9 μm in size without any distinct nucleoid. Only reproductive forms were observed, many of them carrying hexahedral viral capsids which were 60 nm in diameter and which had specific bodies of homogeneous material of average electron density, 0.2–0.5 μm in size. No killer effect was revealed on bacterium-free *P. calkinsi* (Fokin and Sabaneyeva 1993).

Another type of the cytoplasmic bacteria in *P. calkinsi* enclosed with symbiontophorous vacuoles was found and investigated (Fokin and Sabaneyeva 1993;

Fokin et al. 2000). There were two morphologically different forms of the Gram-negative symbiont. One form of the bacterium was rod-like, 1.5–2.0 μm × 0.3–0.4 μm in size, and was always located within the inner cytoplasmic area; sometimes cisternae of the endoplasmic reticulum with or without ribosomes were situated close to some of the bacteria or even decorated them completely (in this case the symbiontophorous vacuole disappeared). Some of the bacteria showed small pili-like structures on their surface. The second form of the bacterium was found in the cortex area of the cells. These symbionts were quite similar to the first form in shape and size, but they never showed any pili structures and were never surrounded by host membranes. FISH reaction revealed these symbionts as alphaproteobacteria (Fokin et al. 2000).

One more bacterium was revealed in a different stock of *P. calkinsi* (Fokin et al. 2000). This symbiont was found all over the cytoplasm as the reproductive form only. This is a rod-shaped Gram-negative bacterium, measuring 1.0–1.5 μm × 0.25–0.30 μm. The infected ciliate did not reveal any killer effect; however, moderate infection capacity of the bacterium was discovered experimentally. FISH reaction revealed this symbiont as an alphaproteobacterium as well (Fokin et al. 2000).

Two symbionts have been revealed so far in the macronucleus of *P. calkinsi* (Fokin and Sabaneyeva 1993). The first were found as rod particles 1.0–2.0 μm × 0.3–0.4 μm in size. A zone of higher electron-density was usually located at the tip of the symbiont. It seems the bacteria do not affect the host nucleus and do not have any infectious or killer capacity. Simultaneous infection of the macronucleus and the cytoplasm was revealed for this symbiont. Another macronuclear endocytobiont in *P. calkinsi* was described as *Holospora curvata*. It is considered in the chapter by Fokin and Görtz in this volume.

## 5.3 Bacteria in *Paramecium caudatum*

A symbiont occurring in the macronucleus of *P. caudatum* was studied by Estève (1978). It confers a killer trait on its host. When it was investigated cytologically, a greatly enlarged macronucleus was observed to contain numerous kappa-like bacteria, some of which contained R bodies. Electron micrographs of this bacterium showed spherical phages inside the R bodies. Schmidt et al. (1987) assumed that the symbiont is identical to *C. caryophilus* described by them.

## 5.4 Bacteria in *Paramecium duboscqui*

A symbiont found in the perinuclear space of *Paramecium duboscqui* was described by Fokin (1988). It was found within the space of the macronuclear envelope of *P. duboscqui*. The symbiont is 0.3 μm wide and 0.7–1.4 μm long, looks spindle-shaped, and is Gram-negative. No killing activity was observed when symbiont bearers were tested against nonsymbiont bearers.

## 5.5 Bacteria in *Paramecium multimicronucleatum*

Jenkins (1970) described a Gram-negative bacterium living within bulbous distensions of the outer membrane of the nuclear envelopes of both the micronucleus and the macronucleus of a strain of *Paramecium multimicronucleatum*. It is a very short rod, sometimes appearing nearly coccoid, approximately 0.35 μm in diameter, with longer forms reaching 0.7 μm in length. It was named "epsilon."

Another quite unusual motile bacterium was briefly described from the macronucleus of the species (Vishnyakov and Rodionova 1999). The bacteria have numerous long flagella, which are arranged all around the surface of the symbiont cell. The symbiont is variable in size and shape: 1.3–2.5 μm × 0.4–1.5 μm, oval or rod-like.

## 5.6 Bacteria in *Paramecium nephridiatum*

A symbiont of the mate-killer type was discovered in the cyloplasm of *P. nephridiatum*, a ciliate living in brackish water at the coast of Barents Sea, Russia (Fokin et al. 1987a; Fokin 1989c). At the beginning the host species was misidentified as *P. woodruffi* (Fokin et al. 1999; Fokin 2004). The symbiont is a Gram-negative rod, 0.2–0.8 μm wide and 0.6–2.5 μm long. It has no flagella. It was found to contain hexagonal viroid particles (Fokin et al. 1987a; Fokin 1989c) (Fig. 7).

Three more bacteria were identified in *P. nephridiatum* collected at the coast of the White Sea, Russia (Fokin 1989b) (Fig. 8). The first one was represented by small rod-like Gram-negative symbionts 0.8–2.0 μm × 0.25–0.30 μm in size lying naked in the cytoplasm; the second one was a spirillum-like bacterium 3.5–7.0 μm × 1.2–1.4 μm in size enclosed in an individual symbiontophorous vacuole.

Another type of complex symbiosis was also revealed for these paramecia. In one stock of *P. nephridiatum* different bacterial symbionts were found in the cytoplasm, perinuclear space of both types of nuclei, and in the macronucleus simultaneously (Fokin 1989a). Perinuclear bacteria found in the stock resembled those described by Jenkins (1970) in *P. multimicronucleatum*. There are small (0.6–1.0 μm × 0.25 μm) spindlelike Gram-negative symbionts living in the perinuclear space of the macronuclei and micronuclei. They are packed in the perinuclear space in a rosettelike manner. The macronuclear symbiont has been described as *H. curvata* (Fokin and Sabaneyeva 1993). This bacterium is discussed in the chapter by Fokin and Görtz in this volume.

## 5.7 Bacteria in *Paramecium putrinum*

Till now only one bacterial symbiont has been detected in the cytoplasm of this *Paramecium* (Fokin et al. 2000). These are coccoid shaped small Gram-negative bacteria about 0.5–1.0 μm × 0.25–0.35 μm in size. They are not encircled by the host

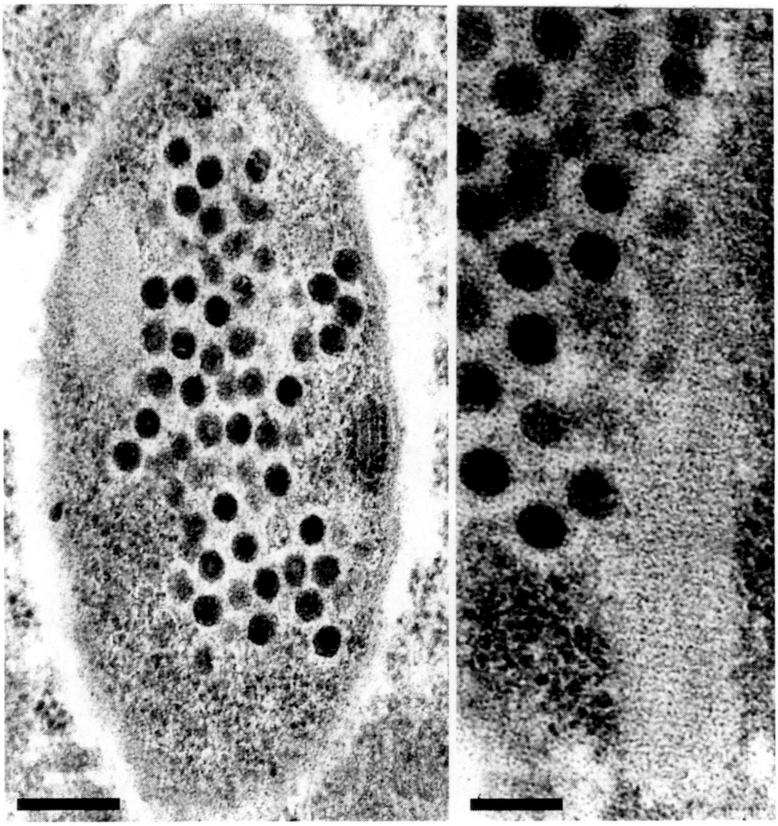

**Fig. 7** Mate-killer symbionts in *P. nephridiatum*. *Left*: Bacterium with many viroid particles in the cytoplasm of *P. nephridiatum*. *Right*: Paracrystalloid-like structure close to the area of phage capsids. *Bars* 0.15 μm (*left*) and 0.1 μm (*right*)

membrane. Most of the symbiont population occupied the central part of the host cell. The FISH reaction revealed this symbiont to belong to *Alphaproteobacteria* as well (Fokin et al. 2000). Another infection of *Paramecium putrinum* concerns a *Holospora* sp. in the macronucleus. It is discussed in the chapter by Fokin and Görtz in this volume.

## 5.8 Bacteria in *Paramecium pentaurelia*

A bacterial symbiont (Fig. 9) that has not been investigated was observed in the cytoplasm of *P. pentaurelia*. *P. pentaurelia* is one of the species of the *P. aurelia* group that has not been studied as much as the others, namely, *P. biaurelia* and *P. tetraurelia*. The previous investigation (Fokin and Ossipov 1986) and current

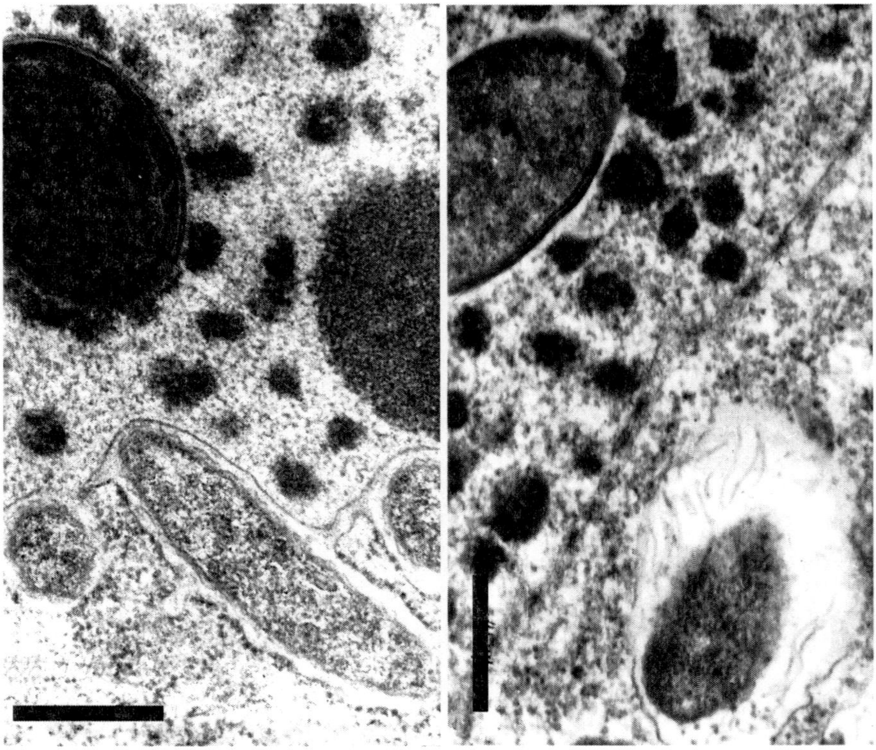

**Fig. 8** Bacteria in the perinuclear space of the macronucleus of *P. nephridiatum*. The macronucleus is infected with *H. bacillata* too. *Bar* = 0.5 μm. *Down*: "Pseudolyticum minutus" in the cytoplasm of *P. nephridiatum* infected with *Holospora bacillata*. Numerous flagella are visible. *Bar* 0.9 μm

observation of an endocytobiont may therefore indicate that *P. pentaurelia* may nevertheless be a potential host of several bacteria.

## 5.9 Bacteria in *Paramecium sexaurelia*

In *P. sexaurelia* isolated from an aquarium with tropical fish, a bacterial endosymbiont was observed by Görtz (1981) (Fig. 10). The bacterium is slightly curved with a diameter of 0.5–0.8 μm and is up to 25 μm long. In the cytoplasm, the symbiont appears closely associated with food vacuoles.

The bacterium was observed again in a *P. sexaurelia* isolated by Fokin (Przybos and Fokin 1997) from a pond in the Wilhelma (zoological and botanical garden, Stuttgart, Germany).

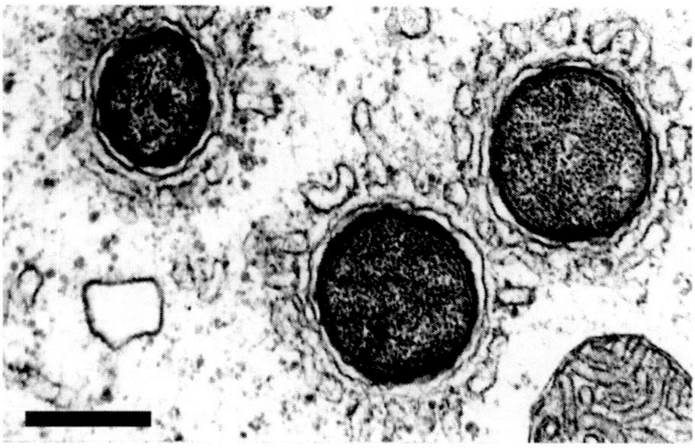

**Fig. 9** Bacteria in the cytoplasm of *Paramecium pentaurelia*. The symbionts are enclosed within a symbiontophorous vacuole decorated by endoplasmic reticulum cisternae. *Bar* 0.3 μm

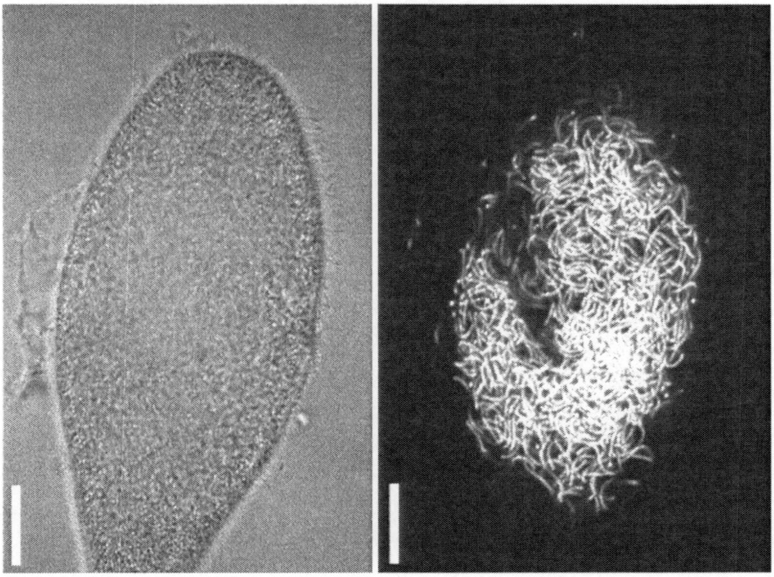

**Fig. 10** Undescribed bacteria in *P. sexaurelia*. *Left*: Fixed cell, micrograph taken with bright field microscopy. Bacteria are visible. *Right*: The same cell after hybridization with a fluorescently labeled eubacterial probe. There are many brightly fluorescing bacteria. Fluorescence microscopy. *Bar* 30 μm

# 6 Phylogeny of Bacteria in *Paramecium*

Phylogenetic analysis of intracellular bacteria was hampered by culture problems: classical methods of classification and determination could not be applied, as the symbionts could not be cultured in vitro. G + C contents, mobility, and morphological and biological data had been used for descriptions as no other features were available.

Only with the availability of DNA sequencing plus FISH did phylogenetic analysis of intracellular bacteria become possible (Amann et al. 1991; Fritsche et al. 1993). Meanwhile in ciliates and other protozoa a number of symbionts have been classified.

The symbionts of *Paramecium* analyzed phylogenetically all belong to *Proteobacteria*, most of them to *Alphaproteobacteria* . We must realize, however, that sequencing data are not available for most symbionts of *Paramecium* and other ciliates. Together with symbionts of other protozoa, namely, small free-living amoebas, symbionts are found in a much wider spectrum of bacterial taxa, including even *Chamydiales* and *Verrucomicrobia* (Görtz and Michel 2003).

Bacteria may immediately separate from their free-living relatives on the occasion of invading a eukaryotic cell. Following colonization of a host cell, evolution of the new endocytobionts may proceed faster than that of the free-living bacteria. Good evidence for this difference was found by Vannini et al. (2007). Comparing symbiotic and free-living *Polynucleobacter* isolates, Springer et al. (1996) found for 16S rRNA genes and internal transcribed spacer sequences a higher divergence in two groups of endosymbionts. Certainly, the population of host cells is a clearly defined niche, not only because of the host cell membrane as a "geographical" barrier, but also because of the special metabolic situation. Thus, populations of intracellular bacteria may be more separated from each other (e.g., in different host populations or different host species) and from free-living relatives than are free-living bacteria.

Evidence for horizontal gene transfer shows the limits of this separation. Transfer of antibiotic resistance genes and infectivity island is not restricted to human pathogens as has been shown for energy-shuttle systems. Energy-shuttle proteins are found in mitochondria, plastids, and energy parasites (intracellular pathogens/symbionts). Linka et al. (2003) found that NTT genes (ATP/ADT translocase genes) in *H. obtusa* and *C. caryophilus* show substantial sequence identity with their counterparts in chloroplasts and intracellular bacterial pathogens of humans. NTT gene homologs in chloroplasts from plants, and green, red, stramenopile, and glaucocystophyte algae proved to be monophyletic. NTT systems of *C. caryophilus* and *H. obtusa* are only distantly related to each other, although these two species are close relatives in 16S rRNA trees. ATP/ADP translocase genes were revealed to originate from a chlamydial ancestor (Schmitz-Esser et al. 2004), rickettisas as well as plants having obtained ATP/ADT translocases by horizontal gene transfer.

On the basis of 16S RNA genes *H. obtusa*, *C. caryophilus*, and *C. macronucleorum* are found within *Alphaproteobacteria* (Amann et al. 1991; Springer et al. 1993; Schrallhammer et al. 2006). Unexpectedly, *C. taenispiralis*, the old kappa particles and first identified killer symbionts of *Paramecium* (Preer 1948, Preer et al. 1974; Preer and Preer 1982), belongs to *Gammaproteobacteria* (Beier et al. 2002). *C. taeniospiralis* was found to form a new evolutionary lineage within this subgroup, a sister group being the family *Francisellaceae* (87% 16S rRNA sequence similarity). The findings show that the killer trait and the ability of forming R bodies cannot be used as phylogenetic markers. *C. caryophilus* was found to be closely related to some *Acanthamoeba* endosymbionts. The sequence information suggests that the progenitor of *C. caryophilus* lived within acanthamoebae prior to the infection of paramecia (Beier et al. 2002).

Phylogenetic analyses merely based upon 16S rRNA-sequence comparison may not give complete details of relationships. For this, inclusion of further genes into the investigations may be necessary. Phylogenetic analyses with multiple protein sequences revealed that *H. obtusa* is positioned basally to the *Rickettsia–Ehrlichia–Wolbachia* assemblage of intracellular bacteria (Lang et al. 2005). The authors postulate that *H. obtusa* is the closest bacterial relative of mitochondria known to date. Still our knowledge of the phylogenetic diversity of bacteria in *Paramecium* is fragmentary. However, though few, the data are encouraging the search for the positions of other endocytobionts in the tree of organisms.

## 7 Conclusions

*Paramecium* is certainly one of the best studied protozoa. This holds true especially for the research field of bacterial symbioses with 50 or more species found, most of them not properly described. The wealth of symbioses observed in *Paramecium* is due to this fact that *Paramecium* has been more investigated than most other protozoa. Nevertheless, as we have tried to show in this chapter, our knowledge of bacterial symbionts in *Paramecium* is still fragmentary. Now, with new methods from molecular biology, biochemistry, and biophysics, phylogenetic details, communication and interaction between symbiont and host, and other phenomena are being investigated, opening new fields of research. As in human pathogens, genetic elements for symbiotic/parasitic traits such as energy shuttle appear to be transferred horizontally. The diversity of bacteria in *Paramecium* and other protists turns out to be fascinatingly rich. We are becoming aware of the ecological consequences of bacterial symbioses in *Paramecium*, certainly being an ecological model system for symbioses in other protozoa too. *Paramecium* like other protozoa may even harbor potentially pathogenic bacteria. Old questions such as metabolic interdependencies and invasion mechanisms may now be addressed and possibly answered in the near future, and we may even get attractive tools for molecular biology and cell biology from the research of bacterial symbioses in *Paramecium*. The finding of many new symbionts in *Paramecium* in the last few years makes it obvious that we still know only some of the bacterial symbionts, making the search for new symbioses necessary.

## References

Allen RD (1988) Cytology. In: Görtz H-D (ed) *Paramecium*. Springer, Berlin, pp 4–40

Amann R, Springer N, Ludwig W, Görtz H-D, Schleifer K-H (1991) Identification in situ and phylogeny of uncultured bacterial endosymbionts. Nature 351:161–164

Beale GH (1954) The genetics of *Paramecium aurelia*. Cambridge University Press, Cambridge

Beale GH, Preer JR Jr (2008) *Paramecium*: Genetics and epigenetics. Taylor and Francis, Boca Raton

Beale GH, Jurand A, Preer JR Jr (1969)The classes of endosymbiont of *Paramecium aurelia*. J Cell Sci 5:65–91

Beier CL, Horn M, Michel, R, Schweikert, M, Görtz H-D, Wagner M (2002) The genus *Caedibacter* comprises endobionts of *Paramecium* spp. related to the *Rickettsiales* (*Alphaproteobacteria*) and to *Francisella tularensis* (Gammaproteobacteria). Appl Environ Microbiol 68:6043–6050

Berger JD (1988) The cell cycle and regulation of cell mass and macronuclear DNA content. In: Görtz H-D (ed) *Paramecium*. Springer, Berlin, pp 97–119

Blackburn EH (1991) Structure and function of telomeres. Nature 350:569–573

Boss AO-L, Borchsenius ON, Ossipov DV (1987) *Pseudolyticum multiflagellatum* n. g., n. sp. – a new symbiotic bacterium in the cytoplasm of *Paramecium caudatum* (Ciliata, Protozoa). Cytologia (Sankt-Petersburg) 29:94–99 (in Russian with English summary)

Chen TT (1955) Paramecin 34, a killer substance produced by *Paramecium bursaria*. Proc Soc Exp Biol Med 88:541–543

Chen TT (1956) Varieties and mating types in *Paramecium bursaria*. II. Variety and mating types found in China. J Exp Zool 132:255–268

Cohen J, Beisson J (1988) The cytoskeleton. In: Görtz H-D (ed) *Paramecium*. Springer, Berlin, pp 363–392

Dieckmann J (1977) An infectious bacterium in the cytoplasm of *Paramecium caudatum*. In Abstract Volume. International Congress of Protozoology, Warzawa, p 77

Dohra H, Fujishima M (1999) Effect of antibiotics on the early infection process of a macronuclear endosymbiotic bacterium *Holospora obtusa* of *Paramecium caudatum*. FEMS Microbiol Lett 179:473–477

Dohra H, Fujishima M, Ishikawa H (1998) Structure and expression of a GroE-homologous operon of a macronucleus-specific symbiont *Holospora obtusa* of the ciliate *Paramecium caudatum*. J Eukaryot Microbiol 45:471–479

DuBois M, Prescott DM (1995) Scrambling of the actin 1 gene in two *Oxytricha species*. Proc Natl Acad Sci U S A 92:3888–3892

Estève J-C (1978) Une population de type "killer" chez *Paramecium caudatum* (Ehrenberg). Protistologica 14:201–207

Euzeby JP (1997) Revised nomenclature of specific or subspecific epithets that do not agree in gender with generic names that end in -bacter. Int J Syst Bacteriol 47:584

Fels D, Kaltz O (2006) Temperature-dependent transmission and latency of *Holospora undulata*, a micronucleus-specific parasite of the ciliate *Paramecium caudatum*. Proc R Soc Lond B 273:1031–1038

Ferrantini F, Fokin SI, Vannini C, Modeo L, Verni F, Petroni G (2007) Ciliates as natural hosts of many novel *Rickettsia*-like bacteria. Protistology 5:28–29

Fok AK, Allen RD (1988) The lysosome system. In: Görtz H-D (ed) *Paramecium*. Springer, Berlin, pp 301–324

Fokin SI (1988) A bacterial symbiont of the macronucleus perinuclear space in the ciliate *Paramecium duboscqui*. Cytologia 30:632–635

Fokin SI (1989a) Bacterial endobionts of the ciliate *Paramecium woodruffi*. II. Endobionts of the perinuclear space. Cytology (Sankt-Petersburg) 31:845–850 (in Russian with English summary)

Fokin SI (1989b) Bacterial endobionts of the ciliate *Paramecium woodruffi*. III. Endobionts of the cytoplasm. Cytology (Sankt-Petersburg) 31:964–969 (in Russian with English summary)

Fokin SI (1989c) Mate-killer effect caused by endobionts of the ciliate *Paramecium woodruffi*. Cytology (Sankt-Petersburg) 31:1085–1089 (in Russian with English summary)

Fokin SI (1997) Morphological diversity of the micronuclei in *Paramecium*. Arch Protistenkd 148:375–387

Fokin SI (2004) Bacterial endocytobionts of Ciliophora and their interactions with the host cell. Int Rev Cytol 236:181–249

Fokin SI, Görtz H-D (1993) *Caedibacter macronucleorum* sp. nov., a bacterium inhabiting the macronucleus of *Paramecium duboscqui*. Arch Protistenkd 143:319–324

Fokin SI, Karpov SA (1995) Bacterial endocytobionts inhabiting the perinuclear space of Protista. Endocytobiosis Cell Res 11:81–94

Fokin SI, Ossipov DV (1986) *Pseudocaedibacter glomeratus* sp. n. – a cytoplasmic symbiont of the ciliate *Paramecium pentaurelia*. Cytologia 28:1000–1004

Fokin SI, Boss AO-L, Ossipov DV (1987a)Virus-containing cytoplasmic symbiont of the ciliate *Paramecium woodruffi*. Cytologia 29:1303–1306

Fokin SI, Sabaneyeva EV (1993) Bacterial endocytobionts of the ciliate *Paramecium calkinsi*. Eur J Protistol 29:390–395

Fokin SI, Ossipov DV, Skoblo II, Rautian MS, Sabaneyeva EV (1987b) *Nonospora macronucleata* g.n., sp. n. – a vegetative nucleus symbiont of the ciliate *Paramecium caudatum*. Cytology (Sankt-Petersburg) 29:963–970 (in Russian with English summary)

Fokin SI, Stoeck T, Schmidt HJ (1999) Rediscovery of *Paramecium nephridiatum* Gelei, 1925 and its characteristics. J Euk Microbiol 46:416–426

Fokin SI, Sabaneyeva EV, Borkchsenius ON, Schweikert M, Görtz H-D (2000) *Paramecium calkinsi* and *Paramecium putrinum* (Ciliaphora, Protista) harboring alpha-subgroup bacteria in the cytoplasm. Protoplasma 213:176–183

Fokin SI, Skovorodkin IN, Schweikert M, Görtz H-D (2004) Co-infection of the macronucleus of *Paramecium caudatum* by free-living bacteria together with the infectious *Holospora obtusa*. J Eukaryot Microbiol 51:417–424

Freiburg M (1985) Isolation and characterization of macronuclei of *Paramecium caudatum* infected with the macronucleus-specific bacterium *Holospora obtusa*. J Cell Sci 73:389–398

Fritsche TR, Horn M, Seyedirashti S, Gautom RK, Schleifer K-H, Wagner M (1993) In Situ detection of novel bacterial endosymbionts of *Acanthamoeba* spp. phylogenetically related to members of the order *Rickettsiales*. Appl Environ Microbiol 65:206–212

Fujishima M, Fujita M (1985) Infection and maintenance of *Holospora obtusa*, a macronuclear-specific bacterium of the ciliate *Paramecium caudatum*. J Cell Sci 76:179–187

Gibson I (1974) The endosymbionts of *Paramecium*. Crit Rev Microbiol 3: 243–273

Gibson I, Beale GH (1961) Genie basis of the mate-killer trait in *Paramecium aurelia*, stock 540. Genet Res 2:82–91

Godiska R, Aufderheide KJ, Gilley D, Hendrie P, Fitzwater T, Preer LB, Polisky, B, Preer JR Jr (1987) Transformation of Paramecium by microinjection of a cloned serotype gene. Proc Natl Acad Sci U S A 84:7590–7594

Görtz H-D (1981) Ein neues symbiontisches Bakterium in *Paramecium sexaurelia*. Verh Dtsch Zool Ges 74:227

Görtz HD (1982) Infection of *Paramecium bursaria* with bacteria and a yeast. J Cell Sci 58:445–453

Görtz H-D (1983) Endonuclear symbionts in Ciliates. Int Rev Cytol Suppl 14:145–176

Görtz H-D (1987) Different endocytobionts simultaneously colonizing ciliate cells. Ann N Y Acad Sci 503:261–268

Görtz, H-D (2007) Symbiotic associations between ciliates and prokaryotes. In: Dworkin M, Falkow S, Rosenberg E, Schleifer KH, Stackebrandt E (eds) The orokaryotes, 3rd edn, vol 1. Springer, East Lansing, pp 403–438

Görtz HD (2008) Towards an understanding of the distribution, dynamics and ecological significance of bacterial symbioses in protists. Denisia 23:307–311

Görtz H-D, Freiburg M (1984) Bacterial symbionts in the micronucleus of *Paramecium bursaria*. Endocytobiosis Cell Res 1:37–46

Görtz H-D, Fujishima M (1983) Conjugation and meiosis of *Paramecium caudatum* infected with the micronucleus-specific bacterium *Holospora elegans*. Eur J Cell Biol 32:86–91

Görtz, H-D, Michel R (2003) Bacterial symbionts in protozoa in aqueous environments – potential pathogens? In: Greenblatt C, Spigelman M (eds) Emerging pathogens. Oxford University Press, Cambridge, pp 25–37

Görtz H-D, Schmidt HJ (2005) Family III. Holosporaceae fam.nov. In: Garrity et al., (eds) Bergey's manual of systematic bacteriology, part C, 2nd edn. Springer, New York, pp 146–160

Görtz H-D, Wiemann M (1989) Route of infection of the bacteria *Holospora elegans* and *Holospora obtusa* into the nuclei of *Paramecium caudatum*. Eur J Protistol 24:101–109

Görtz H-D, Rosati G, Schweikert M, Schrallhammer M, Omura G, Suzaki T (2009) In: White JF Jr, Torres MS (eds) Defensive mutualism in microbial symbiosis. Taylor and Francis, Boca Raton

Hausmann K, Hülsmann N, Radek R (2003) Protistology, 3rd edn. Schweizerbart, Stuttgart

Heckmann K, Görtz H-D (1991) Prokaryotic symbionts of ciliates. In: Balows A, Trüper HG, Dworkin M, Harder W, Schleifer K-D (eds) The prokaryotes, 2nd edn. Springer, Berlin, pp 3865–3890

Heckmann K, Schmidt HJ (1987) *Polynucleobacter necessarius* gen. nov., sp. nov., an obligately endosymbiotic bacterium living in the cyotplasm of *Euplotes aediculatus*. Int J Syst Bacteriol 37:456–457

Holtzman HE (1959) A kappa-like particle in a non-killer stock of *Paramecium aurelia*, syngen 5. J Protozool 6(Suppl):26

Horn M, Fritsche TR, Gautom RK, Schleifer K-H, Wagner M (1999) Novel bacterial endosymbionts of Acanthamoeba spp. Related to the *Paramecium caudatum* symbiont *Caedibacter caryophilus*. Environ Microbiol 1:357–367

Jenkins RA (1970) The fine structure of a nuclear envelope associated endosymbiont of *Paramecium*. J Gen Microbiol 61: 355–359

Jurand A, Rudman BM, Preer JR Jr (1971) Prelethal effects of killing action by stock 7 of *Paramecium aurelia*. J Exp Zool 177:365–388

Katz LA (2001) Evolution of nuclear dualism ciliates: a reanalysis in light of molecular data. Int J Syst Evol Microbiol 51:1587–1592

Kusch J, Görtz H-D (2006) Towards an understanding of the killer trait: *Caedibacter* endocytobionts in *Paramecium*. Prog Mol Subcell Biol 41:61–76

Kusch J, Stremmel M, Breiner H-W, Adams M, Schweikert M, Schmidt H.J. (2000) The toxic symbiont *Caedibacter caryophila* in the cytoplasm of *Paramecium novaurelia*. Microb Ecol 40:330–335

Landis WG (1981) The ecology, role of the killer trait, and interactions of five species of the *Paramecium aurelia* complex inhabiting the littoral zone. Can J Zool 59:1734–1743

Landis WG (1988) Ecology. In: Görtz H-D (ed) *Paramecium*. Springer, Berlin, pp. 419–436

Lang BF, Brinkmann H, Koski LB, Fujishima M, Görtz H-D, Burger G (2005) On the origin of mitochondria and *Rickettsia*-related eukaryotic endosymbionts. Jpn J Protozool 38:171–183

Linka N, Hurka H, Lang BF, Burger G, Winkler HH, Stamme C, Urbany C, Seil I, Kusch J, Neuhaus HE (2003) Phylogenetic relationships of non-mitochondrial nucleotide transport proteins in bacteria and eukaryotes. Gene 306:27–35

Lohse K, Gutierrez A, Kaltz O (2006) Experimental evolution of resistance in *Paramecium caudatum* against the bacterial parasite *Holospora undulata*. Evol Int J Org Evol 60:1177–1186

Meyer GG, Lipps HJ (1980) Chromatin elimination in the hypotrichous ciliate *Stylonychia mytilus*. Chromosoma 77:284–297

Nanney DL (1980) Experimental ciliatology. Wiley, New York

Nidelet T, Kaltz O (2007) Direct and correlated responses to selection in a host-parasite system: testing for the emergence of genotype specificity. Evol Int J Org Evol 61:1803–1811

Nobili R (1961) Alcune considerazioni sulla liberazione e sull' azione delle particelle kappa in individui amacronucleati die *Paramecium aurelia*, stock 51, syngen 4. Atti Soc Tosc Sci Nat Ser B:158–172

Nowacki M, Vijayan V, Zhou Y, Schotanus K, Doak T G, Landweber LF (2008) RNA-mediated epigenetic programming of a genome-rearrangement pathway Nature 451:153–158

Ossipiv DV (1981) Problems of nuclear heteromorphism in the unicellular organisms. Nauka, Leningrad (in Russian with English summary)

Ossipiv DV, Borchsenius ON, Skoblo II, Lebedeva NA (1994) Two new species of symbiotic bacteria in the cytoplasm of the ciliate *Paramecium bursaria*. Cytology (Sankt-Petersburg) 36:411–426 (in Russian with English summary)

Paulin JJ (1996) Morphology and cytology of ciliates. In: Hausmann K, Bradbury PC (eds) Ciliates. Cells as organisms. Fischer, Stuttgart, pp 1–40

Preer JR Jr (1948) A sudy of some properties of the cytoplasmic factor "kappa" in *P. aurelia*, variety 2. Genetics 33:349–404

Preer JR Jr (1977) The killer system in *Paramecium* – kappa and its viruses. Microbiology 1977:576–578

Preer JR, Preer LB (1982) Revival of names of protozoan endosymbionts and proposal of *Holospora caryophila* nom.nov. Int J Syst Bacteriol 32:140–141

Preer JR, Preer LB (1984) Endosymbionts of protozoa. In: Krieg NR (ed) Bergey's manual of systematic bacteriology, vol. 1. Williams and Wilkins, Baltimore, London, pp 795–813

Preer JR, Preer LB, Jurand A (1974) Kappa and other endosymbionts in *Paramecium aurelia*. Bacteriol Rev 38:113–163

Prescott DM (1994) The DNA of ciliated protozoa. Microbiol Mol Biol Rev 58:233–267

Prescott DM, Greslin AF (2005) Scrambled actin I gene in the micronucleus of *Oxytricha nova*. Dev Genet 13:66–74

Przybos E, Fokin SI (1997) Species of the *Paramecium aurelia* complex Sonneborn in Germany. Arch Protistenkd 148:167–172

Quackenbush RL (1978) Genetic relationships among bacterial endosymbionts of *Paramecium aurelia*: deoxyribonucleotide sequence relationships among members of *Caedobacter*. J Gen Microbiol 108:181–187

Quackenbush RL (1982) In: Validation of the publication of new names and new combinations previously effectively published outside the IJSB. List no. 8. Int J Syst Bacteriol 32:266–268

Quackenbush RL (1988) Endosymbionts of killer paramecia. In: Görtz H-D (ed) *Paramecium*. Springer, Berlin, pp 406–418

Raikov IB (1996) Nuclei of ciliates. In: Hausmann K, Bradbury PC (eds) Ciliates. Cells as organisms. Fischer, Stuttgart, pp 221–242

Schmidt HJ (1982) Isolation of omikron-endosymbionts from mass cultures of *Euplotes aediculatus* and characterization of their DNA. Exp Cell Res 140:417–425

Schmidt HJ (1984) Studies on protein synthesis in kappa particles. J Gen Microbiol 130:1517–1523

Schmidt HJ, Görtz H-D, Quackenbush RL (1987) *Caedibacter caryophila* sp nov, a killer symbiont inhabiting the macronucleus of *Paramecium caudatum*. Int J Syst Bacteriol 37:459–462

Schmitz-Esser S, Link N, Collingro A, Beier CL, Neuhaus HE, Wagner M, Horn M (2004) ATP/ADP translocases: a common feature of obligate intracellular amoebal symbiont related to Chlamydiae and Rickettsiae. J Bacteriol 186:683–691

Schneller MV, Sonneborn TM, Mueller JA (1959) The genetic control of kappa-like particles in *Paramecium aurelia*. Genetics 44:533–534

Schrallhammer M, Fokin SI, Schleifer K-H, Petroni G (2006) Characterization of the obligate endosymbiont "*Caedibacter macronucleorum*" Fokin and Görtz, 1993 and of its host *Paramecium duboscqui* strain Ku4-8. J Eukaryot Microbiol 53:499–506

Schweikert M, Meyer B (2001) Characterization of intracellular bacteria in the freshwater dinoflagellate *Peridinium cinctum*. Protoplasma 217:177–184

Sitte P (1993) Symbiotenic evolution of complex cells and complex plastids. Eur J Protistol 29:131–143

Skoblo II, Borchsenius ON, Lebedeva NA, Ossipov DV (1985) A new species of symbiotic bacteria of *Paramecium bursaria* (Ciliophora, Protozoa). Cytology (Sankt-Petersburg) 27:1292–1297 (in Russian with English summary)

Soldo AT (1963) Axenic culture of *Paramecium* - some observations on the growth behavior and nutritional requirements of a particle-bearing strain of *Paramecium aurelia* 299 lambda. Ann N Y Acad Sci. 108:380–388

Soldo AT (1987) *Parauronema acutum* and its xenosomes: a model system. J Protozool 34:447–451

Soldo AT, Godoy GA (1973a) Observations on the production of folic acid by symbiont lambda particles of *Paramecium aurelia* stock 299. J Protozool 20:502

Soldo AT, Godoy GA (1973b) Molecular complexity of *Paramecium* symbiont lambda deoxyribonucleic acid: evidence for the presence of a multicopy genome. J Mol Biol 73:93–108

Sonneborn TM (1938) Mating types in *Paramecium aurelia*: diverse conditions for mating in different stocks; occurrence, number and interrelations of the type. Proc Am Philos Soc 79:411–434

Sonneborn TM (1943) Gene and cytoplasm. I. The determination and inheritance of the killer character in variety 4 of *P. aurelia*. II. The bearing of determination and inheritance of charcters in *P. aurelia* on problems of cytoplasmic inheritance, pneumococcus transformations, mutations and development. Proc Nat Acad Sci U S A 29:329–343

Springer N, Ludwig W, Amann R, Schmidt HJ, Görtz H-D, Schleifer K-H (1993) Occurrence of fragmented 16S rRNA in an obligate bacterial endosymbiont of *Paramecium caudatum*. Proc Natl Acad Sci U S A 90:9892–9895

Springer N, Amann R, Ludwig W, Schleifer KH, Schmidt HJ (1996) *Polynucleobacter necessarius*, an obligate Material endosymbiont of the hypotrichous ciliate *Euplotes aediculatus*, is a member of the ß-subclass of proteobacteria. FEMS Microbiol Lett 135:333–336

Steinbrück G (1983) Overamplification of genes in macronuclei of hypotrichous ciliates. Chromosoma 88:156–163

Steinbrück G (1986) Molecular reorganization during nuclear differentiation in ciliates. In: Hennig W (ed) Results and problems in cell differentiation. Springer, Berlin, pp 13:105–174

Vannini C, Hahn M, Petroni G, Verni F, Lucchesi S, Rosati G (2005) Recent adaptation to symbiotic lifestyle: the Polynucleobacter – Euplotes system. J Eukaryot Microbiol 52:38S–43S

Vannini C, Pöckl M, Petroni G, Wu QL, Lang E, Stackebrandt E, Schrallhammer M, Richardson P, Hahn, MW (2007) Endosymbiosis in statu nascendi: close phylogenetic relationship between obligately endosymbiotic and obligately free-living *Polynucleobacter* strains (Betaproteobacteria). Environ Microbiol 9:347–359

Vishnyakov A, Rodionova G (1999) Motile intranuclear symbionts of ciliate *Paramecium multimicronucleatum*. In: Wagner et al (eds) From symbiosis to eukaryotism. Endocytobiology VII. University of Geneva, Geneva, pp 169–177

Wichterman R (1986) The biology of *Paramecium*, 2nd edn. Plenum, New York

Yao M-C (2008) RNA rules. Nature 451:131–132

Yu X-J, Walker DH (2005) Genus I. *Rickettsia*. In: Garrity et al. (eds) Bergey's manual of systematic bacteriology, part C, 2nd edn. Springer, New York, pp 96–114

# Diversity of *Holospora* Bacteria in *Paramecium* and Their Characterization

**Sergei I. Fokin and Hans-Dieter Görtz**

**Contents**



S.I. Fokin (✉)
Department of Zoology Invertebrate, Biological and Soil Faculty, St. Petersburg State University, Universitetskaya emb. 7/9, 199034, St. Petersburg, Russia; Department of Biology, Pisa University, Via A. Volta, 4, 56126, Pisa, Italy
e-mail: sifokin@hotmail.com

H.-D. Görtz
Department of Zoology, Biologisches Institut, Universität Stuttgart, Pfaffenwaldring 57, 70569, Stuttgart, Germany

M. Fujishima (ed.), *Endosymbionts in Paramecium*, Microbiology Monographs 12, DOI: 10.1007/978-3-540-92677-1_7, © Springer-Verlag Berlin Heidelberg 2009

**Abstract** *Paramecium* spp. may be a host for about 60 species of different bacteria, among of them ten *Holospora* spp. are investigated further. They are infectious alphaproteobacteria, and populate the nuclear apparatus of six ciliates of the genus. The complex life cycle of *Holospora* bacteria and their host and nuclear specificity suggest that there is a system of diverse interactions with the host both at the stage of infection and during infection maintenance in the nuclei. *Holospora* is a very good experimental subject for widely directed investigations in the fields of protistology, symbiontology, parasitology, and cell biology. The considerable number of *Holospora* spp. known to date, the possibility of laboratory infection of aposymbiotic paramecia, and the possibility of experimental modeling of combinations (in which endobionts and hosts possess different morphological and genetic features) make this symbiotic system very convenient for the study of several important problems – especially in experimental cell biology such as intracellular signaling and the development of intracellular communication. Biodiversity, the main morphological and biological features of the bacteria, as well as the ecological significance and phylogenetical position are discussed according to the respective extensive studies of the authors. Historical steps of *Holospora* investigations are reconstructed for the first time.

## 1 Introduction

As the majority of the previous chapters dealt with eukaryotic symbionts of *Paramecium* – zoochlorellae, the intention of this introduction is to present some general information connected with bacterial endocytobionts (symbionts) of ciliates and their interactions within the system. In the broader meaning of the term "symbiosis" it is a close association of individuals of different species (De Bary 1879). In this sense endocytobionts of *Paramecium* could be any organisms living inside the protist's cell. In fact, the majority of them are different bacteria. This symbiosis phenomenon was perceived for a long time as accidental and transient. However, as far back as the nineteenth–early twentieth century some scientists realized the possible implications for general biology of the symbiosis phenomenon in Protista and initiated the idea of symbiogenesis as a way for a cell's evolution (Khahina 1979; Margulis 1981; Fokin 2005, 2007). The value of symbiotic studies of protists comes from the fact that they are cell-organisms and the results obtained from them afford an opportunity for broader generalizations in the field of cell and general biology.

The prokaryotic symbionts of ciliates are much more abundant than eukaryotic ones – algae. Now it is clear that symbiosis between bacteria and ciliates is a very frequent phenomenon (Görtz 1988, 1996, 1998, 2008; Fokin 1993, 2004a; Görtz and Schmidt 2005). Three groups of bacterial endosymbioses can be outlined according to the presumed depth of adaptation in the "bacteria–ciliate cell" system and, therefore, the stability of the system: accidental, permanent, and highly infectious. Assessment of the intimate relationships between the host and the symbiont is usually problematic. It is often difficult to say if the bacterium is a parasite, a commensal, or a true symbiont in the sense of a mutualist. With this in mind, the classification suggested a quarter of a century ago (Görtz 1983) seems relevant, even though it does not always describe the actual nature of the endocytobiotic system. *Holospora* bacteria belong to the last group – highly infectious endocytobionts. Interest in the prokaryotic symbionts of ciliates after their discovery in the middle of the nineteenth century was stimulated again only a century later by the studies of killer traits in paramecia (Sonneborn 1938; Preer 1948). Improvement of microscopic techniques, the invention of the electron microscope, and especially the revival and development of the symbiogenesis theory (Margulis 1981) promoted large-scale investigations of the phenomenon of endocytobiosis in protozoans. The special term "endocytobiology" was even coined (Schwemmler 1980). The phylum Ciliophora is one of the taxa especially rich in bacterial symbionts, above all this is true for *Paramecium* spp. (Preer et al. 1974; Görtz 1983; Fokin 1993, 2004a, 2007; Görtz and Schmidt 2005).

Certain features of Ciliophora (e.g., almost exclusively phagotrophic type of feeding) are probably the prerequisites for the origin of the symbiosis phenomenon. As already stressed, relationships between the partners in ciliophoran symbiotic systems have not always been investigated in detail. It is apparent that they may be represented by all gradations between parasitism and mutualism. In the case of holosporas, it should probably be identified as true parasitism (Görtz 1996).

In the last 40 years symbiosis in ciliates, especially endocytobiosis in *Paramecium* spp., has received considerable attention. The problems studied regarding the ciliates include the diversity of symbionts, adaptations and interactions between partners in symbiotic systems, ecological and evolutionary importance of the symbiosis, and mechanisms of infection. The model host subjects of the investigations are *Paramecium caudaum* and *P. aurelia* (Preer et al. 1974; Fujishima and Görtz 1983; Preer and Preer 1984; Görtz 1986, 1988, 1996, 1998, 2008; Heckmann and Görtz 1991; Görtz and Schmidt 2005; Fokin 1993, 2004a, 2007; Fujishima et al. 2005).

The complex life cycle of infectious *Holospora* bacteria and their host specificity and nuclear specificity suggest that there is a system of diverse interactions with the host both at the stage of infection and during infection maintenance in the nuclei (Fujishima and Fujita 1985; Fokin 1993, 1998, 2004a; Görtz 1996; Görtz and Brigge 1998; Fokin et al. 2003a). Most infection stages appear to be under the double control of the symbiont and the host. They may be described as a cascade of mutual recognition of molecules on the surface of activated infectious forms and the cell structures of the ciliate, culminating (if the partners are complementary) in a stable infection of the nuclei. The results of cooperation of the system elements during endocytobiosis establishment may be assessed at the morphological, biochemical, and physiological levels.

The present chapter aims to throw light chiefly on the history of *Holospora* investigations, on the diversity of *Holospora*–bacteria endocytobionts of *Paramecium*, a "classical" subject of modern symbiontology, and on different adaptations established between the partners in the system.

## 2 History of *Holospora*'s Discovery and Steps in Its Investigation

The official beginning of *Holospora* research should be considered to be 1890, the date when Mardukhey Wolf-Vladimir Hafkin (Hafkine or Khawkine in French) found and described those intranuclear bacteria in *P. caudatum*.

Hafkin (1860–1930) was a graduate of the Odessa (Novorossiyskiy) University, Russia (1884). Ten years later, he became a well-known bacteriologist – in Paris in 1892 Hafkin made the first acting vaccine against cholera, (which was effectively used in India between 1893 and1895), and then a vaccine against the plague (1896–1898), which also saved thousands and thousands of lives in India. In 1899 in Bombay (now Mumbai, India) Hafkin created the special antiplague laboratory, and nowadays the institute is named after him (Popovsky 1963).

Actually, Hafkin started his scientific career as a protistologist working with *Astasia*, *Euglena*, and *Paramecium* during his university course (1879–1884) and spent some years after as a technical assistant at the Zoological Museum of Odessa University. Then in 1889 he was invited to be a junior librarian in the Pasteur Institute (Paris). In parallel with this official position, Hafkin started to work in the laboratory of the future Nobel Prize laureate Ilya Metchnikoff, who was his professor at Odessa as well (Fokin 2004c; 2007; Fig. 1). It was there that Hafkin discovered *Holospora* bacteria in *Paramecium*. As a matter of fact, infected ciliates were isolated from a laboratory aquarium; thus the native origin of *Holospora* is not known (Hafkine 1890). Funnily enough, to the best of our knowledge, holosporas have never been found in France again (Fokin 2004b).

This was the first accurate description of *Holospora* spp. as paramecia parasites using the Linnaean binary nomenclature. Hafkin succeeded in finding three different species of the bacteria in the same population of *P. caudatum* which he named *Holospora obtusa* – species having infected the macronucleus of the ciliate, *H. undulata*, and *H. elegans* – parasites of the micronucleus (Figs. 2h-j, 3b-d, g, h). In relation to the *Holospora* morphology and the main steps of its life cycle, some physiological observations were presented in the article. However, the researcher did not pay much attention to the host's taxonomy – in all of Hafkin's publications the ciliate was discriminated as *P. aurelia*, which was quite a common mistake at that time. Hafkin's materials were used in the lectures of Metchnikoff published as a book with the title *Lectures on comparative pathology of inflammation* (Metchnikoff 1892). There it was stressed that *Holospora* should be a special group of microorganisms not completely able to be characterized as "normal" bacteria. Hafkin did not succeed in achieving experimental infection with *Holospora*. The discussion of this point in Metchnikoff's lectures was quite logical: he proposed that the parasites should come into the nucleus somehow escaping digestion in the cytoplasm. However, no mechanism for this infection process was proposed at that time. As Hafkin very soon shifted to the field of practical bacteriology, *Holospora* research stopped for a decade and a half.

Some indications about rod-like structures inside the nuclear apparatus of paramecia can be found even in earlier literature (Fig. 2), which was mentioned already by

**Fig. 1** Metchnikoff's laboratory members in the Pasteur Institute library, 1890. From the *left* (seated): O. Metchnikoffa, A. Zamshin, I. Metchnikoff, W. Wagner; (standing): W.M. Hafkin, Th. Trapeznikoff, P. Gabritchewskiy, N. Blagoweschenskiy, N. Protopopoff, V. Polowzeff. In the *left corner*: W.M. Hafkin, 1910. The originals are in the Museum of the Pasteur Institute, Paris, France

Hafkin: Müller (1856), Claparede and Lachmann (1858–1861), Balbiani (1861), and Bütschli (1876). Limitations of biological ideas of that time – the view of ciliates as perfect (multicellular) organisms (Ehrenberg 1838) – and not enough good microscopic techniques initially precluded the correct evaluation of the phenomenon discovered. Scientists treated the nuclear apparatus of the ciliate, according to Ehrenberg's idea, as sexual glands and the bacteria inside, correspondingly, as spermatozoids (Maggi 1874; Petchenko 1910, 1911; Wichterman 1953; Fokin 2004a). However, thanks to Balbiani (1861) and Bütschli (1876), who compiled the first survey devoted to parasites of ciliates, this misunderstanding was soon corrected (Bütschli 1887–1889). Moreover, Balbiani (1861) indicated *Holospora*-like bacteria not only in *P. aurelia* (in fact in *P. caudatum*) as Bütschli (1876) did, but also in *P. bursaria*, with special indication that in that ciliate it should be a different species of "vibrions" (Balbiani 1861).

After Hafkin's publication, some data connected with *Holospora* could be seen in the articles that came from the Zootomical Laboratory of Pavel Mitrophanow (1857–1925), at the University of Warsaw (at that time part of Russia). There groups of young scientists investigated the nuclear apparatus of ciliates, mainly *P. caudatum*, and by chance used an infected population of the ciliate for this investigation (Kudelski 1898; Khainsky 1903; Petschenko 1903). They, probably, simply

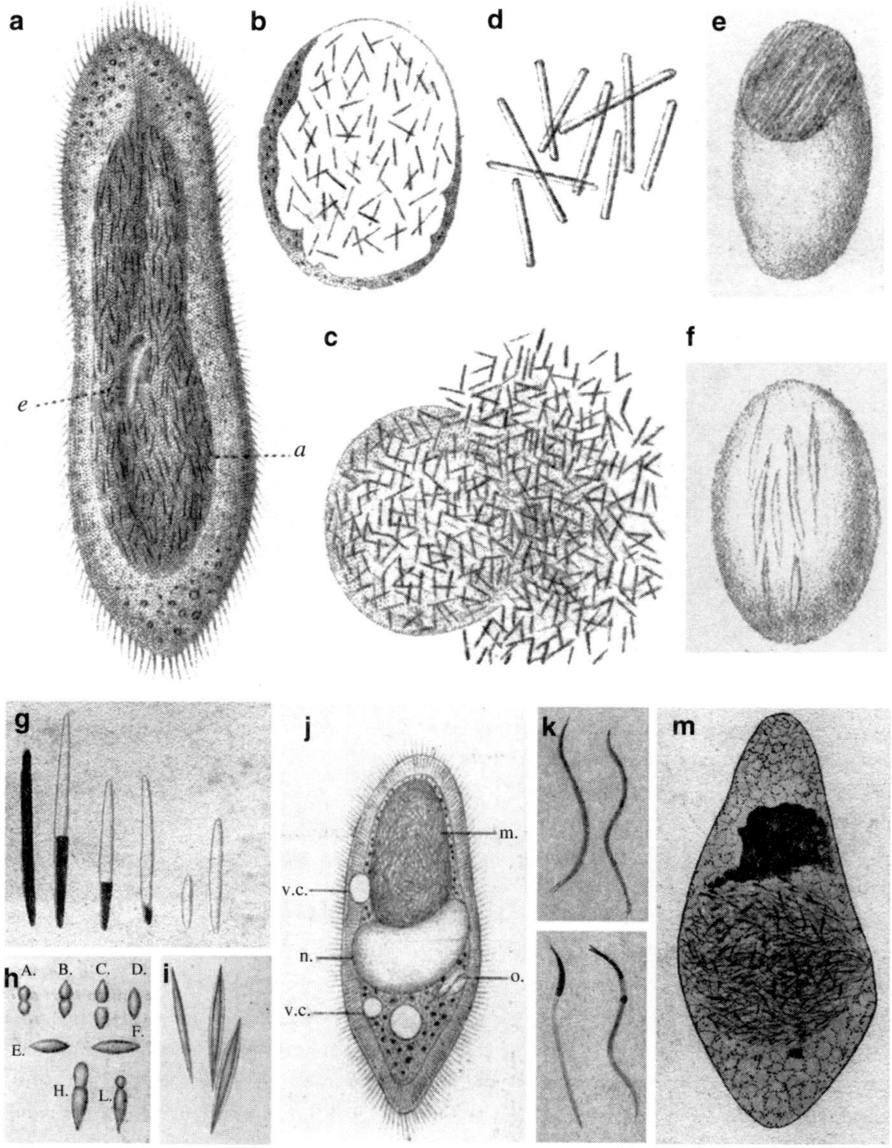


**Fig. 2** Published images of different holosporas. The middle of the nineteenth to the beginning of the twentieth century. **a–d** *Holospora obtusa* (Balbiani 1861). **e**, **f** Probably *H. elegans* and *H. obtusa* (Claparede and Lachmann 1858–1861). **g** *H. obtusa* (Bütschli 1876). **h**, **j** reproductive and infectious forms of *H. undulata*. **i** Infectious forms of *H. elegans* (Hafkine 1890). **k-m** *H. undulata* (*Drepanospira mülleri*). (From Petschenko 1911)

followed their professor's idea about morphological dynamics and plasticity of the macronucleus, which Mitrophanow discussed at the same time using the same materials (Mitrophanow 1903). Thus, some pictures of *H. obtusa* infection were treated

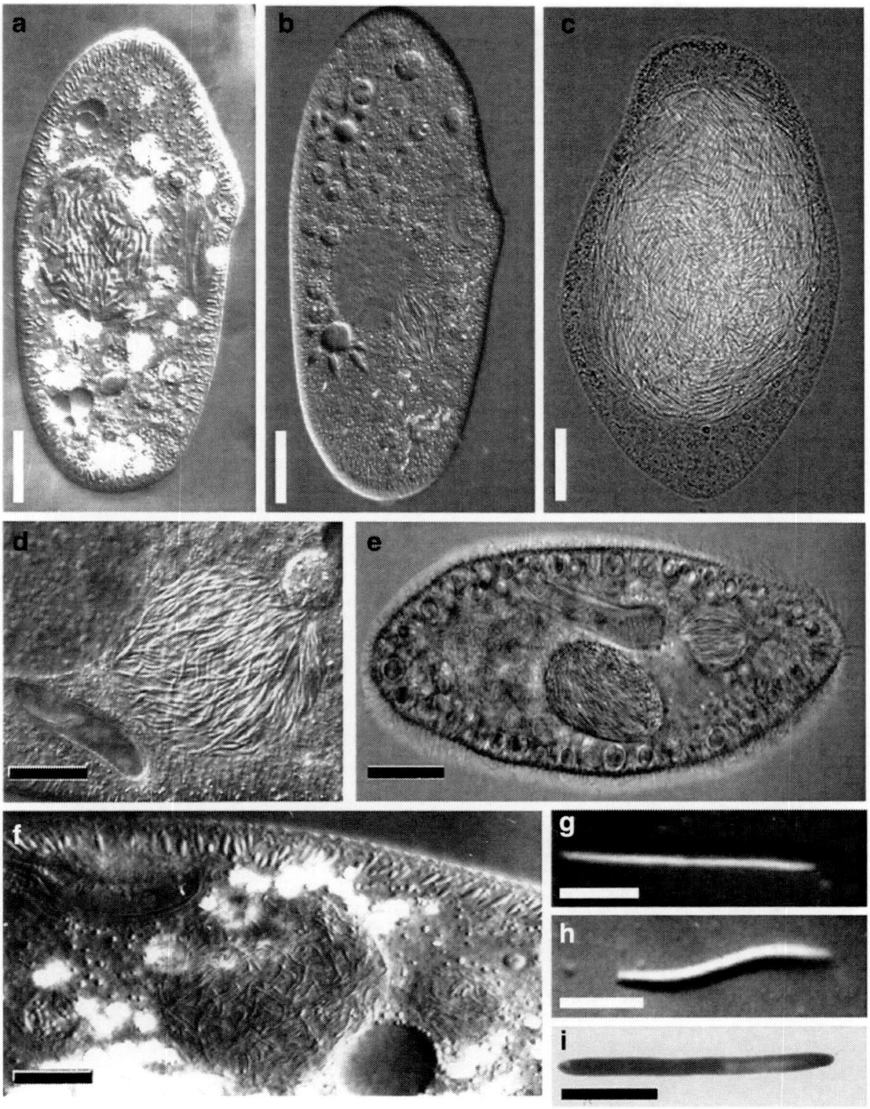


**Fig. 3** Different paramecia infected with *Holospora* bacteria. **a** *Paramecium putrinum* with *Holospora* sp in the macronucleus. **b** *P. caudatum* with *H. elegans* in the micronucleus. **c** *P. caudatum* with *H. obtusa* (hyperinfection) in the macronucleus. **d** *P. caudatum* with *H. undulata* in the micronucleus. **e** *P. bursaria* with *H. curviuscula* in both nuclei. **f** *P. biaurelia* with *H. caryophila* in the macronucleus. Infectious forms of *H. obtusa* (**g**), *H. undulata* (**h**), and *H. bacillata* (**i**). Living cells, differential interference contrast (DIC). *Bars* 15 μm (**a**), 35 μm (**b**, **c**), 15 μm (**d**, **f**), 25 μm (**e**), and 5 μm (**g–i**)

in all of those publications as particular stages of the macronucleus life cycle (crystalloids or chromosomes), even though all of the authors knew Hafkin article (Mitrophanow 1903; Petschenko 1910).

At least one of the students – Boris Petschenko (1882–1936) – proposed later that it could be either the macronuclear structure itself, its derivatives in the cytoplasm, or, finally, a kind of parasite (Petschenko 1910). Petschenko, a graduate of Warsaw University (1904) and later a member of Jagellonian University in Krakow, Poland (1905–1915, 1922–1936), paid more attention to this phenomenon (Petschenko 1910, 1911). Again, as in the case of Hafkin, in the Zootomical Laboratory of Warsaw University, an aquarium population of *P. caudatum* was used for the research. Apparently it was infected by at least two bacteria – *H. obtusa* and another one (Petschenko 1910), which was described by Petschenko 1 year later as *Drepanospira mülleri* (Petschenko 1911) (Fig. 2k-m).

This last species was studied more precisely by Petschenko from a native population of *P. caudatum* collected near Krakow (Petschenko 1911). Whilst definitely fitting the description of *H. undulata*, according to Petschenko, this bacterium had the ability to move, which does not remind us of any holosporas, and, consequently, we cannot explain this observation. Also the author thought that *D. mülleri* is located inside vacuoles in the cytoplasm, but the figures given indicate that it was, in fact, the micronucleus at the different stages of infection; in the mean time Petschenko did not find the micronucleus in infected ciliates (Petschenko 1910, 1911) (Fig. 2m). However, the author did not exclude the possibility of mix up – bacteria in the micronucleus and in the cytoplasmic vesicles (Petschenko 1910). Strangely enough, he did not pay any attention to the motility of his subject, but discussed the appearance of some nuclear elements in stained *D. mülleri*, which Hafkin never indicated for *H. undulata* (Hafkine 1890). Now we know that *H. undulata*, of course, has nucleoides, but it is doubtful that Petschenko saw them. It is more likely that he indicated just parts of the bacterial periplasm.

Then for a long time we cannot find any interest in the investigation of *Holospora* bacteria in the protistological literature. The reason for this was, probably, simply an absence of new findings of the infection. In 1928, *Archiv für Protistenkunde* (the first international protistological journal) published the article by Fiveiskaja (1928), again from Russia, entitled "Influence of intranuclear parasites of ciliates on its metabolism." A population of *P. caudatum* kept for student practice in the laboratory of the Biological Institute of Irkutsk University appeared to be infected with *H. obtusa*. Fiveiskaja, about whom we have not yet been able to find any personal data, was a student of Vladimir Schewiakoff (1859–1930), a well-known Russian protozoologist. Schewiakoff in turn was a student of Otto Bütschli from 1885 to 1889 and for several years worked as his assistant (1891–1894) at Heidelberg University (Fokin 2004d). Apparently, Schewiakoff knew this area of investigation as he was deeply interested in ciliate morphology and was personally acquainted with Metschnikoff and, possibly, also Hafkin. This study was not continued in Irkutsk because when Schewiakoff died in October 1930 his laboratory was closed.

This time could be considered as the end of the first, mainly phenomenological and simply descriptive, stage of *Holospora* bacteria investigations. Only a small number of short reports about nuclear infections in *P. caudatum* and *P. bursaria* were published later in the USA (Wichterman 1940, 1945). This fact indicated the wide spreading of these bacteria in natural ciliate populations.

In spite of the great interest in prokaryotic symbionts of ciliates that occurred after the Second World War, which was stimulated by the studies of killer traits in paramecia (Sonneborn 1943; Preer 1948) and the first review on symbiosis in protists (Kirby 1941a), new *Holospora* research findings were published only a quarter of a century later (Preer 1969).

Preer's article indicated a new level of investigation of *Holospora* spp. This included not only observation of living and stained infected paramecia, but also ultrastructural analysis of the bacterium using transmission electron microscopy, and a set of experiments connected with the microorganism's infectivity and transmission via host generations were included in the investigation (Preer 1969). Two life forms of the bacterium – infectious and reproductive, host specificity (*P. biaurelia*), and adaptation to the host life cycle indicated by the author became fundamental features for *Holospora* spp. However, at the beginning the microorganism was called simply "alpha" with the reference that *H. undulata* and *D. mülleri* are similar to it in appearance. Later the species was renamed first *Cytophaga caryophila* (Preer et al. 1974) and then *H. caryophila* (Preer and Preer 1982) (Fig. 2f).

In 1972, when in the Laboratory of Invertebrate Zoology of the Biological Institute of Leningrad State University (Old Peterhof, Russia) *H. undulata* was rediscovered (Ossipov and Ivakhnyuk 1972) – the first time as omega bacteria, permanent investigations in the field began at this institution. In a few years a large collection of different living stocks of *P. caudatum* infected with *H. undulata* and *H. obtusa* (iota bacteria at that time) was established in the laboratory. With use of this collection wide investigations about different aspects of the *P. caudatum–Holospora* system were performed: host and compartment specificity; ultrastructural studies on the infection process; bacterium's life cycle; geographical distribution and some new *Holospora* spp. descriptions. The first review article of the group was published very soon and in 1981 the main results were accumulated in the book *Problems of heteromorphism of nuclei in the unicellular organisms* (Ossipov 1981). Almost at the same time, Russian and American scientists revised the *Holospora* genus (Gromov and Ossipov 1981; Preer and Preer 1982).

Later, *Holospora* research was continued at four Russian institutions at St. Petersburg (former Leningrad, Russia): Laboratory of Invertebrate Zoology and Laboratory of Protists Karyology (Biological Institute of St. Petersburg State University), Department of Cytology (Biological and Soil Faculty of St. Petersburg State University), and the Laboratory of Unicellular Organisms (Institute of Cytology, Russian Academy of Sciences).

At the end of the 1970s *Holospora* research began in Germany (University of Münster), where Görtz and Dieckmann (1980) redescribed *H. elegans* (Fig. 3b). Very soon via cooperation with Masahiro Fujishima, a member of Yamaguch University (Fujishima and Görtz 1983; Görtz and Fujishima 1983), *Holospora* investigations spread to Japan as well. In 1993 the center of *Holospora* investigations in Germany shifted to the Biological Institute of Stuttgart University. In the meantime, this field of protistology was not continued in the USA and the UK, where only a couple of papers appeared on this topic (Barhey and Gibson 1984; Gibson et al. 1986).

From the beginning of the 1990s the *Paramecium–Holospora* system became an important model subject and a subject of international scientific cooperation in many aspects: the diversity of symbionts, adaptations and interactions between partners in the symbiotic systems, ecological and evolutionary importance of the symbiosis, mechanisms of infection, and many others by using a modern set of approaches such as light microscopy, transmission electron microscopy, and scanning electron microscopy; microsurgical transplantation, UV microirradiation, biochemistry, immunocytochemistry genetics, and molecular biology (Görtz et al. 1990, 1993; Görtz 1996; Görtz and Schmidt 2005; Fujishima et al. 1990, 1991, 1997; Fokin 1993; 2004a; Rautian et al. 1993; Fokin et al. 1996, 2003b, 2005, 2006; Timofeyeva and Rautian 1997; Skovorodkin et al. 2001; Nakamura et al. 2004; Iwatani et al. 2005; Lang et al. 2005; Sabaneyeva et al. 2005, 2007; Fema et al. 2008).

In Russia, Germany, and Japan a number of PhD studies were made in relation to *Paramecium–Holospora* investigations (in Russia – three DSc dissertations were written as well). In the 1990s and at the beginning of this century research on *Holospora* profited from the funding of international cooperation between Russian, German, Canadian, Italian, and French groups. In 1994 international cooperation on the topic was formerly established through annual meetings: 1994 (Stuttgart, Germany), 1995 (Yamaguchi, Japan), and 1996 (St. Petersburg, Russia). Then, in 2003 in St. Petersburg another international seminar on the study of *Holospora* was held, with scientists from Canada, France, Germany, Japan, and Russia participating; new projects on *Holospora* research were stimulated. The group of Lang and Burger has meanwhile sequenced 50% of the *H. obtusa* genome as the basis for advanced phylogenetic and other studies (Lang et al. 2005). A couple of years before, *Holospora* research started at the University of Pierre and Marie Curie, Paris, France (Kaltz and Koella 2003). Now French colleagues shifted "holospora activities" to the University of Montpellier. From 2005 some *Holospora* investigations were initiated at Pisa University, Italy (Fokin et al. 2006; Ferrantini et al. 2007). Although, the groups mentioned are not numerous, the majority of them are quite active. Thus, we can hope that this interesting and promising field of protistology and cell biology will be developed step by step in the future.

## 3 *Holospora* Species in Different Species of the Genus *Paramecium*

Paramecia are among the best studied ciliates. To date, the genus includes about 20 species, among which there are apparently several complexes of biological species (Fokin et al. 2004a). However, most of them cannot be a host for *Holospora* bacteria.

### 3.1 *Holospora* spp. Known from Paramecia

Since Hafkin's discovery of *Holospora* (1890) ten species of the bacteria have been identified from paramecia, but only four of them were validly published (Hafkine 1890;

Gromov and Ossipov 1981; Preer and Preer 1982; Görtz 1983, 1986, 2008; Fokin 1989; Fokin et al. 1996; Table 1). They are immobile, host- and compartment-specific infectious bacteria, invading only nuclei of certain *Paramecium* spp. As a rule, *Holospora* were noted in the macronucleus and more rarely in the generative nuclei (Table 1, Fig. 3a-f). So far the bacteria have been reported from only the large (low polyploid or heteroploid) micronucleus of *P. caudatum* and *P. bursaria* (Hafkine 1890; Borchsenius et al. 1983; Fokin 1991). All species show a developmental cycle with a reproductive form and an infectious form (Figs. 4, 5). Only the infectious form is infectious, and it shows a unique cellular organization (Görtz 1983; Fujishima and Hoshide 1988; Figs. 4b, c, 5i). The bacteria cannot live outside the host cell, not even on very complicated artificial media.

***H. undulata*** (ex Hafkine 1890) Gromov and Ossipov 1981. This occurs as short reproductive forms, spindle-shaped, about 2.5 μm long; infectious forms are up to 15 μm long, spiral-shaped, and 0.8–1.0 μm wide. The ends of infectious forms are pointed, but not equal. It lives in the micronucleus of *P. caudatum*. It was found in Europe, North America, and Asia, and is relatively common.

***H. obtusa*** (ex Hafkine 1890) Gromov and Ossipov 1981. This occurs as short reproductive fusiform rods about 3.0 μm long; infectious forms are up to 20 μm long, 0.8–1.0 μm wide. The ends of infectious forms are rounded. It is found in *P. caudatum* in the macronucleus. It was found in Europe, North America, and Asia, and is common.

***H. elegans*** (ex Hafkine 1890) Preer and Preer 1982. This occurs as short reproductive forms, spindle-shaped, about 3.0 μm long; infectious forms are up to 12 μm long with equally tapered ends, and are 0.8–1.0 μm wide. It lives in the micronucleus of *P. caudatum*. It was found in Europe, and is relatively rare.

***H. recta*** Fokin 1991. Short reproductive forms are spindle-shaped, about 2.5 μm long; infectious forms are up to 13 μm long, 0.7–1.0 μm wide, straight, with one end rounded, and one end tapered. It lives in the micronucleus of *P. caudatum*. It was found in Europe, and is rare.

***H. acuminata*** Ossipov et al. 1980. Reproductive forms are short fusiform rods about 2.5 μm long; infectious forms are 5.0–8.0 μm long, 0.5–0.6 μm wide, straight, and both ends are equally tapered. It lives in the micronucleus of *P. bursaria*. It was found in Europe, North America, and Asia, and is quite common.

***H. curviuscula*** Borchsenius et al. 1983. Reproductive forms are short, spindle-shaped, about 2–5 μm long; infectious forms are 6.0–10.0 μm long, 0.4–0.5 μm wide, slightly curved rods with tapered ends. It lives in the macronucleus (sometimes also in the micronucleus) of *P. bursaria*. It was found in Europe and Asia, and is not rare.

***H. caryophila*** (Preer 1969) Preer and Preer 1982. Reproductive forms are thin, fusiform rods, about 2.0 μm long; spiral infectious forms are 5–6 μm long, 0.2–0.3 μm wide, with both ends equally tapered. It was known previously as alpha (Preer 1969). It lives in the macronucleus of *P. biaurelia* (main host) or in *P. caudatum* (coinfection with *H. obtusa*). It was found in Europe and Asia, and is not rare.

***H. bacillata*** Fokin 1989. Reproductive fusiform rods about 3.0–5.0 μm long; infectious forms are about 15.0 μm long, 0.7–0.8 μm wide, straight, with both ends

**Table 1** Biodiversity and peculiarities of *Holospora* species from *Paramecium* hosts

| Host | Name | Morphology | | | Localization | Connecting piece formation | Distribution | References |
|---|---|---|---|---|---|---|---|---|
| | | Reproductive form (μm) | Infectious form (μm) | Shape of infectious form | | | | |
| *P. caudatum* | *H. undulata* | 2.0–2.5 × 0.8–1.0 | 10.0–15.0 × 0.8–1.0 | Spiral, ends not equally tapered | Micronucleus | + | Russia (European and Asian parts), Ukraine, Estonia, Germany, USA, France, Poland, | Hafkine (1890), Gromov and Ossipov (1981), Görtz and Schmidt (2005) |
| | *H. obtusa* | 2.0–3.0 × 0.8–1.0 | 12.0–20.0 × 0.8–1.0 | Straight, ends rounded | Macronucleus | + | Russia (European and Asian parts), Ukraine, Estonia, Germany, USA, Poland, France, Japan | Hafkine (1890), Gromov and Ossipov (1981), Görtz and Schmidt (2005) |
| | *H. elegans* | 2.0–3.0 × 0.8–1.0 | 10.0–12.0 × 0.8–1.0 | Straight, both ends tapered | Micronucleus | + | Russia (European part), Germany, France, Italy | Hafkine (1890), Gromov and Ossipov (1981), Görtz and Schmidt (2005) |
| | *H. recta* | 2.0–3.5 × 0.7–1.0 | 10.0–13.0 × 0.8–1.0 | Straight, one end tapered, the other one rounded | Micronucleus | + | Russia (European part) | Fokin (1991), Görtz and Schmidt (2005) |
| | *H. caryophila* | 1.0–2.0 × 0.2–0.3 | 5.0–6.0 × 0.2–0.3 | Spiral, both ends tapered | Macronucleus | – | Russia (European part), Germany | Görtz (1987), Fokin (1993), Görtz and Schmidt (2005) |

| | | | | | | | | |
|---|---|---|---|---|---|---|---|---|
| *P. biaurelia* | *H. caryophila* | 1.5–2.5 × 0.2–0.3 | 5.0–6.0 × 0.2–0.3 | Spiral, ends not equally tapered | Macronucleus | – | Russia (European and Asian parts), Germany, Italy | Preer (1969), Preer and Preer (1982), Görtz and Schmidt (2005) |
| *P. nephridiatum* | *H. bacillata* | 1.5–5.0 × 0.7–0.8 | 10.0–17.0 × 0.7–0.8 | Straight, both ends equally rounded | Macronucleus | – | Russia (north of European part) | Fokin (1989), Görtz and Schmidt (2005) |
| *P. calkinsi* | *H. bacillata* | 1.5–5.0 × 0.7–0.8 | 10.0–17.0 × 0.7–0.8 | Straight, both ends equally rounded | Macronucleus | – | Russia (north of European part) | Fokin and Sabaneyeva (1993), Görtz and Schmidt (2005) |
| | *H. curvata* | 2.0–3.0 × 0.7–0.9 | 10.0–15.0 × 0.7–0.9 | Slightly curved, ends not equally rounded | Macronucleus | – | Russia (North of European part) | Fokin and Sabaneyeva (1993); Görtz and Schmidt (2005) |
| *P. bursaria* | *H. acuminata* | 2.0–2.5 × 0.5–0.6 | 5.0–8.0 × 0.5–0.6 | Straight, both ends equally tapered | Micronucleus | + | Russia (European and Asian parts), Germany, Armenia, USA | Hafkine (1890), Gromov and Ossipov (1981), Görtz and Schmidt (2005) |
| | *H. curviuscula* | 4.0–5.0 × 0.4–0.5 | 6.0–10.0 × 0.7–0.8 | Straight, both ends equally tapered | Macronucleus | + | Russia (European and Asian parts), Germany, France | Hafkine (1890), Gromov and Ossipov (1981), Görtz and Schmidt (2005) |
| *P. putrinum* | *Holospora* sp. | 1.5–2.5 × 0.6–0.7 | 5.0–17.0 × 0.7–0.9 | Straight, both ends equally rounded | Macronucleus | – | Germany | Fokin et al. (1996, 1999b), Görtz and Schmidt (2005) |

rounded. At first the host was wrongly discriminated as *P. woodruffi* (Fokin 1989; Fokin et al. 1999a). It lives in the macronucleus of *P. nephridiatum* but can be found in or experimentally introduced into *P. calkinsi* as well (Fokin and Sabaneyeva 1993, 1997). It was found in northern Europe, and is not rare.

***H. curvata*** Fokin and Sabaneyeva 1993. Reproductive fusiform rods about 2.0–3.0 μm long; infectious forms are up to 20 μm long, 0.7–0.9 μm wide, slightly curved, with both ends rounded. It lives in the macronucleus of *P. calkinsi*. It was found in northern Europe, and is rare.

***Holospora*** sp. Fokin et al. 1996. Reproductive fusiform rods about 3.0 μm long; infectious forms are up to 17 μm long, 0.6–0.9 μm wide, rod-like or slightly curved, with both ends rounded. It lives in the macronucleus of *P. putrinum*, and has no special name (Fokin et al. 1996, 1999b). It was found in Europe, and is rare.

To date, holosporas have not been detected in a number of possible hosts: *P. duboscqui, P. woodruffi, P. polycaryum, P. jenningsi, and P. multimicronucleatum.* There are quite common paramecia inhabiting both brackish water (the first two species) and freshwater. All of them could have some other bacterial symbionts and *P. duboscqui, P. multimicronucleatum*, and *P. polycaryum* can be experimentally infected with *H. bacillata* (the first species) and *H. obtusa* (the last two species) but do not maintain the infection (Fokin 2000; Fokin et al. 2005).

During the last few years at least two bacteria which definitely belong to the *Holospora* genus were found in the *Frontonia* genus (Fokin et al. 2006; Ferrantini et al. 2007).

## 3.2 Host and Nuclear Specificity for Different Holosporas

The specificity of infection in *Holospora* bacteria is expressed at two levels: host and nuclear specificity (Ossipov 1973; Fujishima and Fujita 1985; Görtz 1986, 2008; Fokin 1993, 2000; Fokin et al. 1993). Despite this connection, one should remember that infection (entry into the definitive compartment) and maintenance (stable infection) are two different phenomena (Fujishima and Fujita 1985; Fokin 1993). *Holospora* spp. are generally accepted to be species-specific (Görtz and Schmidt 2005). While this is often true, *H. obtusa*, *H. caryophila*, *H. bacillata*, and *H. curvata*, in the experiments and the latter three species sometimes in nature, may infect several *Paramecium* spp. (Barhey and Gibson 1984; Fujishima and Fujita 1985; Görtz 1986; Fokin 1993, 2000; Fokin et al. 1993).

Similar reservations hold true for nuclear specificity. In the genus *Holospora* there are stable macronuclear species (*H. obtusa*, *H. caryophila*, *H. bacillata*, and *H. curvata*) and stable micronuclear ones (*H. undulata, H. elegans, H. recta*, and *H. acuminata*) (Fokin 1993, 2004a; Fokin et al. 1996), yet exceptions also occur. *H. curviuscula*, an endonucleobiont of the macronucleus of *P. bursaria*, often infects both host nuclei (Table 1, Fig. 3e) and occasionally so does *H. acuminata*, normally infecting the micronucleus of this paramecium (Borchsenius et al. 1983, 1990; Skoblo and Lebedeva 1986; Fokin 2004a). Moreover, in the experiment *H. obtusa,*

*H. undulata, H. elegans*, and *H. recta* may also "by mistake" penetrate and even start to develop in a nonspecific nucleus (Görtz and Dieckmann 1980; Fokin 1991; Fokin and Skovorodkin 1991a, b). Such mistakes are infrequent yet reproducible.

## 3.3 Morphological Peculiarities

The population of *Holospora* bacteria in stable infected nuclei includes the infectious and the vegetative (reproductive) stages: infectious and reproductive forms (Fig. 4) The reproductive forms of *Holospora* bacteria are, as a rule, represented by small rod-like cells (1–3 μm) with a homogeneous cytoplasm, whereas the infectious forms are often much larger (10–20 μm) and are differentiated into the cytoplasmic, the periplasmic, and the end (tip) zone (Görtz and Dieckmann 1980;

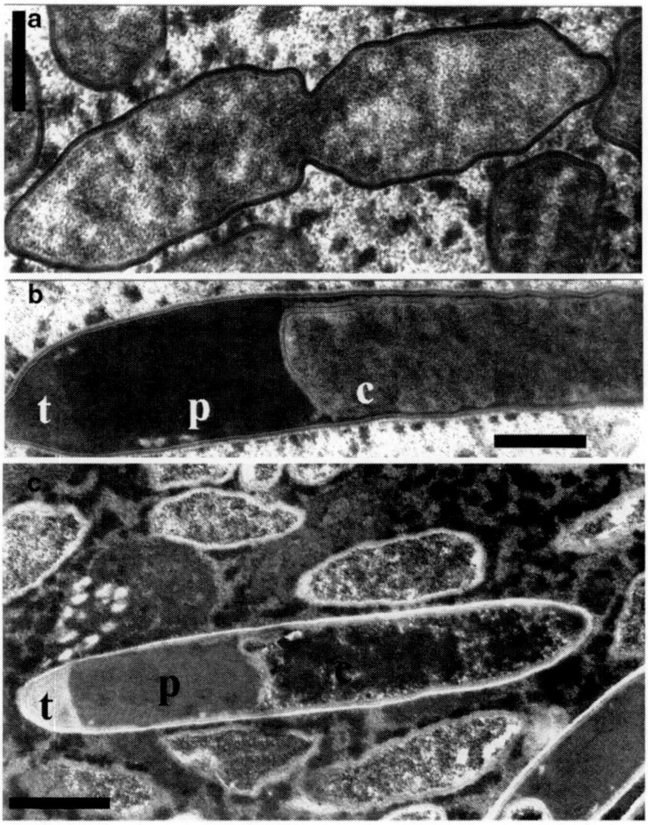


**Fig. 4** Infectious and reproductive forms of different holosporas. **a** Reproductive forms and **b** infectious form of *H. bacillata* in *P. nephridiatum*. **c** *Holospora* sp. population in the macronucleus of *P. putrinum*. *t* recognition tip, *p* periplasmic part, *c* bacterial cytoplasm part. *Bars* 0.5 μm (**a**, **b**) and 1.0 μm (**c**)

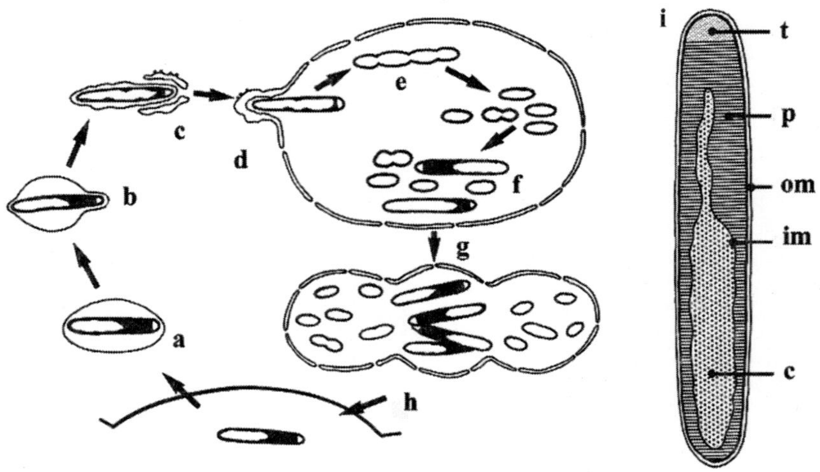


**Fig. 5** Life cycle (**a-h**) and infectious form differentiation (**i**) of *Holospora* bacteria. **a** Step of ingestion into a food (digestive) vacuole of the *Paramecium* host cell. **b** Steps of activation and release from a food vacuole to the cytoplasm. **c** Transportation of a bacterium within a vesicle encircled by endoplasmic reticulum cisternae to the nucleus. **d** Fusing process with the nuclear envelope. **e** Fragmentation leading to development of reproductive forms. **f** Process of differentiation of some reproductive forms into infectious ones. **g** Infected nucleus division with separation of new infectious forms into the connecting piece. **h** Release of the infectious forms to the external medium. *im* inner membrane, *om* outer membrane, other abbreviations are the same as for Fig. 4. (From Görtz 1996)

Fujishima and Hoshide 1988; Görtz et al. 1989; Fig. 5). The infectious forms of holosporas have a species-specific shape - straight rod-like, slightly or distinctively curved with different combinations of the cell ends, rounded, or acute (Görtz and Dieckmann 1980; Fujishima and Hoshide 1988; Görtz et al. 1989; Fokin 1991; Figs. 2d, g, i, k, l, 3g-i). These particular features of infectious forms help to discriminate the species of microorganisms.

*Holospora* infection of paramecia often brings about a considerable increase in the size of the macronucleus and especially of the micronucleus, and sometimes a manifold hypertrophy of the latter (Ossipov and Ivakhnyuk 1972; Ossipov et al. 1975; Görtz and Dieckmann 1980; Fokin 1988; Figs. 2a, m, Fig. 3b-d). In this case infected cells are rather easy to identify under low magnifications of a microscope (or even under a dissection microscope).

## 3.4 Frequency of the Infection in the Natural Population

Data on the presence and dynamics of *Holospora* infection in natural ciliate populations are scarce (Görtz 1983,1986, 2008; Wichterman 1986; Ossipov et al. 1989; Fokin 2004a, b). Bacteria from the genus have so far been discovered only in the Northern Hemisphere (Eurasia and North America), chiefly in regions with a temperate climate (Fokin 1993, 2004b; Hori and Fujishima 2003; Fokin et al. 2006). In some populations of the host (*P. caudatum*) the infection was recorded steadily for many years running, reaching more than 30%. However, usually infection does not rise to over 5% of the host population. In many populations it was noted only once, despite extensive further searches for infected cells (Ossipov et al. 1984; Fokin 2004a, b). For instance, holosporas in paramecia were found just twice in Italy as well as in Japan despite a huge number of samples (Preer 1969; Fokin et al. 2003a, b, 2006; Fokin 2004b; M. Fujishima, personal communication).

Interestingly, *H. obtusa*, *H. undulata*, and *H. caryophila* have a much broader distribution, and are observed much more frequently than *H. elegans* and *H. recta*. *Holospora* sp. (in the macronucleus of *P. putrinum*) was recovered from nature only once (Fokin et al. 1996, 1999b; Görtz 2008). It appears that the temporal dynamics of *Holospora* distribution is epidemic: in some years the number of infected populations increases and then infected cells become rare for a long time. The same holds true for frequencies of occurrence of some other intracellular bacteria: *Caedibacter*, *Nonospora*, *Pseudolyticum* (Fokin 2004b). Of course, it should be kept in mind that a lot of territories (especially exotic ones) are still very poorly investigated not only concerning *Holospora* bacteria infections but also ciliates themselves (Fokin et al. 2004a).

It should also be noted that detection of *Holospora* infections in natural populations (samples) poses certain difficulties, since nuclear infection is very often represented by small reproductive forms, which do not induce a great hypertrophy of the nuclei and are therefore difficult to reveal in total amounts. Their detection calls for screening of living cells under high microscopic magnification and the use of differential interference contrast. Again, the percentage of infected cells could be, as usual, quite low.

## 3.5 Stability of the Infection in Laboratory Cultures

The stability of the endocytobiotic systems of *Paramecium–Holospora* in laboratory cultures has not been specially studied. The data available were mostly obtained from some of *Paramecium* spp., established in laboratory culture in an infected state or infected experimentally. It is evident that the stability of intracellular bacteria maintenance is generally influenced by the characteristics of the symbiotic systems: the degree of mutual adaptation of the host cell and the bacteria, their complementarity (in case of experimental infection) and culturing conditions. These factors, separately or in combination, may greatly affect the stability of the *Paramecium-Holospora* system, taken from nature or established in the laboratory.

Cultivation temperature and feeding regime are probably the most important factors determining the stability of the symbiosis under laboratory conditions. Disintegration of endocytobiosis (loss of bacterial symbionts from the cell) at 25°C is often caused by discrepancy in the division rate of the host and that of the bacteria (Fokin 2004a). Even short-term starvation of *P. caudatum* cells infected by *H. obtusa* or *H. undulata* at room temperature often leads to the formation of numerous infectious forms in the infected nuclei and to a considerable decrease, or even complete arrest, of the host's division. The formation of infectious forms stops altogether at low temperature (4–8°C) (Fokin 2004a). From our own experience of keeping living *Paramecium* stocks infected with different holosporas, we can indicate that *H. caryophila* infection is much more stable than others. This symbiotic system can be maintained in the laboratory for years (in our collection it was 16 years).

# 4 Life Cycle of *Holospora* Bacteria

*Holospora* bacteria undergo a developmental cycle with specialized infectious and reproductive forms, which carry out transmission of the infection from generation to generation of the host (vertically) and spread it in the host population (horizontally).

## 4.1 General Description

The main life strategies of symbionts should ensure stabilization in the host cell, maintenance of infection in the line of generations and, in the case of parasitic bacteria, invasion of new hosts. In accordance with these tasks, the life cycles may be more or less complex and the bacteria should possess certain adaptations. The general pattern of the *Holospora* life cycle has been repeatedly described in the literature (Ossipov and Ivakhnyuk 1972; Görtz and Dieckmann 1980; Ossipov 1981; Görtz 1983; Görtz and Schmidt 2005; Fig. 5a-h).

Highly infectious symbiotic microorganisms belonging to the genus *Holospora* have a quite typical parasitic life cycle (Ossipov 1981; Görtz 1986, 1988; Fokin 2004a). This life cycle includes several steps connected with infectious and reproductive forms of the bacteria (Fig. 5a-h). Infectious forms that have been ingested can infect aposymbiotic cells. The principal stages of the invasion process are phagocytosis (infectious forms can be seen in the food vacuoles), release of infectious forms from phagosomes into the host's cytoplasm, transportation via paramecia cytoplasm with a help of host membranous and actin cytoskeleton systems, and penetration of infectious forms into the nucleus and their stabilization in this definitive compartment (Görtz 1983; Fokin and Skovorodkin 1991b; Kawai and Fujishima 2000; Fokin et al. 2003a; Sabaneyeva et al. 2005, 2007; Figs. 5, 6). In the freshly infected nucleus, infectious forms turn into short fragments (reproductive forms) as a result of their restriction (multiple division) after 24–48 h. The latter start vegetative reproduction, soon occupying the nucleus (48–120 h). From then on (120–168 h) some reproductive forms increase in size and differentiate into infectious forms, which no longer divide (Fig. 5a-h). Depending on the temperature and feeding regime of the host, infectious forms may constitute a minor or a considerable part of the intranuclear population. Depending on the *Holospora* species, infectious forms are

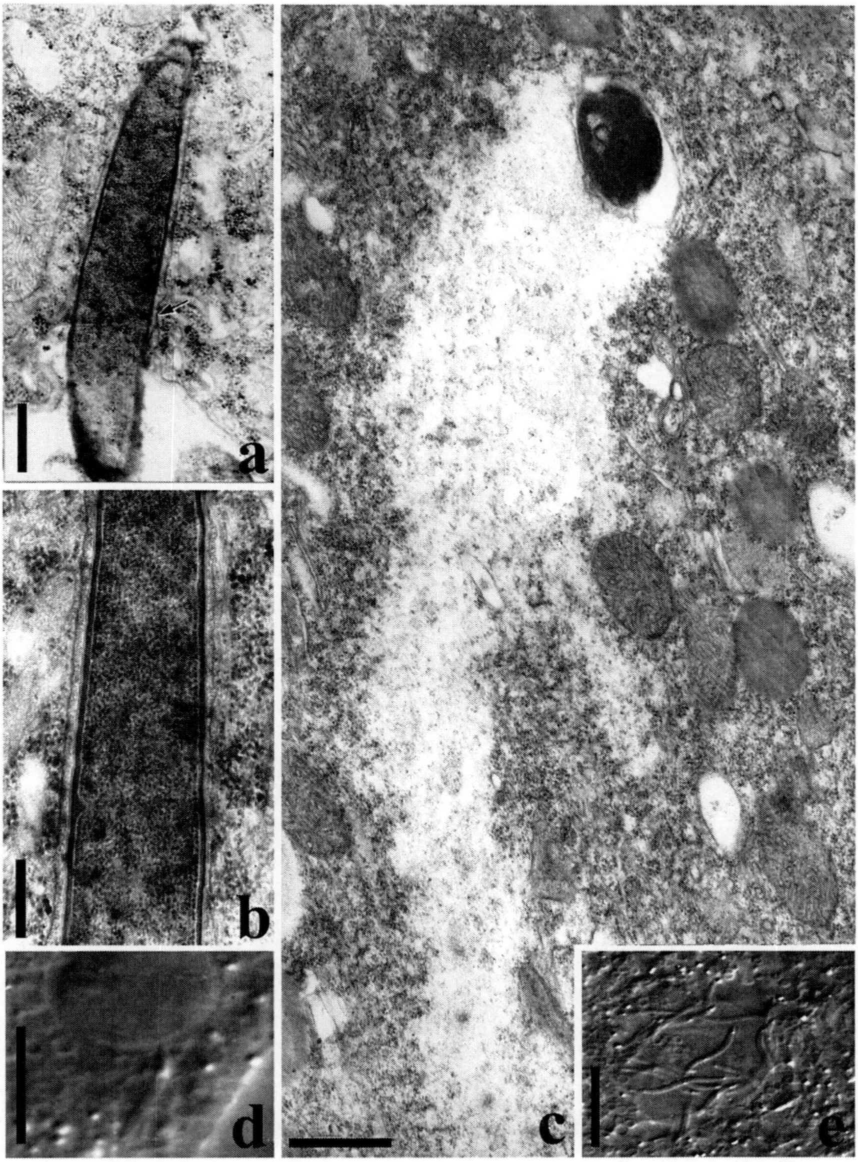

**Fig. 6** The first steps of *P. caudatum* infection by *H. undulata* during experimental infection. **a** Escape from the food vacuole. **b** Escaping bacteria decorated by the food vacuole membrane. **c** Actin-like filaments are associated with the host membrane enclosing an infectious form of the parasite in the host cytoplasm. **d** Interaction of the infectious form with the envelope of the micronucleus. **e** The micronucleus full of infectious forms. *Bars* 1.0 μm (**a**, **c**), 0.5 μm (**b**), and 10 μm (**d**, **e**)

released either after every division of the infected nuclei or regardless of the nuclear cycle phase of the host (Fokin and Sabaneyeva 1997; Figs. 7, 8). If the released infectious forms are eaten by aposymbiotic paramecia, the cycle closes (Fig. 5).

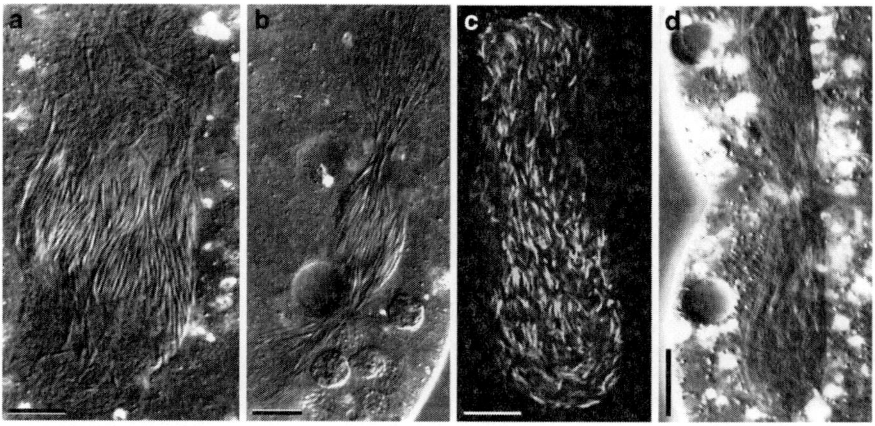


**Fig. 7** Distribution of infections and reproductive forms of holosporas during the process of host cell division in different paramecia. **a** *H. obtusa* in *P. caudatum;* beginning of the connecting piece with formation of infectious forms. **b** *H. recta* in *P. caudatum*; the connecting piece with infectious forms has already formed. **c** *H. caryophila* in *P. biaurelia* and **d** *Holospora* sp. in *P. putrinum*; no connecting piece is formed. Living cells, DIC (**a**, **b**, **d**); fluorescent in situ hybridization (FISH) preparation, fluorescence microscopy (**c**). *Bar* 10 μm

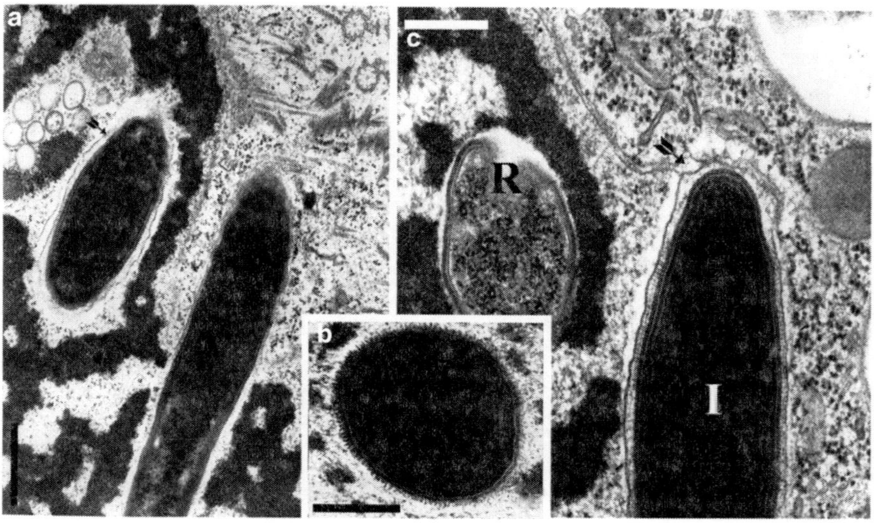


**Fig. 8** Release of the infectious forms of *Holospora* sp. from the macroucleus of *P. putrinum.* **a** Released bacteria (*right*) communicated with the nuclear envelope. *Left*: One already decorated by a membrane closely associated with the fibrous layer (*small arrow*) on the surface of the infectious form. **b** Infectious form with a distinctive fibrous layer before membrane formation. **c** Fusion of the infectious form surrounded by the membrane with the inner membrane of the nucleus envelope (*arrow*). The reproductive form (*left*) does not have a fibrous layer around it. *Bars* 1.0 μm (**a**) and 0.5 μm (**b**, **c**)

## 4.2 Way up to the Target Nucleus

During this stage (Fig. 5a-f) bacteria should be phagocyted by the host, pass the activation stage after the phagosomes coalesce with the acidosomes and leave the phagosome bounded by its membrane, reach the nucleus via the cytoplasm as part of a membrane-fibrillar complex with support of the host actin cytoskeleton, penetrate the karyoplasm after interaction with the nuclear envelope (Fig. 6) and, finally, start formation of the reproductive forms. The details of the process are discussed in the chapter by Fujishima in this volume.

It is noteworthy that bacteria infect the micronucleus only in *P. caudatum, P. bursaria*, and (in the experiment) *P. putrinum* (Fokin et al. 1996; Fokin 2000), i.e., paramecia with large low-polyploid generative nuclei (Ossipov 1981). Karyoplasm infections in small (1–3 μm) and, probably, diploid vesicular or endosomal micronuclei (Fokin 1997) are unknown so far. Whether this is due to the size of such micronuclei, which are too small to be possible "targets," and the special packing of chromatin in them or is due to lack of nutrients for maintenance of symbionts is still unclear. Some infection experiments have shown that for the nuclear recognition the size of the target is not an important feature (Fokin and Skovorodkin 1991a, b; Fokin, 1993). The details of the process are discussed in the chapter by Fujishima in this volume.

## 4.3 Reproduction in the Nucleus

The cell cycle and the character of the interactions with hosts in *Holospora* suggest stable parasitism (Görtz 1988, 1996). Infection maintenance in the host nucleus and differentiation of a part of the population of *Holospora* reproductive forms into infectious forms also has a cooperative character, requiring participation of both the bacteria and the paramecium. In this system there is no direct control of the host over the number of reproductive forms in the nucleus if infected ciliates live under favorable feeding and temperature conditions, but under such conditions few infectious forms are normally formed anyway. Rapid division of reproductive forms in the life cycle of the host does not cause hypertrophy of the nucleus. During starvation of paramecia, however, unlimited division of reproductive forms may destroy the nucleus, and it is then that some reproductive forms cease to divide and differentiate into infectious forms (Figs. 4, 5a-f) – a process in many respects opposite to the changes that infectious forms undergo during infection (Fujishima et al. 1990; Görtz et al. 1990). The details of the process are discussed in the chapter by Fujishima in this volume.

## 4.4 Release of the Infectious Forms from the Host Cell

The most important link in the *Holospora* life cycle is horizontal transmission: the release of infectious agents (infectious forms) and infection of new hosts. Of course, a symbiotic system can only be considered to be stable if infectious bacteria may undergo a complete life cycle (Fig. 5a-h), which usually takes *Holospora* about 7 days (Fokin 1993). Residing in a closed compartment, endonucleobionts are separated from the environment by several barriers (the nuclear envelope, the cytoplasm, the cortex of the host cell), so special adaptations to the release of infectious forms (to be discussed later) should evolve. Depending on the *Holospora* species, infectious forms are released either after every division of the infected nuclei via a special "connecting piece" – an equatorial part of the dividing nucleus where the majority of infectious forms collect (Wiemann and Görtz 1989) (Figs. 5a-g, 7a, b) – or regardless of the nuclear cycle phase of the host – when infectious forms in a group or individually can leave the infected nucleus (Fokin and Sabaneyeva 1997) (Figs. 7c, d, 8). According to this character, all holosporas can be divided into two groups (Fokin et al. 1996). When the released infectious forms are eaten by a new aposymbiotic host cell, the cycle will be complete (Fig. 5a-h).

## 4.5 *Holospora* as a Vector for Other Bacteria

Infectious *Holospora* bacteria probably play a very special vector role in increasing the diversity of nuclear symbionts. Findings of noninfectious bacteria in the nucleus following its experimental infection by *Holospora* (Görtz and Feiburg 1984; Fokin and Skovorodkin 1991a; Fokin and Karpov 1995) suggest a possible coinfection of infectious and noninfectious endobionts (Görtz 1986; Fokin 1993). Experimental testing demonstrated that *H. obtusa* could introduce nonspecific (food) bacteria into the macronucleus (Fig. 9), ensuring a high percentage of macronuclear infection by *H. undulata* (normally infecting the micronucleus) and *H. caryophila* (normally infecting the macronucleus of *P. biaurelia*), as well as by *Nonospora macronucleata*, which has a low infectivity (Fokin et al. 2004b). Analogous "additional" infection (by food bacteria) was observed when *H. obtusa* infected a nonspecific host, *P. multimicronucleatum* (Fokin et al. 2003b). The appearance in the macronucleus of viable food bacteria (*Enterobacter aerogenes* and *Klebsiella pneumoniae*), which normally never leave the phagosomes and are digested there, indicates, incidentally, that *Holospora* do leave the food vacuoles before the latter fuse with the lysosomes (Fig. 9a).

Analysis of the diversity of noninfectious bacteria in the paramecia nuclei shows that they are indeed found first of all in the cells of potentially *Holospora*-infected species (specific or nonspecific hosts of holosporas): *P. bursaria, P. duboscqui, P. nephridiatum, P. calkinsi, P. caudatum*, and *P. multimicronucleatum* (Fokin 2000).

# 5 Two *Holospora* Groups and Some Biological Features of Them

From a morphological point of view, the *Holospora* genus looks quite solid as its representatives manifest a similar life cycle and morphology. However, according to some biological features as well as molecular fluorescent in situ hybridization (FISH) labeling results, they can be split into two groups (Table 1, Fig. 10).

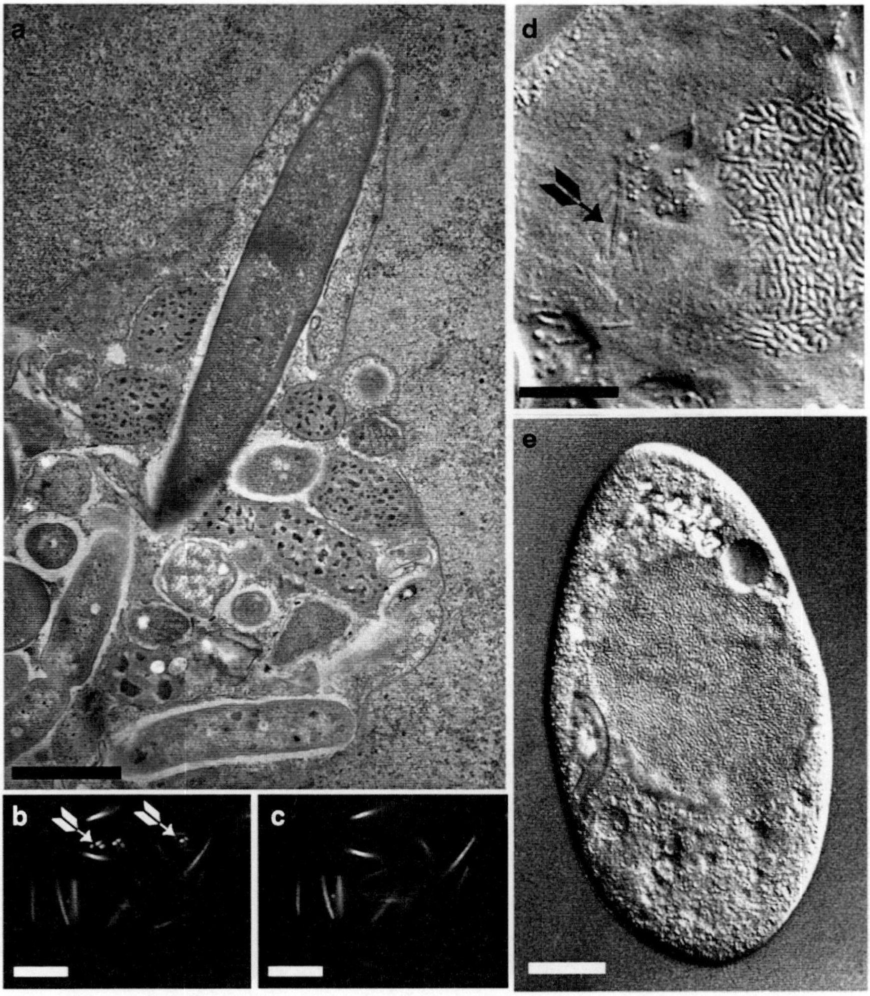

**Fig. 9** Coinfection of *Holospora obtusa* with some food bacteria. **a** Infectious form of *H. obtusa* leaving a phagosome full of different bacteria. **b**, **c** The same freshly infected macronucleus with a number of infectious forms after double FISH with *Holospora*- and eubacteria-specific probes (**b**) and with a *Holospora*-specific probe only (**c**); some food bacteria are visible (*small arrows*). **d** Infected macronucleus after 24 h. A number of food bacteria occupy part of the macronucleus, few holosporas (*arrow*). **e** Ciliate with hyperinfection of the macronucleus by food bacteria. *Bars* 1.0 μm (**a**), 12 μm (**b**, **c**), 15 μm (**d**), and 25 μm (**e**)

## 5.1 Groups Distinguishable by 16S Ribosomal DNA Probes

Since most intracellular bacteria cannot be maintained outside the host cell, they are impossible to characterize by metabolism and growth features, as it is customary with free-living bacteria thus, the identification with the use of traditional microbiological methods is not possible.

At present the phylogenetic position and relations of microorganisms can be established by molecular-biological methods. However, complete nucleotide sequences of the 16S ribosomal DNA (rDNA) gene have so far been obtained only for a few ciliate endocytobionts (Schrallhammer et al. 2006). The attribution of some bacteria to this or that group can now also be determined (or corroborated) by the results of FISH with corresponding oligonucleotide probes (Amman et al. 1991, 1995; Görtz and Brigge 1998). Amman et al. (1991) designed a fluorochromated oligonucleotide probe according to a sequence of the 16S rDNA of *H. obtusa*. With use of this probe it was possible to label *H. obtusa* as well as *H. elegans* and *H. undulata*, and the probe was therefore regarded as genus-specific for *Holospora* (Fokin et al. 1996). According to the FISH results, a number of *Holospora* bacteria from the macronucleus of different paramecia – *P. biaurelia* (*H. caryophila*), *P. nephridiatum* (*H. bacillata*), *P. putrinum* (*Holospora* sp.) - do not belong *Alphaproteobacteria* as "classical" *Holospora*, though all their morphological and most biological characters correspond to those of the holosporas (Fokin et al. 1996; Fokin 2004a; Görtz and Schmidt 2005; Table 1, Fig. 3). Thus, FISH data indicate that the *Holospora* genus may at least belong into two phylogenetic groups (Fig. 10).

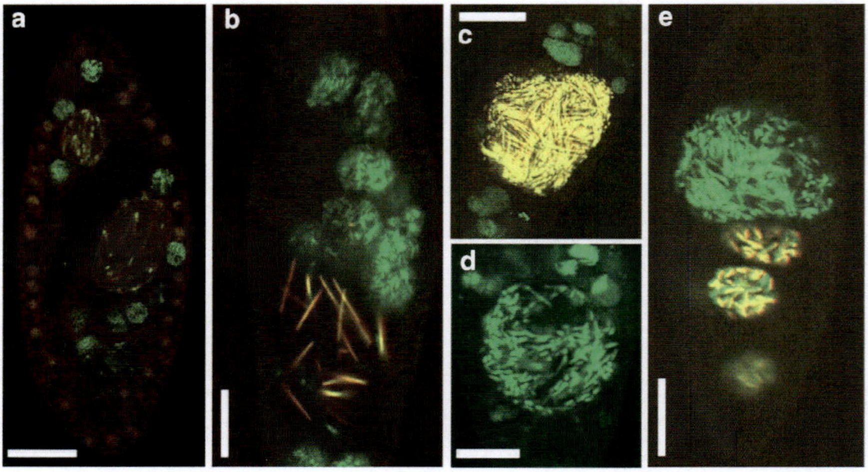


**Fig. 10** FISH of several *Holospora* spp. with two specific oligonucleotide probes. **a** Double labeling of *P. bursaria* infected with *H. curviuscula*. **b** Double labeling of *P. caudatum* infected with *H. obtusa* after 1 h of experimental infection. **c** Part of a *P. caudatum* cell heavily infected with *H. obtusa*, double labeling. **d** Part of a *P. putrinum* cell with *Holospora* sp. in the macronucleus after double labeling. *Holospora*-specific and eubacteria-specific probes were used. No *Holospora* signal is present. **e** *P. biaurelia* infected with *H. caryophila* after experimental feeding with alphaproteobacteria, double labeling with alpha-specific and eubacteria-specific probes. No alpha signal is present in the macronucleus, only in food vacuoles. *Bars* 30 μm (**a**, **d**), 20 μm (**b**, **e**), and 30 μm (**c**)

## 5.2 Behavior of Different *Holospora* Species During Nuclear Division of the Host

The release of infectious forms and the infection of new hosts are a major objective of parasitic bacteria. *Holospora* "solve" this problem in two different ways, which invited the suggestion of a polyphyletic nature of the genus (Fokin et al. 1996). In some holosporas – *H. obtusa, H. undulata, H. elegans, H. recta, H. acuminata*, and *H. curviuscula* – during division of the infected nucleus (the macronucleus or the micronucleus) a connecting piece, where infectious forms mostly stay, is formed in the equatorial region of the division figure, with reproductive forms being regularly segregated into the daughter nuclei (Görtz 1986; Fokin 1993, 2004a; Fokin et al. 1996; Table 1; Fig. 7a, b). Later the connecting piece degrades and numerous infectious forms contained in it are packed in vacuoles and are transported to the cytoproct to be released outside. This process takes about 1 h. The dynamics, light optical, and ultrastructural details of the release of infectious forms from the macronucleus and the micronucleus were studied, respectively, in *H. obtusa* and *H. elegans* (Wiemann 1989; Wiemann and Görtz 1989).

No such connecting piece is formed during the division of the infected macronucleus in the case of infection by *H. caryophila* (*P. biaurelia* and *P. caudatum*), *H. bacillata* (*P. nephridiatum* and *P. calkinsi*), *H. curvata* (*P. calkinsi*), and *Holospora* sp. (*P. putrinum*) (Fig. 7c, d). Their infectious and reproductive forms are randomly distributed between the daughter nuclei (Fokin and Sabaneyeva 1997; Fokin 2004a; Fig. 7c, d). In the latter three species this process is associated with a backward translocation of individual infectious forms via the nuclear envelope: first into the perinuclear space and then, in vacuoles formed from the outer nuclear membrane, into the cytoplasm (Fokin and Sabaneyeva 1997; Fig. 8). Interaction between the infectious forms released from the nucleus and the membranes of the nuclear envelope appears to be mediated by a fine fibrillar layer and then the membrane, exposed on the surface of the infectious forms, and lacking in the reproductive forms of these bacteria (Fig. 8). For about 1 h these vacuoles undergo cyclosis in the host's cytoplasm and then their contents (infectious forms) are released via the cytoproct (Fokin and Sabaneyeva 1997).

The mode of release of infectious forms from the host nuclei may have changed successively in the course of the evolution of the system: from single bacteria released into the cytoplasm (*H. bacillata, H. curvata*, and *Holospora* sp.) to their complexes (*H. caryophila*) and, finally, to the exploitation of the host's nuclear division apparatus for effective segregation of infectious forms with their subsequent extrusion into the cytoplasm at each division of the paramecia (other holosporas).

## 5.3 Behavior of Different *Holospora* Species During the Sexual Process of the Host Cell

Especially interesting are the strategies of intracellular bacteria from paramecia aimed at survival during the sexual propagation (conjugation) of the host's life cycle when, in the course of the nuclear reorganization, the old macronucleus degrades and the micronucleus, having undergone meiosis and a number of post-meiotic divisions, gives rise to a new nuclear apparatus (Ossipov 1981; Hiwatashi and Mikami 1989). It had long been thought that all the infected cells reversibly lose the ability to enter the sexual process (Ossipov 1981), which is analogous, to a certain extent, to the phenomenon of parasitic castration in higher organisms (Baudoin 1975). Later it was shown that both *H. elegans* (the micronucleus infection) and *H. obtusa* (the macronucleus infection) did not always prevent the host from entering into conjugation (Görtz and Fujishima 1983; Fokin 1998), but that exconjugants either lost bacteria together with the old macronucleus during nuclear reorganization (*H. obtusa* infection) or were nonviable (infection with *H. elegans*) (Görtz and Fujishima 1983; Fokin 1998).

It is now clear that endonucleobionts of paramecia may be retained not only when the host's sexual process is blocked but also if they modify the normal course of nuclear reorganization in exconjugants, which leads to regeneration of the old macronucleus or formation of a new infected heterokaryotic macronucleus. This is the case for *H. caryophila* (Fokin 1998). It can manifest even one more unique survival strategy during the host's sexual process. Infectious forms of *H. caryophila* may enter the cytoplasm from the fragments of the old infected macronucleus and infect the anlagen of the new somatic nucleus (Preer 1969). Since autogamy is characteristic of this paramecium species and nuclear reorganization occurs approximately once in 20 days, such adaptations favor the infection maintenance in the ciliate population.

## 5.4 Phylogenetic Positions and Taxonomy of the Endocytobionts

According to the modern view, *Holospora* bacteria belong to the family *Holosporaceae* fam. nov. Görtz and Schmidt 2005 inside the class *Alphaproteobacteria* (Görtz and Schmidt 2005). *Holospora* were the first intracellular bacteria in *Paramecium* for which the phylogenetic position was determined (Amman et al. 1991). *H. obtusa*, *H. elegans*, and *H. undulata* belong to *Alphaproteobacteria*. The closest relative among other symbionts in ciliates was found to be *Caedibacter caryophilus* and "*C. macronucleorum*" as well as some endocytobionts of *Acanthoamoeba* (Schrallhammer et al. 2006), the closest relatives among other bacteria found up to now being *Rickettsia* and *Ehrlichia* (Amman et al. 1991; Springer et al. 1993; Schrallhammer et al. 2006) (Fig. 11). In the search for bacterial relatives of mitochondria Lang et al. (2005) found good indications for *H. obtusa* being the closest known relative of mitochondria.

It is tempting to regard the striking biology (developmental cycle, host specificity for *Paramecium*, etc.) and the unique morphology of the infectious form as homologous features proving the close relationship and monophyletic origin of these bacteria. However, because of some observations, this may be doubted (Fokin et al. 1996). The behavior of the infectious forms of certain *Holospora* species to assemble in the connecting piece of the dividing host nucleus is certainly highly advanced and must

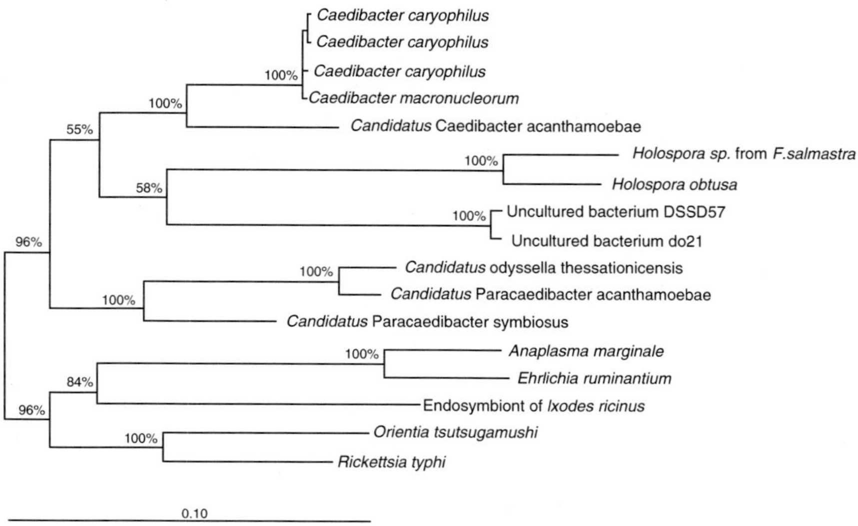


**Fig. 11** Maximum parsimony phylogenetic tree based on almost complete16S ribosomal DNA gene sequences of *Holosporacea* and related rickettsias. The tree was calculated using a filter that retains 1,321 unambiguously aligned positions. The numbers represent bootstraps values (percent) of 100 pseudoreplicates. The *scale bar* corresponds to ten nucleotide substitutions per 100 positions

be regarded as an apomorphic feature. This behavior ensures that the infectious forms are specifically collected and released by the host cell. *H. caryophila*, *H. bacillata, H. curvata, and Holospora* sp. (from *P. putrinum*) do not share this feature with the other holosporas (Figs. 7, 8). They either are more primitive than the other holosporas, as a quantitative separation of infectious forms and reproductive forms is not yet formed, or are phylogenetically not closely related. In conclusion, at the time two species groups may be distinguished in the genus *Holospora* (Table 1, Fig. 10).

# 6 Host Cell Resistance Against *Holospora* Infection

Some holosporas are quite widely distributed in nature, but *Paramecium*-infected populations are not numerous despite the fact that *P. caudatum, P. bursaria*, and *P. aurelia* at least are cosmopolitan and abundant on all continents (Fokin et al. 2004a). Since *Holospora* spp. are highly infectious bacteria we could expect a higher rate of infection, but in reality this is not the case.

## 6.1 General Remarks

The *Holospora* infection rate in paramecia populations is usually quite low (see Sect. 3.4). Analysis of such a rate for *P. caudatum* on the territory of the former USSR, based on long-standing observations, yields a total figure of about 5%, even less for some regions (Ossipov et al. 1984, 1989; Fokin 2004b). Considering the high infectivity of *Holospora* and its wide distribution, this appears to indicate that potential hosts of these bacterial endonucleobionts are often resistant to infection. Even in the case of experimental infection it is not always 100% of the cells that maintain it, despite the fact that early in the experiments all the cells were infected (Görtz and Dieckmann 1980; Fokin and Skovorodkin 1991a, b). Ciliates, apparently, possess some defensive mechanisms against bacterial infection (Görtz 1986; Fokin and Skovorodkin 1997).

The investigations revealed that some clones and possibly even populations of the host were unsusceptible to infection. The number of such clones was high for *H. undulata* (22%) and relatively low for *H. obtusa* (9%) (Ossipov 1973; Rautian et al. 1990), though for another *H. obtusa* strain and another sampling of *P. caudatum* clones the proportion of resistant host cell clones was much higher: 41% (Fujishima and Fujita 1985) and even 60% (Fokin et al. 2003b). Apparently the reasons for such resistance could be different and infection could be blocked at various steps. For three species of potential hosts - *P. bursaria*, *P. multimicronucleatum*, and *P. caudatum* – resistance ability against *Holospora* infection was investigated experimentally.

## 6.2 Experimental Data

To define when exactly the infection was blocked, early infection stages (from the phagosome to entry to the nucleus) were successively analyzed and newly infected nuclei were observed over several days following the start of experimental infection. It has been shown that failure to infect paramecia with holosporas or to maintain stable infection in the nucleus may be associated with disruptions at most early infection stages. Cells of some *P. caudatum* clones obtained from nature cannot ingest infectious forms of *Holospora*, probably owing to food selectivity (Fokin and Skovorodkin 1991a; Skoblo et al. 1996; Fokin et al. 2003b). The infection process has also been found to be blocked at the phagosomal stage (failure of bacteria to enter the cytoplasm) and at the cytoplasmic stage (failure of bacteria to enter the nucleus) (Skoblo et al. 1996; Fokin et al. 2003b). Even penetration into the nucleus (the macronucleus or the micronucleus) does not guarantee the establishment of the symbiotic system, since bacteria may be rapidly lysed in the karyoplasm and/or may not produce reproductive forms (Skoblo et al. 1990; Fokin and Skovorodkin 1991a, b; Skoblo et al. 1996; Fokin et al. 2003b).

Rapid lysis of infectious forms of *H. undulata* and *H. acuminata* in the micronucleus and sometimes in the macronucleus, noted in some resistant paramecium clones, is a good example of interaction in a noncomplementary system (Skovorodkin and Fokin 1990; Ossipov et al. 1993; Fokin and Skovorodkin 1997; Fig. 12). The mechanisms of this phenomenon have not been studied. The lysis may be brought about by activation of the host enzymes: the rapidity of the process (3–5 h for *H. undulata*) indicates this option. However, if this is the case, it is not clear how the host protects the structures of its own karyoplasm.

In experiments involving genetic transformation of the lysis-resistant clone and the creation of heterokaryons bearing different combinations of nuclei

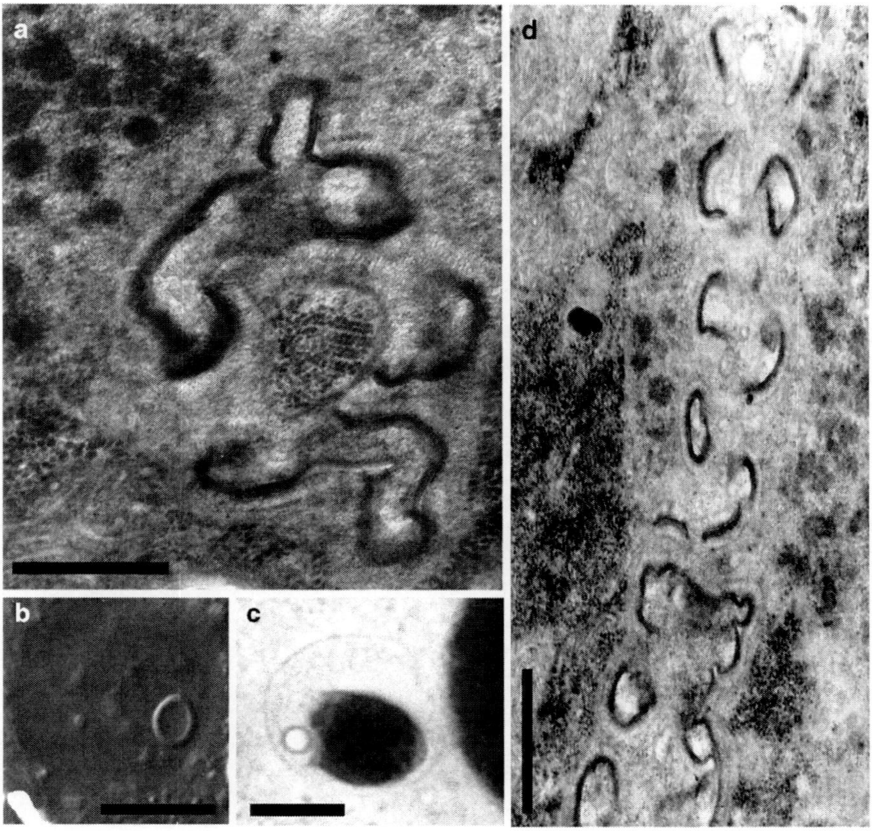


**Fig. 12** Bacterial lysis inside the nuclei in *P. caudatum* and morphology of the resistant micronucleus. **a** *H. obtusa* lysis in the macronucleus. **b** Living and **c** Feulgen-stained resistant micronucleus. **d** Fragments of *H. undulata* in the resistant micronucleus. *Bars* 1.0 μm (**a**, **d**), and 5 μm (**b**, **c**)

from strains resistant or sensitive to infection, it has been shown that the character of lysis is determined by the genome of the macronucleus, not the micronucleus, where the process occurs (Rautian et al. 1996; Fokin and Skovorodkin 1997).

Another type of host reaction to infection was revealed when it was attempted to infect with *H. obtusa* the macronucleus of *P. multimicronucleatum*, a nonspecific host of this bacterium (Fujishima and Fujita 1985; Fokin et al. 2005). In general, *P. multimicronucleatum* can be infected with infectious forms within 2 h, but it mainly loses the microorganisms from the macronucleus after 9–16 h. They are released from the host nucleus into the cytoplasm and then outside. This process in the infected cells can be stopped by low temperature or nocodazole treatment. This experiment indicates that some microtubules in the nuclear matrix probably play a role in the ciliate defense against intranuclear bacteria (Fokin et al. 2005).

## 7 Simultaneous Infection of the Host with Different *Holospora* and Other Bacteria

It is not uncommon that the ciliate cell contains simultaneous infections by several endocytobionts at the same time, in different compartments or, less frequently, in the same one (Preer 1969; Görtz 1987; Fokin et al. 2000). Examples of such complex endocytobioses can be found in special reviews (Görtz 1986; Fokin 1993). These systems are usually unstable under laboratory conditions, apparently owing to relationships between the symbionts or between the symbionts and the host. Different bacteria may be antagonistic, especially when they are occupying the same compartment (Fig. 13). For instance, *H. caryophila* in the macronucleus of *P. caudatum* supersedes *H. obtusa* in the case of common infection; *N. macronucleata* supersedes *H. obtusa* (Fokin et al. 1987; Fokin 1988); in the case of common experimental infection of the micronucleus by *H. undulata* and *H. recta*, only the cells with either the former or the latter symbionts are soon observed in the infected paramecia culture, with the common infection disappearing (Fokin 1991). Most probably, the antagonism expresses itself in modification of conditions in the cell compartment by one symbiont, so that they become unsuitable for other bacterial species. It may be also expressed in a direct interspecific interaction, by production of metabolites or competition for some metabolites of the host (Görtz 1987).

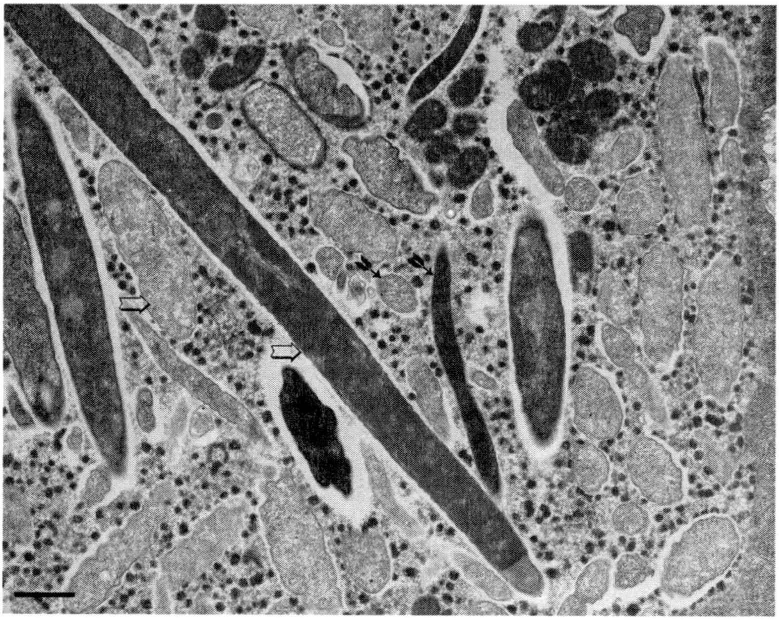

**Fig. 13** *H. obtusa* and *H. caryophila* simultaneously infected the macronucleus of *P. caudatum*. Infectious and reproductive forms of *H. obtusa* (*arrows*), infectious and reproductive forms of *H. caryophila* (*small arrows*). *Bar* 1.0 μm

*Holospora* infection of one of the paramecia nuclei specifically changes the state of the other one. For instance, at the vegetative stage of *H. obtusa* development in the macronucleus, the generative nucleus may be effectively infected by *H. undulata*, but as soon as numerous infectious forms appear in the macronucleus, the micronucleus become unsusceptible to *H. undulata* infection (Fokin and Skovorodkin 1991a, b). A similar resistance of the macronucleus is observed when *H. undulata* in the micronucleus passes from the reproductive to the infectious stage. Later the double infection, as a rule, disappears (Fokin 1991, 1993). On the other hand, there are cases when only the cells already infected by certain bacteria may be infected by additional endocytobionts (Preer 1969; Gibson 1973; Barhey and Gibson 1984).

## 7.1 *Holospora* and *Holospora*-Like Endosymbiotic Bacteria in Other Ciliates

As already mentioned, ten *Holospora* spp. have been found so far in paramecia. Very recently two more examples of bacteria definitely belonging to the genus were discovered in the *Frontonia* genus (Fokin et al. 2006; Ferrantini et al. 2007; Fig. 14a). Although *Paramecium* and *Frontonia* are the sister genera, we can postulate now that *Holospora* are not exclusively endonucleobionts of *Paramecium* spp. as was always indicated before (Görtz and Schmidt 2005). Thus, in the future it can be expected that some more holosporas in different groups of ciliates will be found. Indeed, several other bacteria, which have a life cycle similar to *Holospora* represented by two morphological forms, have been mentioned from different ciliates

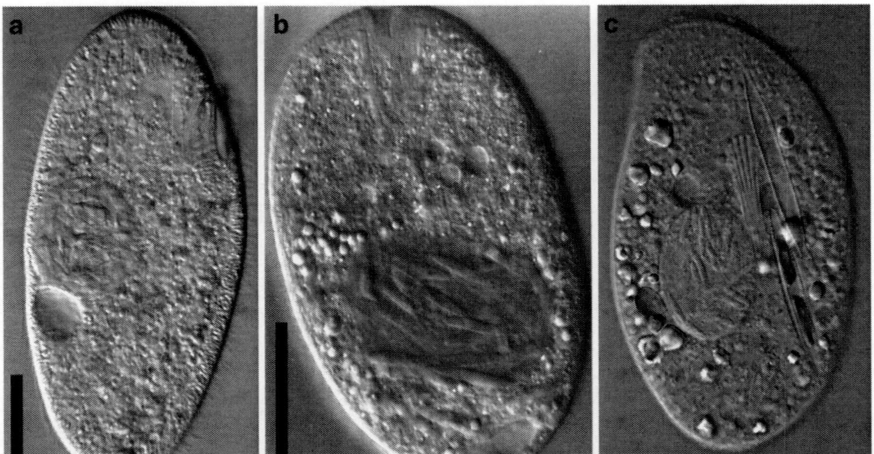


**Fig. 14** *Holospora* and *Holospora*-like bacteria in other ciliates. **a** *Holospora* sp. in the macronucleus of *Frontonia salmastra*. **b** *Holospora*-like bacteria in the macronucleus of *Prorodon teres*. **c** *Holospora*-like bacteria in the macronucleus of *Trithigmastoma cucullulis*. *Bar* 35 μm

(Fig. 14). All these bacteria appear to be infectious, though experimental evidence on the matter exists only for one of them (Görtz and Maier 1991).

There are macronuclear symbionts of *Metopus caudatus* (Jankowski 1964), *Prorodon teres* (Stein 1867; Fig. 14b), *Stentor multiformis* (Görtz and Wiemann 1987), *S. polymorphus* (Balbiani 1892; Fokin 2004a) *S. roeselii* (Stein 1867), *Trithigmostoma cucullulus* (Görtz and Maier 1991; Fig. 14c), *Vorticella* sp. (Kirby 1941b), and *Zoothamnium pelagicum* (Laval 1970). For the majority of the ciliates *Holospora*-like bacteria were just mentioned without serious further investigation or were even shown as figures without any information (*Trichodina pediculus*). According to the morphology of the infectious form, it is possible that not all of them are holosporas since sometimes the distribution of the periplasmic part and the tip part morphology deviated quite a bit from that of "classical" holosporas (Görtz and Wiemann 1987; Görtz and Maier 1991; Fokin et al. 2006)

## 7.2 Ecological and Evolutionary Significance of *Holospora* Infection

We know especially little about the actual situation regarding the presence and maintenance of bacterial infection in ciliates in natural bodies of water. The low percentage of bacterial infection in most ciliates revealed so far appears to indicate that most endocytobionts are parasites or commensals, and not mutualistic symbionts, otherwise, their broad distribution in the population as a result of positive selection would have been observed. This is apparently true for *Holospora*, which should be considered as parasites of paramecia.

Living in the host cell and feeding at its expense, parasitic bacteria should negatively influence the growth of ciliate populations. The presence in some populations of genotypes that do not support infection and stabilization of bacteria in the cell may be considered as a kind of defensive mechanism limiting the opportunities for such endocytobiosis. For example, in some regions of Japan the share of such populations of *P. caudatum* may reach 40–60% (Fokin et al. 2003b). However, in places where holosporas are widespread (Europe), the bacteria may influence the population structure. An increased proportion of amicronuclear (genetically dead) cells and cells with the polymorphic micronucleus and cortical anomalies, inability of infected paramecia to enter the sexual process, are the usual "payment" for maintaining *H. undulata* as well as, probably, *H. elegans* and *H. recta* infections in the population (Ossipov 1981; Fokin 1985, 2004a; Görtz 1986, 2008).

The results of laboratory investigations cannot always be extrapolated for natural ciliate populations and yet our knowledge is mostly based on such works. Experiments show that infectious endocytobionts may increase the presence of other bacteria in ciliate cells by means of coinfection and the diversity of endocytobionts should thus be higher in the populations that are, or were, infected by holosporas. Thus, holosporas can be regarded as native vectors and a factor of increasing endocytobiont diversity.

The host specificity of *Holospora* spp., ten of which in nature infect six *Paramecium* species, invites the supposition about coevolution of these genera. Experimental cross-infections of eight *Paramecium* spp. by five species of *Holospora* revealed groups infected by similar sets of bacteria, pointing to relatedness of paramecia in these groups. The good agreement of this classification with phylogenetic groups of *Paramecium* revealed by 18S rDNA analysis of the same set of species (Fokin 2000; Fokin et al. 2004) testifies to the sensitivity of this "symbiotic" method.

A recent model of the evolution of mixed transmission of parasitic genotypes has shown that transmission proceeds either predominantly vertically or horizontally (Lipsitch et al. 1996). In contrast to this prediction, the example of *Holospora* points to adaptive phenotypic plasticity as an alternative to genetic specialization in dealing with ecological variability. This parasite has a mixed, phenotypically plastic strategy, ranging from exclusive expression of vertical transmission to predominant expression of horizontal transmission, which possibly evolved as an adaptation to variable host growth conditions (Kaltz and Koella 2003; Fels and Kaltz 2006; Restif and Kaltz 2006).

An experimental study of coevolution in the microcosm of several *P. caudatum* clones infected with *H. undulata* made recently revealed an increase in host resistance, but not in parasite infectivity (Lohse et al. 2006). Cross-infection experiments showed significant host clone-parasite isolate interactions, and the evolved host tended to be more resistant to its own (local) parasites than to *H. undulata* isolated from other host clones. The authors postulated de novo evolution of host resistance, associated with both direct and indirect costs, and thus illustrated how interactions with holosporas can lead to the genetic divergence of initially identical ciliate populations (Lohse et al. 2006).

## 8 Concluding Remarks and Further Perspectives

*Holospora* has proven to be an excellent experimental subject for widely directed investigations in the fields of protistology, symbiontology, parasitology, and cell biology. This experimental system opens broad vistas both for investigations of endosymbiosis in ciliates and for the study of symbiosis in general biological aspects. The infection capacity of *Holospora* bacteria allows one not only to maintain but also to recreate the symbiotic system under laboratory conditions. A considerable number of *Holospora* spp. are known to date. This offers the possibility of laboratory infection of aposymbiotic paramecia and of experimental modeling of combinations in which endobionts and hosts possess different morphological and genetic features, making this symbiotic system convenient for the study of important problems of experimental biology such as intracellular signaling and the development of intracellular communication, and also of cellular resistance against bacterial infection.

Holosporas can be regarded as native vectors and then as factors increasing the diversity of endocytobionts. We must be even aware that holosporas may introduce harmful microorganisms into ciliates as well. The latter aspect may be of special practical interest because of the close phylogenetic relationship between some symbionts in *Paramecium* and human pathogens. Thus, further research on the cell biology of holosporas and on the ecology of infected populations is necessary.

## References

Amman RI, Springer N, Ludwig W, Görtz H-D, Scheifer K-H (1991) Identification in situ and phylogeny of uncultured bacteria endosymbionts. Nature 351:161–164

Amman RI, Ludwig W, Scheifer K-H (1995) Phylogenetic identification and in situ detection of individual cells without cultivation. Microb Rev 59:143–169

Balbiani G (1861) Phénomènes sexuels des infusoires. Masson, Paris

Balbiani E-G (1892). Nouvelles recherches expérimentales sur la mérotomie des infusoires ciliés. Ann Microgr 5:1–25

Barhey K, Gibson I (1984) A study on the conditions of infection of *Holospora caryophila*, a macronuclear symbiont of *Paramecium biaurelia*. Micron Microscop Acta 15:261–268

Baudoin M (1975) Host castration as a parasitic strategy. Evolution 29:335–352

Borchsenius ON, Skoblo II, Ossipov DV (1983) *Holospora curviuscula* – a new species of macronuclear symbiotic bacteria of *Paramecium bursaria*. Cytology (Sankt-Petersburg) 25:91–97 (in Russian with English summary)

Borchsenius ON, Skoblo II, Lebedeva NA, Ossipov DV (1990) The accidental penetration and momentary residence of symbiotic bacteria *Holospora acuminata* in the macronucleus of the ciliate *Paramecium bursaria*. Cytology 32:578–583 (in Russian with English summary)

Bütschli O (1876) Studien uber die Ersten Entwicklungsvorgange der Eizelle, die Zelltheilung und die Conjugation der Infusorien. Winter, Frankfurt

Bütschli O (1887–1889) Protozoa. In: HG Bronn's Klassen und Ordnungen des Thierreich in Wort und Bild. III. Abt. Infusoria und System der Radiaraia. Winter, Leipzig, pp 1098–2035

Claparede E, Lachmann J (1858–1861) Études sur les infusoires et les rhizopodes. vols 1 and 2. Kessmann, Geneva

De Bary A (1879) Die Erscheinungen der Symbiose. Trübner, Strassbourg

Ehrenberg CG (1838) Die Infusionsthierchen als vollkommene Organismen. Leopold Voss, Leipzig

Fels D, Kaltz O (2006) Temperature-dependent transmission and latency of *Holospora undulata*, a micronucleus-specific parasite of the ciliate *Paramecium caudatum*. Proc R Soc Lond B 273:1031–1038

Fema A, Dohra H, Fujishima M (2008) Fate of the 63-kDa periplasmic protein of the infectious form of the endonuclear symbiotic bacterium *Holospora obtusa* during the infection process. FEMS Microbiol Lett 280:21–27

Ferrantini F, Fokin SI, Vannini C, Verni F, Petroni G (2007) Characterization of a novel *Holospora*-like symbiont from *Frontonia* (Ciliophora, Oligohymenophorea). J Eukaryot Microbiol 54:75a

Fiveiskaja A (1928) Einfluss der Kernparasiten der Infusorien auf den Stoffwechsel. Arch Protistenkd 65:75–298

Fokin SI (1985) Vegetative function of the micronucleus and cell morphogenesis of *Paramecium caudatum*. I. Changes induced by nuclear symbionts. Cytology 27:322–328 (in Russian with English summary)

Fokin SI (1988) *Paramecium caudatum*: morphological analysis of consequences of infection with different nuclear symbionts. Cytology 30:471–477 (in Russian with English summary)

Fokin SI (1989) Bacterial endobionts of the ciliate *Paramecium woodruffi*. I. Endobionts of the macronucleus. Cytology 31:839–844 (in Russian with English summary)

Fokin SI (1991) *Holospora recta* sp. nov. - a micronucleus-specific endobiont of the ciliate *Paramecium caudatum*. Cytology 33:135–141 (in Russian with English summary)

Fokin SI (1993) Bacterial endobionts of ciliates and their employment in experimental protozoology. Cytology 35:59–91 (in Russian with English summary)

Fokin SI (1997) Morphological diversity of the micronuclei in *Paramecium*. Arch Protistenkd 148:375–387

Fokin SI (1998) Strategies of the macronuclear endocytobionts of *Paramecium* during the sexual process of the host. Symbiosis 225:323–342

Fokin SI (2000) Host specificity of *Holospora* and its relationships with *Paramecium* phylogeny. Jpn J Protozool 33:94

Fokin SI (2004a) Bacterial endocytobionts of Ciliophora and their interactions with the host cell. Int Rev Cytol 236:181–249

Fokin SI (2004b) Some bacterial infections of *Paramecium*. Occurrence and dynamics. Abstr 23 An Meet DGP Innsbruck. Ber Nat Med Ver Innsbruck 27

Fokin SI (2004c) A brief history of ciliate studies (late XVII – the first third of the XX century). Protistology 4:283–297

Fokin SI (2004d) Professor Otto Bütschli und Seine Russischen Schüler. Microkosmos 93:91–99

Fokin SI (2005) Konstantin Sergeevich Merezhkovskiy [CS Mereschkowsky] (1855–1921). In: 100 years of the endosymbiotic theory: from Prokaryotes to eukaryotic organelles. International Symposium Hamburg, pp 6–7

Fokin SI (2007) Phylium Ciliophora Dofline, 1901 (Ciliates). General part. In: Frolov AO, Krilov MV (eds) Protista: handbook on zoology, vol 2. Nauka, St Petersburg, pp 417–494 (in Russian with English summary)

Fokin SI, Karpov SA (1995) Bacterial endocytobionts inhabiting the perinuclear space of Protista. Endocytobiosis Cell Res 11:81–94

Fokin SI, Sabaneyeva EV (1993) Bacterial endocytobionts of the ciliate *Paramecium calkinsi*. Eur J Protistol 29:390–395

Fokin SI, Sabaneyeva EV (1997) Release of endonucleobiotic bacteria *Holospora bacillata* and *Holospora curvata* from the macronucleus of their host cells *Paramecium woodruffi* and *Paramecium calkinsi*. Endocytobiosis Cell Res 12:49–55

Fokin SI, Skovorodkin IN (1991a) *Holospora obtusa* – endonucleobiont of the ciliate *Paramecium caudatum* in search for macronucleus. Cytology 33:101–115 (in Russian with English summary)

Fokin SI, Skovorodkin IN (1991b) *Holospora undulata* – endonucleobiont of the ciliate *Paramecium caudatum* in search for micronucleus. Cytology 3:64–75 (in Russian with English summary)

Fokin SI, Skovorodkin IN (1997) Experimental analysis of the resistance of *Paramecium caudatum* (Ciliophora) against infection by bacterium *Holospora undulata*. Eur J Protistol 33:214–218

Fokin SI, Ossipov DV, Skoblo II, Rautian MS, Sabaneyeva EV (1987) *Nonospora macronucleata* g.n., sp.n. - a vegetative nucleus symbiont of the ciliate *Paramecium caudatum*. Cytology 29:963–970 (in Russian with English summary)

Fokin SI, Skovorodkin IN, Sabaneyeva EV (1993) Specificity of *Holospora* – bacterial endonucleobionts of paramecia (Ciliophora). In: Abstracts of 9th International Congress on Protozoology, Berlin, pp 41

Fokin SI, Brigge T, Brenner J, Görtz H-D (1996) *Holospora* species infecting the nuclei of *Paramecium* appear to belong into two groups of bacteria. Eur J Protistol 32(Suppl1):19–24

Fokin SI, Stoeck T, Schmidt HJ (1999a) Rediscovery of *Paramecium nephridiatum* Gelei, 1925 and its characteristics. J Eukaryot Microbiol 46:416–426

Fokin S, Brigge T, Görtz H-D (1999b) An infectious bacterium inhabiting the macronucleus of *Paramecium putrinum*. J Eukaryot Microbiol 46:11a

Fokin SI, Sabaneyeva EV, Borkchsenius ON, Schweikert M, Görtz H-D (2000) *Paramecium calkinsi* and *Paramecium putrinum* (Ciliaphora, Protista) harboring alpha-subgroup bacteria in the cytoplasm. Protoplasma 213:176–183

Fokin SI, Skovorodkin IN, Sabaneyeva EV (2003a) Initial steps of infection of the ciliate *Paramecium* with bacteria *Holospora* spp. Cytology 45:1227–1233

Fokin SI, Schweikert M, Görtz H-D, Fujishima M (2003b) Bacterial endocytobionts of Ciliophora. Diversity and some interactions with the host. Eur J Protistol 39:475–480

Fokin SI, Przyboś E, Chivilev SM, Beier CL, Horn M, Skotarczak D, Wodecka B, Fujishima M (2004a) Morphological and molecular investigations of *Paramecium schewiakoffi* sp. nov. (Ciliophora, Oligohymenophorea) and current status of taxonomy. Eur J Protistol 40:225–243

Fokin SI, Skovorodkin IN, Schweikert M, Görtz H-D (2004b) Coinfection of the macronucleus of *Paramecium caudatum* by free-living bacteria together with the infectious *Holospora obtusa*. J Eukaryot Microbiol 51:417–424

Fokin SI, Schrallhammer M, Vanini C, Ferrantini F, Petroni G, Görtz H-D (2006) New *Holospora* endocytobionts in some common ciliates. 25th Wiss Tag Deut Gesel Protozool Liebenwalde, pp 28

Fujishima M, Görtz H-D (1983) Infection of macronuclear anlagen of *Paramecium caudatum* with the macronucleus-specific symbiont *Holospora obtusa*. J Cell Sci 64:137–146

Fujishima M, Fujita M (1985) Infection and maintenance of *Holospora obtusa*, a macronucleus-specific bacterium of the ciliate *Paramecium caudatum*. J Cell Sci 76:179–187

Fujishima M, Nagahara K, Kojima Y (1990) Changes in morphology, buoyant density and protein composition in differentiation from the reproductive short form to the infectious long form of *Holospora obtusa*, a macronucleus-specific symbiont of the ciliate *Paramecium caudatum*. Zool Sci 7:849–860

Fujishima M, Nagahara K, Kojima Y, Sayama Y (1991) Sensitivity of the infectious long form of the macronuclear endosymbionts *Holospora obtusa* of the ciliate *Paramecium caudatum* against chemical and physical factors. Eur J Protistol 27:119–126

Fujishima M, Dohra H, Kawai M (1997) Quantitative changes in periplasmic proteins of the macronucleus-specific bacterium *Holospora obtusa* in the infection process of the ciliate *Paramecium caudatum*. J Eukaryot Microbiol 44:636–642

Fujishima M, Hoshide K (1988) Light and electron microscopic observation of *Holospora obtusa*: a macronucleus-specific bacterium of the ciliate *Paramecium caudatum*. Zool Sci 5: 791–799

Fujishima M, Kawai M, Yamamoto R (2005) *Paramecium caudatum* acquires heat-shock resistance in ciliary movement by infection with the endonuclear symbiotic bacterium *Holospora obtusa*. FEMS Microbiol Lett 243:101–105

Gibson I (1973) Transplantation of killer endosymbionts in *Paramecium*. Nature 241:127–129

Gibson I, Bedingfield G, Horne RW, Bird B (1986) Electron microscope study of the structure of *Holospora caryophila*. Micron Microsc Acta 17:247–257

Görtz H-D (1983) Endonuclear symbionts in ciliates. Int Rev Cytol(Suppl) 14:145–176

Görtz H-D (1986) Endonucleobiosis in ciliates Int Rev Cytol 102:169–213

Görtz H-D (1987) Different endocytobionts simultaneously colonizing ciliate cells. Ann N Y Acad Sci 503:261–268

Görtz H-D (1988) Endocytobiosis. In. Görtz H-D (ed) *Paramecium*. Springer Verlag, Berlin, Heidelberg, pp 393–405

Görtz H-D (1996) Symbiosis in Ciliates. In: Hausmann K, Bradbury PS (eds) Ciliates. Cells as organisms, Fischer, Stuttgart, New York, pp 441–462

Görtz H-D (1998) Aquatic symbionts and pathogens–ancient and new. In: Greenblatt CL (ed) Digging for Pathogens. Balaban Publishers, Rehovot, Philadelphia, pp 97–115

Görtz H-D (2008) Towards an understanding of the distribution, dynamics and ecological significance of bacterial symbioses in protists. In: Aescht E, Berger H (eds) The Wilhelm Foissner Festschrift. Atribute to an outstanding protistologist on the occasion of his 60th birthday. Denisia 23:307–311

Görtz H-D, Brigge T (1998) Intracellular bacteria in Protozoa. Naturwissenschaften 85:359–368

Görtz H-D, Dieckmann J (1980) Life cycle and infectivity of *Holospora elegans*, a micronucleus-specific symbiont of *Paramecium caudatum* Ehrenberg. Protistologica 16: 591–603

Görtz H-D, Freiburg M (1984) Bacterial symbionts in the micronucleus of *Paramecium bursaria*. Endocyt Cell Res 1:37–46

Görtz H-D, Fujishima M (1983) Conjugation and meiosis of *Paramecium caudatum* infected with the micronucleus-specific bacterium *Holospora elegans*. Eur J Cell Biol 32:86–91

Görtz H-D, Maier G (1991) A bacterial infection in a ciliate from sewage sludge. Endocytobiosis Cell Res 8:45–52

Görtz H-D, Schmidt HJ (2005) Genus *Holospora*. In: Garrity et al. (eds) Bergey's manual of systematic bacteriology, vol 2, part C, 2nd edn. Springer, New York, pp 149–151

Görtz H-D, Wiemann M (1987) Colonization of the ciliate *Stentor multiformis* by three different endocytobionts. Endocyt Cell Res 4:177–184

Görtz H-D, Ahlers N, Robenek H (1989) Ultrastructure of the infectious and reproductive forms of *Holospora obtusa*, a bacterium infecting the macronucleus of *Paramecium caudatum*. J Gen Microbiol 135:3079–3085

Görtz H-D, Lellig S, Miosga O, Wiemann M (1990) Changes in fine structure and polypeptide pattern during development of *Holospora obtusa*, a bacterium infecting the macronucleus of *Paramecium caudatum*. J Bacteriol 172:5664–5669

Görtz H-D, Bentig J, Freiburg M, Wiemann M, Fokin SI (1993) The infection of the nuclei of *Paramecium* by *Holospora* bacteria. In: Sato S, Ishida M, Ishikawa H (eds) Endocytobiology V. Tübingen University Press, Tübingen, pp 509–514

Gromov BV, Ossipov DV (1981) *Holospora* (ex Hafkine 1890) nom. rev., a genus of bacteria inhabiting the nuclei of paramecia. Int J Syst Bacteriol 31:389–352

Hafkine MW (1890) Maladies infectieuses des paramécies. Ann Inst Pasteur 4:363–379

Heckmann K, Görtz H-D (1991) Prokaryotic symbionts of ciliates. In: Balows A, Truper HG, Dworkin M, Harder W, Schleifer K-H (eds) The prokaryotes. 2nd ed. Springer Verlag, Berlin, New York, pp 3865–3890

Hiwatashi K, Mikami K (1989). Fertilization in *Paramecium*: process of the nuclear reorganization. Int Rev Cytol 114:1–19

Hori M, Fijishima M (2003) The endosymbiotic bacterium *Holospora obtusa* enhances heat-shock gene expression of the *Paramecium caudatum*. J Eukaryot Microbiol 50:293–298

Iwatani K, Dohra H, Lang BF, Burger G, Hori M, Fujishima M (2005) Translocation of an 89-kDa periplasmic protein is associated with *Holospora* infection. Biochem Biophys Res Commun 337:1198–1205

Jankowski AW (1964) Morphology and evolution of Ciliophora. III. Diagnoses and phylogenesis of 53 sapropelebionts, mainly of the oder Heterotrichida. Arch Protistenkd 107:185–294

Kaltz O, Koella JC (2003) Host growth conditions regulate the plasticity of horizontal and vertical transmission in *Holospora undulata*, a bacterial parasite of the protozoan *Paramecium caudatum*. Evolution 57:1489–1497

Kawai M, Fujishima M (2000) Invasion of the macronucleus of *Paramecium caudatum* by the bacterium *Holospora obtusa*: fates of the bacteria and timing of invasion steps. Eur J Protistol 36:46–52

Kirby HJ (1941a) Organisms living on and in Protozoa. In: Calkins GN, Summers FS (eds) Protozoa in biological research. Colombia University Press, New York, pp 1009–1113

Kirby H (1941b) A parasite of the macronucleus of Vorticella. J Parasit 27:311–314

Khahina LN (1979) The problem of symbiogenesis. Nauka, Leningrad (in Russian)

Khainsky (1903) About changing in the nuclear apparatus of paramecia. Zootom Lab Wars Univ Works 30:1–22 (in Russian)

Kudelski A (1898) Metamorphoses of nuclear apparatus in *Paramecium*. Zootom Lab Wars Univ Works 18:1–35 (in Russian)

Lang BF, Brinkmann H, Koski LB, Fujishima M, Görtz H-D, Burger G (2005) On the origin of mitochondria and Rickettsia-related eukaryotic endosymbionts. Jpn J Protozool 38:171–183

Laval M (1970) Presence de bacteries intranucleaires chez *Zoothamnium pelagicum* (Cilie Peritriche) leur role dans la formation des pigments intracytoplasmiques des zoides. Septième Congrès International la Microscopie Électronique, Grenoble, pp 403–404

Lipsitch M, Nowak MA, Ebert D, May RM (1995) The population dynamics of vertically and horizontally transmitted parasites. Proc. R. Soc. Lond. 260: 321–327

Lohse K, Gutierrez A, Kaltz O (2006) Experimental evolution of resistance in *Paramecium caudatum* against the bacterial parasite *Holospora undulata*. Evolution 60:1177–1186

Maggi L (1874) Sulla storia naturale degli esseri interiori (infusorj). Bernardoni, Milan

Margulis L (1981) Symbiosis in cell evolution. Freemann, San Francisco
Metchnikoff E (1892) Leçons sur la pathologie comparée de l'inflammation. Masson, Paris
Mitrophanow P (1903) Nouvelles recherches sur l'appareil nucléaire des Paramécies. Arch Zool Exp Gen Ser 4(1):411–436
Müller J (1856) Einige Beobschtungen an Infusorien. Monatsber Preuss Akad Wiss 389–393
Nakamura Y, Aki M, Aikawa T, Hori M, Fujishima M (2004) Differences in gene expression of the ciliate *Paramecium caudatum* caused by endonuclear symbiosis with *Holospora obtusa*, revealed using differential display reverse transcribed PCR. FEMS Microbiol Lett 240:209–213
Ossipov DV (1973) Specific infectious specificity of the omega-particles, micronuclear symbiotic bacteria of *Paramecium caudatum*. Cytology 15:211–217 (in Russian with English summary)
Ossipiv DV, Skoblo II, Rautian MS (1975) Iota-particles, macronuclear symbiotic bacteria of ciliate *Paramecium caudatum* clone M-115. Acta Protozool 14:263–280
Ossipov DV (1981) Problems of nuclear heteromorphism in the unicellular organisms. Nauka, Leningrad (in Russian with English summary)
Ossipov DV, Ivakhnyuk IS (1972) Omega-particles - micronuclear symbiontic bacteria of *Paramecium caudatum* clone M1–48. Cytology 14:1414–1418 (in Russian with English summary)
Ossipiv DV, Skoblo II, Borchsenius ON, Rautian MS (1980) *Holospora acuminata* – a new species of symbiotic bacterum from the micronucleus of the ciliate *Paramecium bursaria* Focke. Cytology (Sankt-Petersburg) 22:922–929 (in Russian with English summary)
Ossipov DV, Skoblo II, Borchsenius ON, Rautian MS, Fokin SI (1984) Occurence of endosymbionts in natural populations of infusoria of the genus *Paramecium* in various areas of the Soviet Union. In: Andrushaitis GP, Boikova EE, Liepa RA (eds) Ecology of marine and freshwater protozoans. Proceedings of the I-st symposium. Biological Institute of Latvian Academy of Sciences, Salaspils, pp 84–86 (in Russian)
Ossipov DV, Fokin SI, Borchsenius ON, Lebedeva NA, Rautian MS, Skoblo II (1989) Diversity and distribution of the intranuclear symbionts of ciliates of the Soviet Union. Proceedings of the II-d symposium Ecology of marine and freshwater protozoans. Yaroslavl, pp 53 (in Russian)
Ossipov DV, Skoblo II, Borchsenius ON, Lebedeva NA (1993) Interactions between *Paramecium bursaria* (Protozoa, Ciliophora, Hymenostomatida) and their nuclear symbionts. Eur J Protistol 29:61–71
Petschenko B (1903) About changing in the nuclear apparatus of paramecia in natural conditions. Zootom Lab Wars Univ Works 30:1–35 (in Russian)
Petschenko B (1910) Fibrous-crystalloid structures in paramecia body. Zootom Lab Wars Univ Works 38:1–52 (in Russian)
Petschenko B (1911) *Drepanospira mülleri* n.g., n.sp. parasite des paramécıums; contribution a l'étude de la structure des bactéries. Arch Protistenkd 22:248–98
Popovsky M (1963) The fate of doctor Hafkin. East literature, Moscow (in Russian)
Preer JR (1948) A sudy of some properties of the cytoplasmic factor "kappa" in *P. aurelia*, variety 2. Genetics 33:349–404
Preer LB (1969) Alpha, an infectious macronuclear symbiont of *Paramecium aurelia*. J Protozool 16:570–578
Preer JR, Preer LB (1982) Revival of names of protozoan endosymbionts and proposal of *Holospora caryophila* nom.nov. Int J Syst Bacteriol 32:140–141
Preer JR and Preer LB (1984) Endosymbionts of protozoa. In: Krieg NR (ed) Berger's manual of systematic bacteriology. vol. 1. Williams and Wilkins, Baltimore, London, pp 795–813
Preer JR, Preer LB, Jurand A (1974) Kappa and other endosymbionts in *Paramecium aurelia*. Bacteriol Rev 38:113–163
Rautian MS, Skoblo II, Lebedeva NA, Ossipov DV (1990) Complementation in the symbiont–host system formation between the clones of the ciliate *Paramecium caudatum* and the stocks of the bacterium *Holospora obtusa*. Cytology 32:584–590 (in Russian with English summary)
Rautian MS, Skoblo II, Lebedeva NA, Ossipov DV (1993) Genetics of symbiontic interactions between *Paramecium bursaria* and the intranuclear bacterium *Holospora acuminata*, natural genetic variability by infectivity and susceptibility. Acta Protozool 32:165–173

Rautian M, Vishnyakov A, Makarov S, Ossipov D (1996) Transformation of a *Paramecium caudatum* clone resistent to infection by intranuclear symbiotic bacteria of the genus *Holospora*. Eur J Protistol 32(Suppl 1):135–140

Restif O, Kaltz O (2006) Condition-dependent virulence in a horizontally and vertically transmitted bacterial parasite OIKOS 114:148–158

Sabaneyeva EV, Fokin SI, Gavrilova EV, Kornilova ES (2005) Nocodazole inhibits macronuclear infection with *Holospora obtusa* in *Paramecium caudatum*. Protoplasma 226:147–153

Sabaneyeva EV, Benken KA, Derkacheva VE, Skovorodkin IN, Fokin SI (2007) *Holospora* trafficking in *Paramecium*: the role of the host cell cytoskeleton. Protistology 5:68

Schrallhammer M, Fokin SI, Schleifer K-H, Petroni G (2006) Molecular Characterization of the obligate endosymbiont "*Caedibacter macronucleorum*" Fokin and Görtz 1993 and of its host *Paramecium duboscqui* Strain Ku4–8. J Eukaryot Microbiol 53:499–506

Schwemmler W (1980) Endocytobiology: a modern field between symbiosis and cell research. In: Schenk HE, Schwemmler W (eds) Endocytobiology: endosymbiosis and cell biology, a synthesis of recent research. Vol. 1. de Gruyter, Berlin, New York, pp 943–967

Skoblo II, Lebedeva NA (1986) Infection of the nuclear apparatus of *Paramecium bursaria* (Ciliata) by the symbiotic bacterium *Holospora curviuscula*. Cytology 28:367–372 (in Russian with English summary)

Skoblo II, Borchsenius ON, Lebedeva NA, Ossipov DV (1990) Symbiogenic lysis of bacteria *Holospora acuminata* in the generative nucleus of the ciliate *Paramecium bursaria*. Cytology 32:515–519 (in Russian with English summary)

Skoblo II, Lebedeva NA, Rodionova GV (1996) The formation of the *Paramecium-Holospora* symbiotic system: a sudy of the compatibility of different symbiont isolates and host clones. Eur J Protistol 32(Suppl 1):147–152

Skovorodkin IN, Fokin SI (1990) The resistence of some stocks of ciliate *Paramecium caudatum* to the infection by bacterial endonucleobiont *Holospora undulata*. Cytology 33:98–103 (in Russian with English summary)

Skovorodkin IN, Fokin SI, Fujishima M (2001) Fates of the endonuclear symbiotic bacteria *Holospora obtusa* and *Holospora undulata* injected into the macronucleus of *Paramecium caudatum*. Eur J Protistol 37:137–145

Sonneborn TM (1938) Mating types, toxic interactions and heredity in *Paramecium aurelia*. Science 88:503

Sonneborn TM (1943) Gene and cytoplasm. I. The determination and inheritance of the killer character in variety 4 of *Paramecium aurelia*. Proc Nat Acad Sci U S A 90:9892–9895

Springer N, Ludwing W, Amman R, Schmidt HJ, Görtz H-D, Schleifer KH (1993) Occurence of fragmented 16S rRNA in an obligate bacteral endosymbiont of *Paramecium caudatum*. Proc Nat Acad Sci U S A 90:9892–9895

Stein F von (1867) Der Organismus der Infusionsthiere. II. Abth 2 Naturgeschichte der heterotrichen Infusorien. Wilhelm Engelmann, Leipzig

Timofeyeva AS, Rautian MS (1997) *Holospora undulata*: the genome size determined by pulsed-field gel electrophoresis. Cytology 39:634–639 (in Russian with English summary)

Wichterman R (1940) Parasitism in *Paramecium caudatum*. J Parasitol 26:29

Wichterman R (1945) Schizometes parasites in *Paramecium bursaria*. J Parasitol 31:25

Wichterman R (1953) The biology of *Paramecium*. Blakiston, New York

Wichterman R (1986) The biology of *Paramecium*, 2- ed. Plenum Perss, New York, London

Wiemann M (1989) The release of *Holospora obtusa* from *Paramecium caudatum* observed with a new device for extended in vivo microscopy. J Protozool 36:176–179

Wiemann M, Görtz H-D (1989) Release of the endoucleobiotic bacterium *Holospora elegans* from its host cell *Paramecium caudatum*. Eur J Protistol 25:100–108

# Infection and Maintenance of *Holospora* Species in *Paramecium caudatum*

**Masahiro Fujishima**

**Contents**



**Abstract** Infectious forms of *Holospora* species have a cytoplasmic region and a periplasmic lumen with an electron-translucent invasion tip. Bacterial 89-kDa proteins with two actin-binding motifs translocate from the invasion tip lumen to outside the tip when infectious forms are ingested in digestive vacuoles of a *Paramecium* cell. With the invasion tip, the bacteria disrupt the digestive vacuole membrane, appear in the host cytoplasm, migrate to their target nucleus, distinguish host nuclear envelopes of two kinds through affinity between the bacterial lipopolysaccharides of

M. Fujishima (✉)
Department of Environmental Science and Engineering, Graduate School of Science and Engineering, Yamaguchi University, Yoshida 1677-1, Yamaguchi, 753-8512, Japan
e-mail: fujishim@yamaguchi-u.ac.jp

M. Fujishima (ed.), *Endosymbionts in Paramecium*, Microbiology Monographs 12, DOI: 10.1007/978-3-540-92677-1_8, 

the outer membranes and an unknown substance of the target nuclear envelope, and then penetrate the target nuclear envelope with the invasion tip. The first bacterium invades the host nucleus within 10 min after mixing. After endosymbiosis the bacteria alter host gene expression, providing an excellent opportunity for elucidating host-symbiont interactions between eukaryotic and prokaryotic cells and establishment of new associations leading to eukaryotic cell evolution.

## 1 Introduction

The Gram-negative bacterium *Holospora* species are endonuclear symbionts of the ciliate *Paramecium* species (Fokin and Sabaneyeva 1997; Gibson et al. 1986; Preer 1969; Ossipov 1973; Ossipov et al. 1975, 1980; Skoblo and Lebedeva 1986; see Fokin and Görtz, this volume) and belong to *Alphaproteobacteria* (Amann et al. 1991; Lang et al. 2005). *Holospora* species are found in paramecia living in cold areas, such as northern Europe and the Kamchatka Peninsula (Fokin et al. 1996). To date, nine *Holospora* species have been described (Fokin et al. 1996). All show species specificity and nucleus specificity in their habitats. They cannot grow outside the host cell with ordinary culture media. They show two different forms during their life cycle: a short reproductive form (RF, 1.5–2 μm long) and a long infectious form (IF, 10–15 μm long) (Fokin et al. 1996; Fujishima et al. 1990b; Görtz 1980; Görtz et al. 1989; Gromov and Ossipov 1981). The bacterium exists as a short RF cell and divides by binary fission in the host nucleus when the host is growing vegetatively. The RF stops dividing and differentiates into a longer IF cell through intermediate forms when the host cell starves (Fujishima et al. 1990a; Görtz 1983). During this differentiation, the bacterium forms a distinctive structure, half of which contains the cytoplasm; the other half is a periplasmic lumen with an electron-translucent tip (Dohra and Fujishima 1999a; Fujishima and Hoshide 1988; Görtz 1980; Görtz and Wiemann 1989; Görtz et al. 1989; Iwatani et al. 2005). The IF cells engulfed into the host digestive vacuoles (DVs) escape with the electron-translucent tip ahead and penetrate the target nuclear envelope with this special tip when the IF cells are mixed with paramecia (Fujishima and Fujita 1985; Fujishima and Kawai 2004; Görtz and Wiemann 1989). For that reason, this tip is designated as an “invasion tip” (Iwatani et al. 2005). Under a phase-contrast microscope, the cytoplasmic region appears dark, but the periplasmic region appears as a refractile region (Dohra and Fujishima 1999; Görtz and Dieckmann 1980). In the macronucleus-specific *Holospora obtusa* of *Paramecium caudatum*, the IF cells form two distinctive nucleoids during differentiation (Fujishima et al. 1990a). The shape of the nucleoids stained with 4′,6-diamidino-2-phenylindole dihydrochloride (DAPI) is different in strains of *Holospora* (M. Fujishima, unpublished observation). This bacterium also changes the buoyant density and protein compositions (Fujishima et al. 1990a) and surface morphologies of the outer membrane (Fujishima et al. 1990b) during differentiation. When the host divides again, the IF cells of *H. obtusa* are collected in a connecting piece of the dividing nucleus; then they are freed from the dividing nucleus by wrapping with the nuclear membrane. They are eventually expelled from the host cytoproct. On the other hand, the outer membrane of the RF has a stronger affinity to bind the host chromatin than the IF cells, so RF cells remain in the daughter nuclei when the host cell divides (Ehrsam and Görtz 1999; Fokin et al. 1996; Görtz et al. 1992; Wiemann 1989). When the macronucleus is filled with many IFs, the host cells cannot grow and are eventually killed by the bacteria; the IFs are freed from the cells. Consequently, the IFs appear outside the host cell by these two means and can then infect new host cells. A *Paramecium* cell has a limited life span; therefore, *Holospora* species must escape the host to infect younger cells. For that reason, a different nature of the outer membranes of these two forms is indispensable for *Holospora*'s survival.

The phenomenon of bacterial invasion of a target nucleus is designated as "infection." On the other hand, stable multiplication of the infected bacteria is designated as "maintenance" (Fujishima and Fujita 1985). The infection is controlled using a specific binding between *Holospora*'s outer membrane and the target nuclear envelope and by bacterial penetration of the nuclear envelope (Fujishima and Fujita 1985; Iwatani et al. 2005). The maintenance is controlled by the host genotypes (Fujishima and Mizobe 1988). Therefore, infection and maintenance are independently controlled phenomena. The whole infection process occurs in as little as 10 min (Fujishima and Görtz 1983). To date, the only organism having an ability to distinguish a somatic macronucleus from a germinal micronucleus of the host is *Holospora* species. Apparently, these bacteria know some differences between the two kinds of nuclei originating from a common fertilization nucleus. This review specifically examines molecular mechanisms underlying morphological and functional differentiation of cell structures of *Holospora* species in their life cycle, infection of the target nucleus by *Holospora*, and changes of the host cell that occur because of the infection.

## 2 Reproductive Form and Infectious Form of *Holospora*

*Holospora* species change their morphology according to the host physiological conditions. An RF cell is observed in a growing host cell, and is capable of binary fission in the host nucleus, but it has no infectivity. On the other hand, an IF cell is observed in a starved host cell, it has infectivity but has no ability to grow in the host nucleus.

### 2.1 Triggers for Differentiation

Figure 1 presents the life cycle of the *Holospora* species. Actually, *H. obtusa* is a macronucleus-specific bacterium and *H. undulata* is a micronucleus-specific bacterium of *P. caudatum*. Differentiation of the IF from the RF cell of these bacteria is induced by starvation of the host cell (Fujishima et al. 1990a; Görtz 1983). This differentiation is also induced when the host protein synthesis is inhibited with emetine in the presence of the bacterial RNA and protein syntheses (Fujishima 1993). These results suggest that the binary fission of the RF cell needs the host nuclear proteins for growth, and that depletion of this protein pool in the host nucleus ceases the binary fission of the RF cell and induces *Holospora*'s gene expressions needed for differentiation to the IF cell. This hypothesis is supported by evidence obtained by 2D sodium dodecyl sulfate polyacrylamide gel electrophoresis (SDS-PAGE) that more than 65% of protein spots of the RF cells and the IF cells were each form-specific (Fujishima et al. 1990a). Wiemann and Görtz (1991) also showed stage-specific proteins in the life cycle of *H. obtusa*.

On the other hand, differentiation of the RF cell from the infected cell occurs early in the infection process (Fig. 1). Indirect immunofluorescence with a monoclonal antibody (mAb) specific for an outer membrane of *H. obtusa* showed that the first constriction and the first single RF appeared at 32–34 and 34–36 h, respectively, after mixing at 25°C. The mean number of constrictions per long form is about seven (Kawai and Fujishima 2000). For the micronucleus-specific bacterium *H. recta* of *P. caudatum*, the differentiation of the RF cells occurs within 18 h after mixing (Kawai and Fujishima 1996). This differentiation occurs even if the host cells are in the stationary phase of growth. The RF cells can continue several binary fissions in the starved host cell, unlike another differentiation from the RF to the IF cell.

## 2.2 Isolation of *Holospora* from Host Cells

Isolation of *Holospora* from homogenates of the host cells first succeeded in macronucleus-specific *H. caryophila* of *P. biaurelia* using an ECTEOLA ion-exchange column (Preer 1969). Later, *H. obtusa* (Freiburg 1985; Fujishima and Nagahara 1985), micronucleus-specific *H. elegans* (Schmidt et al. 1987), *H. recta* (Kawai and Fujishima 1996), and *H. undulata* (Timofeyeva and Rautian 1997) of *P. caudatum* were isolated from homogenates of the host cells or isolated nuclei using Percoll gradient centrifugation. In the case of *H. obtusa* (Fujishima et al. 1990a), the buoyant density of the RF cell and intermediate forms with various lengths in differentiation from the RF to the IF cell is 1.09 g $ml^{-1}$, and that of the IF cell is 1.16 g $ml^{-1}$. During the differentiation from the RF to the IF cell, dispersed DAPI-positive nucleoids in the RF cell concentrate to two nucleoids in the IF cell (Fig. 1). Similarly, *H. elegans* and *H. undulata* form two nucleoids in the IF cell (M. Fujishima, unpublished observations). In *H. recta*, however, a condensed large nucleoid is formed near the periplasmic region of the IF (Kawai and Fujishima 2000).

Scanning electron microscopy observation of the isolated *H. obtusa* shows that the RF cell and intermediate forms have an entirely rough surface. The IF cell has an entirely smooth surface, except for the invasion tip, which always shows a rough surface (Fujishima et al. 1990b). This invasion tip is depressed easily by surface tension if the bacterium is dried without using a critical-point dryer (Fujishima and Hoshide 1988).

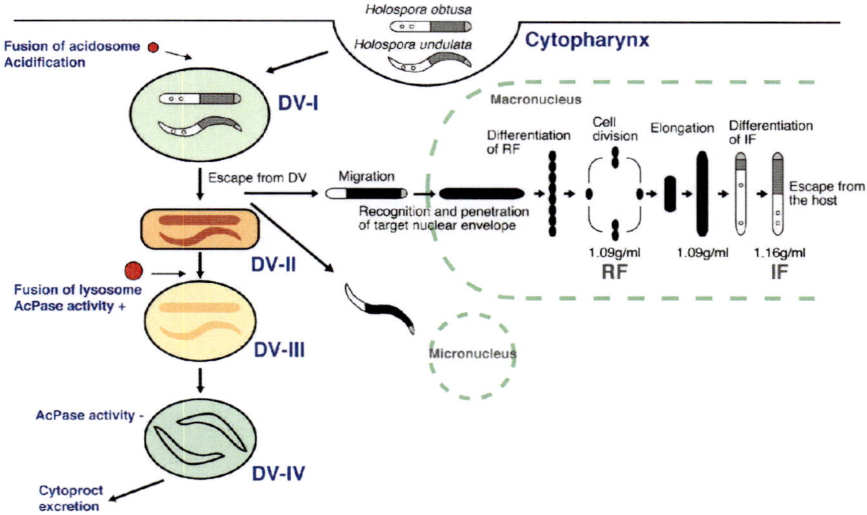


**Fig. 1** The infection process and life cycle of *Holospora* species. Classification of digestive vacuoles (DVs) is followed by the results for *Paramecium multimicronucleatum* (Fok and Allen 1988). The spherical DV-I vacuole differentiates to a condensed and acidified DV-II vacuole by fusion of acidosomes and evagination of the DV membrane to the cytoplasm. Then the vacuole differentiates to a swollen DV-III vacuole by fusion of primary lysosomes. Undigested materials remain in the acid phosphatase-less DV-IV vacuole. The DV-IV vacuole fuses to a cytoproct and discharges the content. Some infectious-form (IF) cells of *Holospora* species escape from the acidified DV without wrapping with the DV membrane. Immediately after escaping from the host DV, the bacteria differentiate to activated forms (Sect. 3.1) in the host cytoplasm, and migrate toward the target nucleus with the help of the host actin (Sect. 3.2). The bacteria distinguish their target nucleus by specific binding between lipopolysaccharides of the outer membrane and the unknown nuclear envelope substance (Sect. 3.3). Then, the bacterium penetrates the target nuclear envelope with an invasion tip leading. After the invasion, the bacterial cytoplasmic region increases and the large periplasmic region decreases to form constrictions for differentiation to the reproductive forms (RFs). During this infection process, the bacterium decreases its buoyant density from 1.16 to 1.09 g ml$^{-1}$. The RF continues to divide by binary fission when the host cell is also growing by binary fission, but the RF halts the binary fission, elongates itself, and differentiates to the IF when the host cell starves or the host's protein synthesis is inhibited. During this differentiation, the bacterium increases the buoyant density, and forms a large periplasmic region, an invasion tip, and two nucleoids. The infectious forms are freed from the cells (Sects. 2.1, 2.5). Acridine orange stains the DV-II vacuole *orange*, the DV-III vacuole *yellow*, and other DVs *yellow-green*.

## 2.3 Cryopreservation of Isolated Infectious Forms

*Holospora* species cannot grow outside the host cell; therefore, if cryopreservation of the isolated infectious forms of *Holospora* species were practiced, it would enable preservation of the bacteria for a long time without the need for laboratory culture procedures. The first cryopreservation was done for *H. obtusa* by freezing in 10 mM phosphate buffer, pH 7.0, containing 10% glyceroi or 10% dimethyl sulfoxideat −85°C for 24 weeks without loss of the infectivity or reproductivity (Fujishima et al. 1991). Later, the cryopreservation was performed for *H. recta* (Kawai and Fujishima 1996), *H. undulata* (Millot and Kaltz 2006), and *H. elegans* (M. Fujishima and M. Kawai, unpublished results).

## 2.4 Form-Specific Antigens

To date, many mAbs and antisera that are specific for RF or IF cells have been raised for *Holospora* species using whole cells, homogenates of the cells, or purified proteins from 2D-SDS-PAGE gels.

### 2.4.1 Infectious-Form-Specific Periplasmic Antigens

During differentiation of an IF cell, *Holospora* forms a large periplasmic region (Görzt et al. 1989; Iwatani et al. 2005). This expansion of the periplasmic region suggests that it contains important molecules that are necessary for the infection. Dohra et al. (1994) developed mAbs IF-3-1 and IF-3-2 against 39- and 15-kDa periplasmic proteins of the IF cell and illustrated qualitative and quantitative changes of these proteins during the infection process (Fujishima et al. 1997). Immunoblotting and immunogold labeling showed that the amounts of both antigens were reduced within 1 h after the bacteria had been mixed with the host cells, but that the amounts of IF-3-2 antigens declined earlier than those of the IF-3-1 antigen. These antigens might have a function to sustain the large periplasmic structure and are a main cause of the change in the bacterial buoyant density in the life cycle.

Dohra et al. (1997) purified a 5.4-kDa periplasm-specific peptide from IF cells of *H. obtusa*, raised an antiserum, and cloned a gene encoding this peptide consisting of 49 amino acids. This protein is synthesized in an intermediate form cell, and is a major protein of the periplasmic region of the IF cell. The molecular mass of this peptide determined by Tricine SDS-PAGE is 11 kDa; therefore, this 5.4-kDa peptide is thought to be identical to the 11.5-kDa polypeptide found by Wiemann and Görtz (1991). The *Escherichia coli* cell transformed with the 5.4-kDa peptide gene was scarcely able to divide (Dohra et al. 1997). Differentiation of the IF cell requires prior cessation of binary fission of the RF cell; therefore, the 5.4-kDa peptide might be involved in the inhibition of division of the RF cell, which switches the cell differentiation to the IF cell.

On the other hand, mAbs specific for a 63-kDa periplasmic protein of *H. obtusa*'s IF cell show that this protein is secreted into the host macronucleus, not only from the bacteria that invaded in early infection but also from the IF cells that differentiated from the RF cells in the macronucleus (Abamo et al. 2008). Indirect immunofluorescence with mAbs shows that the protein is secreted into the macronucleus immediately after the bacterial invasion of the nucleus. The use of inhibitors for the host and the bacterial protein synthesis illustrates that, in infection, not only the preexisting but also a newly synthesized 63-kDa protein is secreted into the macronucleus. The amino acid sequence deduced from a gene encoding the 63-kDa protein shows that this protein is a novel protein and involves the predicted two DNA-binding motifs. Many intracellular pathogens are known to enter and survive within the host cell by exporting proteins to modify the host gene expression (Gilbert et al. 2007; Saeij et al. 2007). Because infection of *H. obtusa* alters the host gene expression (Hori and Fujishima 2003; Nakamura et al. 2004), it is suggested that the 63-kDa protein might change the host gene expression to the advantage of the bacteria, for example, a modulation of the host cell's apoptosis mechanism (Clifton et al. 1998; Tunbridge et al. 2006). The 63-kDa proteins are present in the host macronucleus at their highest levels between 1 and 3 days. This stage is crucial for the bacteria to differentiate to the RF cells for growth. The bacteria are then given the chance to multiply inside the host if the 63-kDa proteins have antiapoptotic activity at this stage.

### 2.4.2 RF-Specific Outer-Membrane Antigen

Isolated RF cells have a rough surface, whereas IF cells have a smooth surface, except for the invasion tip (Fujishima et al. 1990b). Furthermore, unlike in the IF cells, an outer membrane of the RF cell of *H. obtusa* has a strong affinity with the host chromatin (Fokin et al. 1996; Wiemann 1989). Taken together, this evidence suggests that this bacterium changes its outer-membrane components during its life cycle. Although many mAbs specific for the outer membranes of both the RF and the IF cells have been developed in *H. obtusa*, *H. undulata*, *H. elegans*, and *H. recta* (M. Kawai and M. Fujishima, unpublished results), the reactivities of these mAbs are species-specific, and cross-react neither with other *Holospora* species nor with *Klebsiella pneumoniae* food bacteria of the host cell, suggesting that antigenicities of the outer membranes of respective *Holospora* species differ considerably. On the other hand, the outer-membrane-specific mAb R-1-1 for the RF cell of *H. obtusa* was developed (Y. Harayama and M. Fujishima, unpublished results). This mAb does not react with the IF cell, which demonstrates that the bacterium changes its outer membrane's antigenicity in the life cycle. Bacteria that invaded the host macronucleus become reacted with this R-1-1 mAb at about 30 h at 25°C after mixing, prior to forming constrictions to differentiate to the RF cells.

The mAb IR-4-1 reacts with lipopolysaccharides (LPSs) of the outer membranes of both the IF and the RF cells of *H. obtusa* (Kawai and Fujishima 2000). Using this mAb, results showed that the IF cells secrete LPSs originating from the outer membranes into the host macronucleus and that the LPSs decrease the permeability of the nuclear envelope and eventually kill the host cell (M. Kawai and M. Fujishima, unpublished results). On the other hand, the RF cells do not secrete LPSs outside the bacterial cell. Consequently, the IF is harmful for the host, but the RF is not.

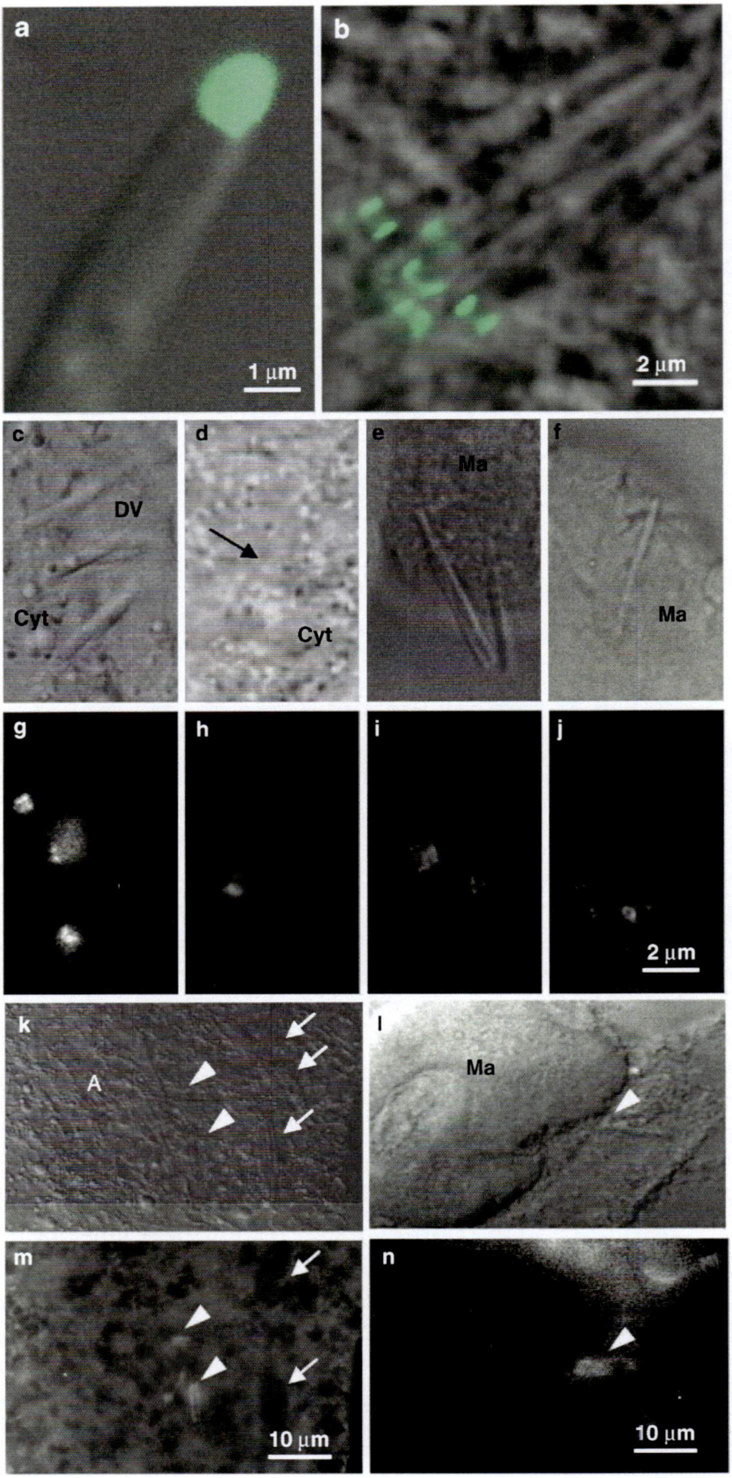
a
1 μm
b
2 μm
c
DV
Cyt
d
Cyt
e
Ma
f
Ma
g
h
i
j
2 μm
k
A
l
Ma
m
10 μm
n
10 μm

## 3 Infection Process of *Holospora*

When IF cells of *Holospora* species are mixed with paramecia, they are soon ingested into the host DVs and escape from there to appear in the host cytoplasm. Then the bacterium moves to its target nucleus, distinguishes the host nuclei of two kinds, and invades the target nucleus. The first bacterium appears in the nucleus within 10 min after mixing at 25°C. Many interesting phenomena are involved in this infection process. When and how does the IF cell escape from the host DV? How can the bacterium approach its target nucleus without a flagellum? How can the bacterium distinguish the two kinds of host nuclei? How can the bacterium invade the target nucleus?

### 3.1 Invasion-Tip-Specific Proteins

As presented in Fig. 1, DVs of *P. caudatum* can be classified into four different stages according to Fok and Allen (1988) in *P. multimicronucleatum*. The IF cells ingested in the DV-I vacuole escape from there while the DV-I vacuole is acidified and becomes a condensed DV-II vacuole (M. Fujishima, unpublished observation). In the presence of vacuolar-type ATPase(V-ATPase) inhibitors, concanamycinA or concanamycin B, both the acidification of the DV and the bacterial escape from the DV are inhibited completely (Fujishima and Kawai 1997). In addition, a V-ATPase-specific mAb labels the DV membrane (M. Fujishima, unpublished observation). These results indicate that the acidification of the host DV is a prerequisite phenomenon for the bacterial escape from the host DV. Bacteria appearing in the host cytoplasm

◄

**Fig. 2** Invasion-tip-specific 89-kDa protein of *Holospora obtusa* and the host *P. caudatum* actin in infection. **a**, **b** Confocal laser scanning microscope image of 89-kDa-protein-specific monoclonal antibody (mAb) IF-4-2. **a** Isolated IF cell treated with 20 mM NaOH for 1 min. Alexa Fluor 488 (AF488) fluorescence of the 89-kDa protein appears inside the invasion tip. **b** IF cells in the DV-I vacuole, 15 min after mixing with paramecia. Not treated with NaOH. Actually, AF488 fluorescence of 89-kDa protein can be labeled without treatment with NaOH, indicating that the 89-kDa protein translocates to outside the invasion tip. **c–j** Indirect immunofluorescence with 89-kDa-protein-specific mAb IF-4-2. **c**, **g** Bacteria escaping from the DV-I vacuole of *P. caudatum*, 15 min after mixing. The 89-kDa protein translocates to the outside of the *Holospora* invasion tip, completely covering it. **d**, **h** Bacterium after evasion into the cytoplasm, 15 min after mixing. *Arrow* bacterium in the cytoplasm. **e**, **f**, **i**, **j** Isolated macronuclei into which the bacteria are just penetrating the nuclear envelope, 30 min after mixing. The AF488 fluorescence of the 89-kDa protein remains at the entry point on the nuclear envelope. **c–f** Differential interference contrast. **g–j** AF488 fluorescence. **k–n** Indirect immunofluorescence with mAb PCactin-1 raised for *P. caudatum* actin. **k**, **m** Fifteen minutes after mixing of paramecia and holosporas. The AF488 fluorescence of actin appears around holosporas in the host cytoplasm (*arrowheads*), but not around the bacteria in the host DVs (*arrows*). **l**, **n** Isolated macronucleus from *P. caudatum*, 30 min after mixing. A *H. obtusa* bacterium is just penetrating the macronuclear envelope (*arrowhead*). Actin fluorescence appears in the same region as the fluorescence of the 89-kDa protein. (**a**, **b** Courtesy of H. Dohra. **c–j** Updated from Iwatani et al. 2005)

are designated as an activated form. This bacterium appears darker than the IF cell under a phase-contrast microscope (Görtz and Wiemann 1989). Kawai and Fujishima (1997) developed the mAb AF-1 specific for 58.8- and 22.9-kDa proteins of the activated form of *H. obtusa*. When the isolated IF cells are treated with acidic buffers, pH 3–4, for 20 min, the IF cells become a morphologically activated form and 58.8- and 22.9-kDa activated-form-specific proteins appear, indicating that a trigger for the differentiation of the activated form is low pH in the host DV.

On the other hand, IF cells always escape from the host DVs with the invasion tip ahead (Görtz and Dieckmann 1980; Iwatani et al. 2005). Furthermore, the activated form migrates to the target nucleus, distinguishes the two kinds of host nuclei, and invades the target nucleus, with this invasion tip leading. Consequently, it seems that this invasion tip is responsible for these phenomena. To elucidate the function of materials inside the invasion tip, an invasion tip about 1 μm long of each IF cell of *H. obtusa* was harvested from 3,438 bacteria using a laser capture micro-dissection system (LM 100; Olympus), subjected to SDS-PAGE, and 89, 76, and 63-kDa bands were detected by silver staining. Then, IF cells of *H. obtusa* ($1 \times 10^8$) were subjected to SDS-PAGE; proteins of 60–90 kDa were electroeluted from a respective gel slice. The eluted proteins were injected interperitoneally into mice, which subsequently developed a hybridoma cell. That cell produced anti-invasion-tip-specific mAb IF-4-2 (Fig.2a) (Iwatani et al. 2005). This mAb reacts with an 89-kDa protein of the invasion tip of *H. obtusa* but not with the RF cells nor with other *Holospora* species examined. Subsequently, proteins from $4 \times 10^9$ IF *cells of H. obtusa* were subjected to 2D-SDS-PAGE and stained with Coomassie brilliant blue; then the 89-kDa antigen spot was excised. The 89-kDa protein in the gel slice was digested using the Cleveland peptide mapping method (Cleveland et al. 1977). Partial amino acid sequences of the fragments were determined and used to design primers for PCR amplification of the gene encoding the 89-kDa protein from genomic DNA. The open reading frame of the gene was 2,253 nt long with a G + C content of 32.5%. The predicted protein of 750 residues has a molecular mass of 85 kDa and isoelectric points of 8.75. The predicted amino acid sequence of the 89-kDa protein shows two actin-binding motifs, one coiled-coil motif, and one cadherin signature, suggesting that the 89-kDa protein is a cytoskeleton-associated protein (Iwatani et al. 2005).

Antibodies only slightly enter the periplasm or cytoplasm of the *Holospora* cell, even if the bacteria are fixed and treated with detergents such as Triton X-100 or Tween 20 (Kawai and Fujishima 2000). Therefore, immunodetection of proteins in the periplasm or the cytoplasm requires prior sectioning of the bacteria (Fujishima et al. 1997; Dohra et al. 1994), brief sonication (Dohra et al. 1998), or treatment with 20 mM NaOH for permeabilization of the outer membrane (Iwatani et al. 2005). Figure 2a portrays that the 89-kDa protein localized in the invasion tip of the isolated IF cell of *H. obtusa* when the bacterium was permeabilized by 20 mM NaOH for 1 min. However, after ingestion into the host DV-I vacuole, even nonpermeabilized IF cells show Alexa Fluor 488 (AF488) fluorescence in the invasion tip region (Fig. 2b), indicating that the antigen has translocated to the outside of the tip in the DV. The bacterium escapes from the host DV with the invasion tip ahead (Fig. 2c, g); the bacterium appearing in the host cytoplasm also retains the 89-kDa proteins outside the tip (Fig. 2d, h). However, upon penetration of the macronuclear envelope the fluorescent material remains at the entry point of the bacterium on the nuclear envelope (Fig. 2e, f, i, j). Viewing the bacteria from different angles (Fig. 2e, i present side views and Fig. 2f, j show a top view) reveals that the 89-kDa proteins surround the outside of the invasion tip, from which it is stripped off during entry into the nucleus. The fluorescent ring disappears from the macronuclear envelope about 24 h after infection.

Immunoelectron microscopy with mAb IR-4-2 failed. However, fibrous structures corresponding to the 89-kDa protein fluorescence are visible between the outer membrane of the invasion tip and the host DV membrane using ordinary transmission electron microscopy (TEM) (Fig. 3). Furthermore, very fine fibrous structures appear outside the protruded DV membrane (Fig. 3a). Recently, indirect immunofluorescence microscopy with a DV-II vacuole membrane-specific mAb has shown that the protruded DV membrane near the invasion tip is disrupted before the escape of the bacterium from the DV (M. Fujishima and C. Morikawa, unpublished results), indicating that the bacterium escapes from the host DV without wrapping by the host's DV membrane or with wrapping by a new membrane which has no epitope to the mAb. Fibrous structures still cover the invasion tip of *H. obtusa* when the bacterium appears in the cytoplasm (Fig. 3b). The fibers are oriented at a 90° angle with respect to the bacterial outer membrane. Furthermore, a thin lining runs parallel to the outer membrane near the invasion tip. These fibrous structures are absent from the posterior end of the bacterium. This bacterium is not wrapped by the host membrane in the cytoplasm. Figure 3c and d portrays bacteria that are just penetrating the macronuclear envelope. It is noteworthy that both the fine transverse and the longitudinal filaments remain at the entry point of the bacterium on the nuclear envelope (Fig. 3d). Consequently, the positions of the fibrous structures observed by TEM correspond exactly to that of the AF488 fluorescence of 89-kDa proteins presented in Fig. 2g–j. Because 76- and 63-kDa bands are also involved in the invasion tip, the 89-kDa protein might be a part of the fibrous structures observed by TEM. The fibrous structures might have a function to facilitate the entry of *Holospora* into the macronucleus by pushing the bacterium through the nuclear envelope or by aligning the bacterium at the correct angle with respect the nuclear membrane. Recently, induction of translocation of the 89-kDa protein to outside the invasion tip by treatment of the IF cells with an acidic buffer has been succeeded (H. Fujise and M. Fujishima, unpublished results).

## 3.2 Actin-Based Motility of *Holospora*

The intracellular parasitic bacteria *Listeria*, *Shigella*, and *Rickettsia* induce polymerization of the host actin as a driving force for movement through the host cytoplasm (Goldberg 2001; Gouin et al. 1999; Heinzen 2003; Heinzen et al. 1999). Actually, *H. obtusa* likely uses a similar mechanism, consistent with reports of a comet-tail-like structure around the periplasm of *Holospora* when it is migrating through the host cytoplasm (Görtz and Wiemann 1989) because the deduced sequence of the 89-kDa

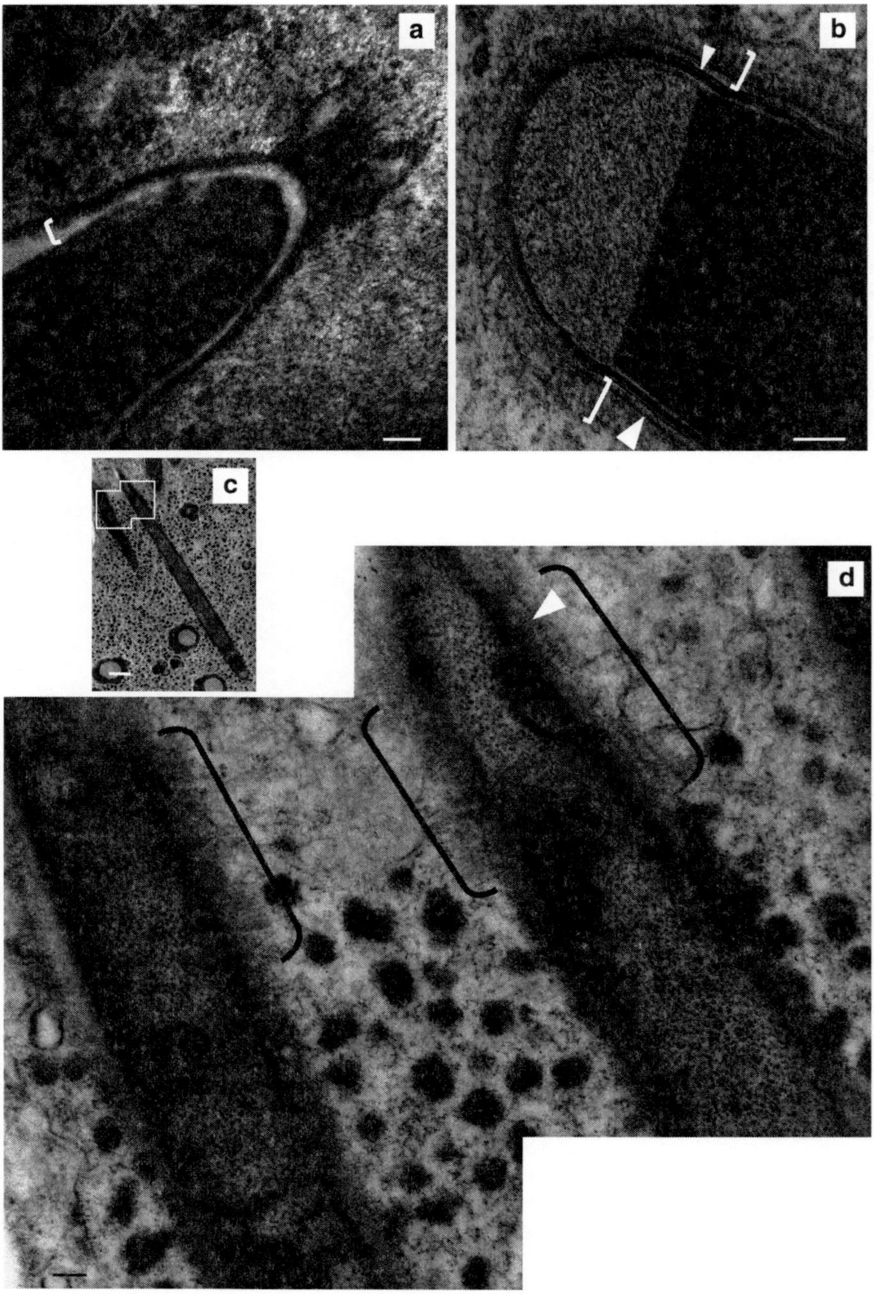


**Fig. 3** Transmission electron microscopy photomicrographs during the *H. obtusa* infection process. **a** Bacterium forming an extrusion of the host DV, 15 min after mixing. Fine fibrous structures are visible between the outer membrane of the invasion tip and the membrane of the DV (*bracket*). **b** Bacterium appears in the host cytoplasm, 30 min after mixing. The thin lining running parallel to

protein has two actin-binding motifs. To prove participation of the host actin for the infection process of *H. obtusa*, the mAb PCactin1 was developed using a synthesized peptide, YEEELKKYKE, which was deduced from the *P. caudatum* actin gene (DDBJ: AB070223) (Y. Naka, Y. Nakamura and M. Fujishima, unpublished results). This actin gene corresponds to the actin1-1 gene of *P. tetraurelia*. Indirect immunofluorescence shows that AF488 fluorescence of the actin appears around *H. obtusa*, when the bacterium escapes from the host DV and appears in the host cytoplasm, but the fluorescence does not appear when the bacteria are in the host DVs (Fig. 2k, m). The fluorescence remains at the entry point when the bacterium penetrates the macronuclear envelope (Fig. 2i, n) (Fujishima et al. 2007; Iwatani et al. 2006). The same result was also obtained using green fluorescent protein-actin transfected paramecia (Sabaneyeva et al. 2007). To confirm whether the host actin has an indispensable role in infection by *Holospora*, an effect of the actin polymerization inhibitor latrunculin B was examined. *P. caudatum* cells were mixed with IF cells of *H. obtusa* or *H. elegans* for 5 min at densities of $5 \times 10^3$ paramecia $ml^{-1}$ and $6 \times 10^6$ bacteria $ml^{-1}$; then the paramecia were transferred to modified Dryl's solution (Dryl 1959: Fujishima et al. 1990b) containing 0.2 mM latrunculin B, and fixed with 100% methanol at 1 h after mixing with *Holospora*. Invasion of their target nuclei by these bacteria was inhibited completely, indicating that actin polymerization is indispensable for these *Holospora* species. These results show that the host actin participates in the bacterial migration in the cytoplasm and in the bacterial invasion of the target nucleus.

Taken together, these results suggest that the fibrous structures surrounding the invasion tip of the infectious *Holospora* cells contain the 89-kDa protein and that this protein plays a crucial role with the host actins in the bacterium's escape from the host DV, in the migration to the macronucleus, and in the invasion of the host's macronucleus. The cartoon in Fig. 4 portrays the dynamics of the 89-kDa protein and the host actin in the early infection process of *H. obtusa*.

## 3.3 Recognition of the Target Nuclear Envelope

Fujishima and Kawai (2004), respectively, developed mAbs IR-4-1 and Rlu-1 specific for the outer membranes of *H. obtusa* and *H. undulata*. Both mAbs labeled outer membranes of the RF and IF cells without pretreatment with NaOH, indicating that the epitopes are exposed outside the outer membranes. When the antigens were extracted from the SDS-PAGE gels, and mixed with freshly isolated macronuclei

the outer membrane is indicated by a *large arrowhead*; the outer membrane of the bacterium's anterior region is shown by a *small arrowhead*. Fine transverse filaments appear (*brackets*). **c**, **d** Bacterium penetrating the macronuclear envelope. **d** High magnification of **c**, 30 min after mixing. The fine fibrous structures (*brackets)* that remain on the macronuclear envelope consist of transverse filaments (*brackets*) and a longitudinal filament (*big arrowhead*) around the outer membrane. *Bars* 100 nm in **a**, **b**, and **d**; 1 μm in **c**. (From Iwatani et al. 2005)

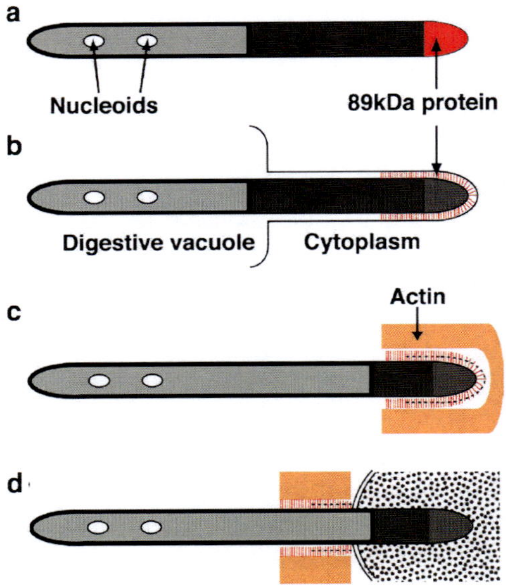


**Fig. 4** Model of 89-kDa-protein translocation and actin-based movement of *Holospora* in infection. **a** An IF cell of *H. obtusa* has an 89-kDa protein in the invasion tip. **b** In the host DV, the vacuole membrane extends and surrounds the bacterium starting from the invasion tip. During this phase, the 89-kDa proteins translocate to the outside of the invasion tip, forming a fine fibrous structure between the bacterial outer membrane and the DV membrane. **c** In the host cytoplasm, the invasion tip region maintains the attachment of the 89-kDa proteins and the fibrous structure, and host actins accumulate around the tip. **d** During invasion into the macronucleus, the 89-kDa proteins, the fibrous structures, and actins are left behind at the entry point on the nuclear envelope. (Updated from Iwatani et al. 2005)

and micronuclei of *P. caudatum*, they bound with nuclear envelopes of their target nuclei (Fujishima and Kawai 2004). Relative molecular masses of the outer-membrane substances were 16 kDa in *H. obtusa* and 13 kDa in *H. undulata*. These antigens are resistant against proteinase K and can be stained neither with Coomassie brilliant blue nor by an ordinary silver stain. However, the antigens were stained with silver for bacterial LPS, indicating that their outer-membrane substances are LPSs (Fujishima and Kawai 2004). Consequently, these results show that the bacterial recognition of the target nuclei is controlled by an affinity between a LPS of the outer membrane of the IF cell and an unknown receptor substance of the target nuclear envelope. These results also demonstrate that *Holospora* arrives at its target nucleus without wrapping with membranes derived from the host cell because the bacterium distinguishes its target nuclear envelope by direct binding between their LPSs and nuclear envelopes. To date, several mAbs specific for the macronuclear envelope have been developed. Among these antigens, a receptor substance for

*Holospora*'s LPS might be present. These LPS molecules in the IF cells of *H. obtusa* and *H. undulata* are exposed not only on the invasion tip, but also on the entire IF cell surface (M. Kawai and M. Fujishima, unpublished results).

It is noteworthy that mAbs IR-4-1 and Rlu-1 label not only the outer membranes of the IF cells, but also those of the RF cells of both *Holospora* species by indirect immunofluorescence. The molecular masses of the antigens in the RF cells are almost identical to those of the IF cells determined by immunoblotting. However, unlike LPSs of the IF cells, those of the RF cells cannot bind to their target nuclear envelopes of the isolated host nuclei, which suggests that LPSs of the outer membranes of the RF cells acquire an ability to bind their target nuclear envelopes through partial modification of the LPS during differentiation into the IF cells.

### 3.4 Penetration of the Target Nuclear Envelope and Invasion of the Nucleus

*Holospora* species always penetrate the target nuclear envelope with the invasion tip leading, but not with the other end, which suggests that the invasion tip has important molecules facilitating penetration of the nuclear envelope. This tip involves at least three polypeptides: 89-, 76-, and 63-kDa proteins. A possible function of the 89-kDa protein is portrayed in Fig. 4. Therefore, the remaining 76- and 63-kDa proteins might have some function in the penetration of the host nuclear envelope.

On the other hand, Ossipov and Podlipaev (1977) and Görtz and Wiemann (1989) reported that *H. obtusa* is transported surrounded by the host membranes in the cytoplasm and the bacterium can enter the macronucleus by membrane fusion between the enveloping membranes and the macronuclear membrane. Similar observations were also reported in *H. undulata* (Podlipaev and Ossipov 1979) and *H. elegans* (Görtz 1983). Around an evagination of the host DV by IF cells, actin-like fibers appear (Görtz and Wiemann 1989; K. Iwatani and M. Fujishima, unpublished observation). Cross sections of this evagination suggest that the bacterium appears in the host cytoplasm wrapped by the host membrane.

## 4 Infectivity of *Holospora* Against the Target Nucleus

*Holospora* species can distinguish host nuclei of two kinds, a vegetative macronucleus and a germinal micronucleus, and invade the alternative nucleus. This infectivity is acquired during differentiation of the IF cell from the RF cell of this bacterium. No other microorganism is known to have this special ability; therefore, molecules responsible for this ability have received the attention of many researchers.

### 4.1 Sensitivity of Bacterial Infectivity to Chemical and Physical Factors

Fujishima et al. (1991) examined the effects of various temperatures, pHs, enzymes, acid and NaOH, alcohols, detergents, and antibiotics on the morphology, infectivity, and reproductivity after infection of the IF cell of *H. obtusa*. For example, the infectivity and reproductive activity are unaffected if the bacteria are kept at 4°C for 10 weeks or at 25°C for 10 days, but the infectivity is easily lost at 30°C for 3 days and 40°C for 5 min. These abilities are not affected by a change in pH over a wide pH range (3–12) for 20 min at 25°C. These results show that acidosomal fusion with the host DV does not inactivate the bacterial infectivity because the pH in the DV at the maximum acidification is about 3 in *P. multimicronucleatum* (Fok et al. 1982). Among various enzymes, 220 μm ml$^{-1}$ α-mannosidase induces the loss of infectivity, suggesting that the bacterium must escape from the DV before lysosomal fusion occurs. Treatments with 0.04% trypsin or 3 mg ml$^{-1}$ proteinase K do not affect infectivity, which suggests that proteins exposed on the outside of an outer membrane of the IF cell reported by Görtz et al. (1992) have no direct function for infectivity. Consequently, the IF cell easily loses its infectivity at moderately high temperature, but it is resistant against various chemicals and enzymes.

### 4.2 *Holospora* Infects Phylogenetically Closed *Paramecium* Species

Fujishima and Fujita (1985) and Fujishima (1986) found that a property of the macronucleus necessary for it to be recognized and infected by *H. obtusa* is commonly provided by *P. caudatum*, *P. multimicronucleatum*, and *P. aurelia* species but not by *P. jenningsi*, *P. bursaria*, *P. trichium* (=*P. putrinum*), *P. duboscqui*, or *P. woodruffi*, although the bacteria can escape from the DVs and appear in the cytoplasm. Later, results showed that *P. calkinsi*, *P. polycaryum*, and *P. nephridiatum* also cannot be infected by *H. obtusa*. For *P. jenningsi*, recently, an infection-capable strain was found: its infectivity is strain-specific (M. Fujishima, unpublished results). Stable maintenance of the infected bacteria in the host nucleus is achieved only in certain strains of *P. caudatum*. All strains of *P. caudatum*, *P. multimicronucleatum*, and *P. aurelia* species examined were infected by *H. obtusa*. Fokin et al. (2004) proposed that the genus *Paramecium* be subdivided into four subgenera—*Chloroparamecium* (*P. bursaria*), *Hekianter* (*P. putrinum*, *P. dubosqui*), *Cypriostomum* (*P. calkinsi*, *P. polycaryum*, *P. nephridiatum*, *P. woodruffi*, *P. pseudotrichium*), and *Paramecium* (*P. caudatum*, *P. aurelia* species, *P. multimicronucleatum*, *P. jenningsi*, *P. wichtermani*, *P. africanum*, *P. jankovskii*, *P. ugandae*, *P. schwiakoffi*) – on the basis of their morphologies and small subunit ribosomal RNA gene sequences. A phylogenetic tree based on cytosol-type *hsp*70 gene sequences of *Paramecium* species by Hori et al. (2006) also supports this proposal. Species-specific infectivity of *H. obtusa* and the phylogenetic tree of *Paramecium* species shows that this bacterium can invade the macronucleus of closely related *Paramecium* species that belong to the subgenus *Paramecium*, but cannot infect species belonging to other subgenera (Fokin 2000; Fujishima 1986; Fujishima and Fujita 1985). Whether the cause of the failure of the infection by *H. obtusa* results from non-affinity between *H. obtusa*'s LPS and their macronuclear envelopes or the bacterium cannot penetrate their nuclear envelopes remains unknown. On the other hand, nuclei of *Didinium nasutum*, *Blepharisma japonicum*, *Pseudourostyla levis*, *Euplotes daidaleos*, *E. patella*, *E. plumipes*, *E. aediculatus*, *E. octocarinatus*, *E. woodruffi*, *E. eurystomus*, and *Tetrahymena thermophila* are not infected by *H. obtusa*. In *B. japonicum*, although the bacteria were ingested into DVs, they could not escape from the DV. In *P. levis, Euplotes* species, and *T. thermophila*, the bacteria can appear in the cytoplasm, but they cannot penetrate the host nuclear membrane (Fujishima and Fujita 1985).

## 4.3 Timing of Appearance of Nuclear Infectability During Nuclear Differentiation

*Holospora* species can infect a specific nucleus, either the macronucleus or the micronucleus, of *Paramecium* species. These nuclei of two kinds originate from a common fertilization nucleus; therefore, it is assumed that each nucleus acquires a property necessary for it to be recognized by the bacterium at a certain time during nuclear differentiation. This ability of *Holospora* provides us with an opportunity to determine the stage at which the post-zygotic nuclei differentiate into macronuclei and micronuclei, with respect to the acquisition of infectability. Fujishima and Görtz (1983) mixed IF cells of *H. obtusa* and exconjugants of *P. caudatum* at various times after exconjugation, and found that the infectability by *H. obtusa* is acquired by four of the eight postzygotic nuclei as soon as the four nuclei differentiate morphologically into macronuclear anlagen. Old macronuclear fragments were also infected. These results indicate that the macronuclear anlagen acquire infectability by *H. obtusa* at almost identical times as the first recognizable change in the macronuclear anlagen. Quite recently, macronuclear-envelope-specific substances recognized by *H. obtusa* species have been detected (K. Tanaka and M. Fujishima, unpublished results).

# 5 Maintenance of Infected *Holospora*

Infection (recognition and penetration of the target nuclear envelope) and maintenance (stable multiplication of infected bacteria) of *Holospora* are independently controlled phenomena. The infection is controlled by affinity between LPSs of the IF cells and unknown substances on their target nuclear envelope and penetration of the nuclear envelope (Fujishima and Fujita 1985; Fujishima and Kawai 2004;

Iwatani et al. 2005). The maintenance is probably controlled by the host genotypes (Fujishima and Mizobe 1988).

## 5.1 Species Specificity and Strain Specificity of Maintenance

As described already, *H. obtusa* can infect *P. caudatum*, *P. multimicronucleatum*, 14 species of *P. aurelia* (Fujishima 1986; Fujishima and Fujita 1985). and *P. jenningsi* (M. Fujishima, unpublished results). However, stable maintenance of the infected bacteria in the host nucleus is achieved only in certain strains of *P. caudatum*. Bacteria that infected the macronuclei of *P. multimicronucleatum* and *P. aurelia* species can differentiate to the RF cells and can multiply at least 2–5 days after infection; however, the bacteria gathered together afterwards, forming a large aggregate in the macronucleus and were eventually extruded synchronously from the host nuclei of the cells. For *P. aurelia* species, multiplication of the infected bacteria is observed only in *P. decaurelia* and *P. quadecaurelia*, but the bacteria also disappeared from the host nuclei in the same manner as in *P. multimicronucleatum* (Fokin et al. 2005; Fujishima 1986; Fujishima and Fujita 1985). A similar phenomenon is reported in *H. acuminata* of *P. bursaria* (Skoblo et al. 1990). For *P. jenningsi*, infected *H. obtusa* were unable to form constrictions to differentiate RF cells and were eventually expelled (M. Fujishima, unpublished observation). Synchronous elimination of the infected bacteria suggests that infection by *H. obtusa* actuates an unknown defense mechanism of the host cell against the infecting *Holospora*. Before disappearance of the bacteria from the nuclei, vacuolar structures of various sizes appear frequently in both or alternate nuclei (Skovorodkin et al. 2001). Similar vacuolar structures are also reported in the micronucleus of a certain strain of *P. caudatum*, which cannot maintain the infected *H. undulata* in the nucleus (Skovorodkin and Fokin 1991). On the other hand, vacuolar structures also appear when *H. obtusa* or *H. undulata* is injected into a macronucleus of *Holospora*-free *P. caudatum* cells (Skovorodkin et al. 2001). Tsukii et al. (1995) reported morphologically similar structures in the macronucleus of *P. caudatum* that were not exposed in any way to holosporas. They named the structures "inclusion bodies," and suggested that they might be produced using a virus or viruslike element. The inclusion bodies were isolated from the macronuclei. Therefore, the relations between this inclusion body and the vacuolar structure can be examined further in the future using antibodies raised against the inclusion bodies.

In *P. caudatum*, to date, all strains tested were infected by *H. obtusa*, but only certain strains can maintain the infected bacteria in the macronucleus (Fujishima and Fujita 1985). It is noteworthy that, to date, all strains belonging to syngen 12 are infected by *H. obtusa*, but were insufficient to maintain the infected bacteria. The fact that stable maintenance of the bacteria is a highly specific property of certain strains suggests that the maintenance of *H. obtusa* is influenced by the host's genotypes, as in the case of kappa and other symbionts of the *P. aurelia* species complex (Preer et al. 1974). Genetic analyses by cross-fertilization show that the ability to maintain the infected bacteria is determined by two loci: M and S. Dominant gene M is needed for maintenance of infected *H. obtusa*, but dominant S suppresses M. Therefore, genotypes must be either MMss or Mmss to maintain the bacteria (Fujishima and Mizobe 1988). On the other hand, the IF cell of a micronucleus-specific *H. acuminata* is known to infect *P. bursaria* only slightly and is only slightly maintained if the previous host cell of the bacteria and a new host cell belong to different syngens. Furthermore, Rautian et al. (1993) also showed that some strains of *P. bursaria* are incapable of infection by *H. acuminata*.

## 5.2 Fate of *Holospora* Injected into a Nontarget Nucleus

Skovolodkin et al. (2001) examined the fates of micronucleus-specific *H. undulata* and macronucles-spcific *H. obtusa* cells that were injected from the donor nucleus to the recipient macronucleus of *P. caudatum* by microinjection. The RF cells of *H. undulata* initially grew in the recipient macronucleus, but were eventually eliminated from the nucleus within 5 days after the injection. On the other hand, the IF cells of both *Holospora* species did not differentiate into the RF cells in the recipient macronucleus. They were then expelled from the nucleus into the cytoplasm. In normal infection through the host DVs, the infected bacteria in the target nucleus soon acquire an affinity for nuclear chromatin (Görtz 1983) and the outlines of the bacteria become obscure under a light microscope. However, the injected IF cells do not become obscure under a light microscope and do not form constrictions for binary fission in the macronucleus. The fact that acidification of the isolated IF cells in vitro induces activated-form-specific morphology and protein synthesis (Kawai and Fujishima 1997) suggests that the IF cell has no ability to differentiate into RF cells unless it infects the nucleus after differentiation into the activated form through the host DV. On the other hand, the RF cells of *H. obtusa* and *H. undulata* were able to divide in the recipient macronucleus, whereas *H. undulata* cells were expelled from the nucleus within 5 days after injection, which signifies that the micronucleus-specific bacterium cannot be maintained in the macronucleus even if the bacterium is injected into the macronucleus of a genetically identical cell.

# 6 Effects of Infection of *Holospora* on the Host

To date, unlike endosymbionts of *Euplotes* species (Fujishima and Heckmann 1983; Heckmann 1983) and *Amoeba proteus* (Jeon and Jeon 1976), *Holospora* species is not a necessary endosymbiont for the host's survival. However, *Holospora*-bearing *Paramecium* cells can acquire heat-shock resistance (Fujishima et al. 2005; Hori et al. 2008; Hori and Fujishima 2003), high salts concentration resistance (Fujishima et al. 2006) and osmotic-shock resistance (Smurov and Fokin 1998); therefore, *Paramecium* cells become adapted to unsuitable environments for their habitations through *Holospora* species.

## 6.1 *Holospora*-Induced Alteration of the Host's Gene Expression

Differential display reverse transcribed PCR showed that *H. obtusa* alters multiple gene expression of the host after establishing endosymbiosis (Nakamura et al. 2004). Studies of functions of the gene products are now progressing. Infection of *H. obtusa* enhances heat-shock gene expression of the host cell (Hori and Fujishima 2003). However, the molecular mechanism for the symbiont-induced alteration of the host's gene expression remains unknown. A periplasmic 63-kDa protein of *H. obtusa* might be one cause for induction of the host's gene alteration (Sect. 2.4.1) (Abamo et al. 2008). On the other hand recent results showed that 16S ribosomal DNA of the micronucleus-specific bacterium *H. elegans* is transferred to the macronucleus of *P. caudatum*. Simultaneously the host cell irreversibly enhances hsp70 gene expression (Hori et al. 2008), which implies that a mobile genetic element might be involved in alteration of the host's gene expression. Changes of the host's gene expression are not only a problem caused by *Holospora*, but are a general phenomenon for endosymbiosis. Therefore, *Paramecium* and *Holospora* might serve as a good model system to elucidate the mechanism of the horizontal gene transfer between the eukaryotic host cells and the symbiotic bacteria.

## 6.2 Advantages and Disadvantages for the Host

*Holospora* species had been considered as a parasite because no benefit has been identified for the host cell over many years. Moreover, the numerous IF cells in the host macronucleus inhibit the host cell's growth and eventually kill it (Sect. 2.4.2). However, sometimes *Holospora*-bearing paramecia can survive for a longer time at 10°C compared with a genetically identical strain of the *Holospora*-free cells (M. Fujishima, unpublished observation). Results showed that the *H. obtusa*-bearing paramecia can acquire temperature-shift resistance if the bacteria in the nucleus are RF cells (Hori et al. 2008; Hori and Fujishima 2003; Fujishima et al. 2005). Dohra et al. (1998) showed that the groEL gene is highly expressed in the RF of *H. obtusa*, even at 25°C, but that of the IF cell is only present in a trace amount. Damage to the host cell from various stresses such as heat and osmotic shocks might be lessened if the groEL homolog is able to protect the essential proteins of the host cell. On the other hand, Görtz and Fujishima (1983) showed that if a micronucleus is infected by *H. elegans*, the nucleus could not give rise to new functional macronuclei after conjugation. Therefore, *H. elegans*-bearing paramecia are genetically dead even if the host cell is able to acquire stress resistance.

# 7 Concluding Remarks and Further Perspectives

Endosymbiosis is a major driving force behind eukaryotic cell evolution leading to acquisition of new intracellular components and cell diversity. A *Paramecium* cell can be infected with *Holospora* species isolated from *Holospora*-bearing cells by mixing them. Various phenomena observed in this infection process provide an excellent opportunity to elucidate cell-to-cell interactions between the eukaryotic and the prokaryoticcells for establishment of endosymbiosis.

*Holospora* species cannot grow outside the host cell, and are not necessary for the host. The host cell can acquire stress resistance through RF cells (Sect. 6.2) and the RF cell does not kill the host. No harmful effects of the RF cell on the host cell are known except for a micronucleus-specific *Holospora* (Sect. 6.2). Therefore, their interaction is mutualistic rather than commensal if the host cell is in a stressful environment. However, when the bacteria are IF cells and the bacteria are numerous, the host cell is killed by LPSs secreted from the IF cells (Sect. 2.4.2), indicating that their interaction is parasitic. For that reason, interaction between these organisms might be in a transitional stage from parasitic to commensal or might represent a mutualistic relation. We can expect to collect *Paramecium* cells from natural environments that have nonharmful IF cells, and can expect that such cells will provide us with an opportunity to understand what changes in parasitic bacteria become a trigger to induce a more intimate relation with the host cell.

**Acknowledgements** The author thanks all coworkers at Yamaguchi University. The author also thanks Yuuki Kodama of Tsukuba University for her help in preparing the figures used in this chapter. This work was supported by a grant-in-aid for Scientific Research (B) (no. 17405020) to M.F. from the Japan Society for the Promotion of Science.

# References

Abamo F, Dohra H, Fujishima M (2008) Fate of the 63-kDa periplasmic protein of the infectious form of the endonuclear symbiotic bacterium *Holospora obtusa* during the infection process. FEMS Microbiol Lett 280:21–27

Amann R, Springer N, Ludwig W, Görtz H-D, Schleifer K-H (1991) Identification in situ and phylogeny of uncultured bacterial endosymbionts. Nature 351:161–164

Cleveland DW, Fischer SG, Kirschner MW, Laemmli UK (1977) Peptide mapping by limited proteolysis in sodium dodecyl sulfate and analysis by gel electrophoresis. J Biol Chem 252:1102–1106

Clifton D, Goss R, Sahni S, van Antwerp D, Baggs R, Marder V, Silverman D, Sporn LA (1998) NF-KB-dependent inhibition of apoptosis is essential for host cell survival during *Rickettsia rickettsii* infection. Microbiology 95:4646–4651

Dohra H, Fujishima M (1999a) Cell structure of the infectious form of *Holospora*, an endonuclear symbiotic bacterium of the ciliate *Paramecium*. Zool Sci 16:93–98

Dohra H, Fujishima M (1999b) Effects of antibiotics on early infection process of a macronuclear endosymbiotic bacterium *Holospora obtusa* of *Paramecium caudatum*. FEMS Microbiol Lett 179:473–477

Dohra H, Fujishima M, Hoshide K (1994) Monoclonal antibodies specific for periplasmic materials of the macronuclear specific bacterium *Holospora obtusa* of the ciliate *Paramecium caudatum*. Eur J Protistol 30:288–294

Dohra H, Yamamoto K, Fujishima M, Ishikawa H (1997) Cloning and sequencing of a periplasmic 5.4 kDa peptide gene of the macronucleus-specific symbiont *Holospora obtusa* of the ciliate *Paramecium caudatum*. Zool Sci 14:69–75

Dohra H, Fujishima M, Ishikawa H (1998) Structure and expression of a groE operon homolog of the macronucleus-specific *Holospora obtusa* of the ciliate *Paramecium caudatum*. J Eukaryot Microbiol 45:71–79

Dryl S (1959) Antigenic transformation in *Paramecium aurelia* after homologous antiserum treatment during autogamy and conjugation. J Protozool 6(Suppl):25

Ehrsam E, Görtz H-D (1999) Surface proteins of the Gram-negative bacterium *Holospora obtusa* bind to macronuclear proteins of its host *Paramecium caudatum*. Eur J Protistol 35:304–308

Fok A, Allen RD (1988) The lysosome system. In: Görtz H-D (ed), *Paramecium*. Springer, Berlin, pp 301–324

Fok A, Lee Y, Allen RD (1982) The correlation of digestive vacuole pH and size with digestive cycle in *Paramecium caudatum*. J Protozool 29:409–412

Fokin SI (2000) Host specificity of *Holospora* and its relationships with *Paramecium* phylogeny. Jpn J Protozool 33:94

Fokin SI, Sabaneyeva EV (1997) Release of endonucleobiotic bacteria *Holospora bacillata* and *Holospora curvata* from the macronucleus of their host cells *Paramecium woodruffi* and *Paramecium calkinsi*. Endocytobiosis Cell Res 12:49–55

Fokin SI, Brigge T, Brenner J, Görtz H-D (1996) *Holospora* species infected the nuclei of *Paramecium* appear to belong into two groups of bacteria. Eur J Protistol 32:19–24

Fokin SI, Przybos E, Chivilev SM, Beier CL, Horn M, Skotarczak B, Wodecka B, Fujishima M (2004) Morphological and molecular investigations of *Paramecium schewiakoffi* sp. nov. (Ciliophora, Oligohymenophorea) and current status of distribution and taxonomy of *Paramecium* spp. Eur J Protistol 40:225–243

Fokin SI, Schweikert S, Fujishima M (2005) Recovery of the ciliate *Paramecium multimicronucleatum* following bacterial infection with *Holospora obtusa*. Eur J Protistol 41:129–138

Freiburg M (1985) Isolation and characterization of macronuclei of *Paramecium caudatum* infected with the macronucleus-specific bacterium *Holospora obtusa*. J Cell Sci 73:389–398

Fujishima M (1986) Further study of the infectivity of *Holospora obtusa*, a macronucleus specific bacterium of the ciliate *Paramecium caudatum*. Acta Protozool 25:345–350

Fujishima M (1993) Control of morphological changes of the endonuclear symbiont *Holospora* of the ciliate *Paramecium*. In: Sato S, Ishida M, Ishikawa H (eds) Endocytobiology V. Tübingen University Press, Tübingen, pp 505–508

Fujishima M, Fujita M (1985) Infection and maintenance of *Holospora obtusa*, a macronucleus-specific bacterium of the ciliate *Paramecium caudatum*. J Cell Sci 76:179–187

Fujishima M, Görtz H-D (1983) Infection of macronuclear anlagen of *Paramecium caudatum* with the macronucleus-specific symbiont *Holospora obtusa*. J Cell Sci 64:137–146

Fujishima M, Heckmann K (1983) Intra- and interspecies transfer of endosymbionts in *Euplotes*. J Exp Zool 230: 339–345

Fujishima M, Hoshide K (1988) Light and electron microscopic observations of *Holospora obtusa*: A macronucleus-specific bacterium of the ciliate *Paramecium caudatum*. Zool Sci 5: 791–799

Fujishima M, Kawai M (1997) Acidification in digestive vacuoles is an early event required for *Holospora* infection of *Paramecium* nucleus. In: Achenk HEA, Herrmann RG, Jeon KW, Müller NE, Schwemmler W (eds) Eukaryotism and symbiosis. Springer, Berlin, pp 367–370

Fujishima M, Kawai M (2004) Endonuclear symbiotic bacteria *Holospora* species distinguish the host nuclear envelopes. Endocytobiosis Cell Res 15:71–76

Fujishima M, Mizobe Y (1988) Host genes controlling maintenance of a macronucleus-specific symbiont *Holospora obtusa* of *Paramecium caudatum*. Zool Sci (Abst):1272

Fujishima M, Nagahara K (1985) Isolation of endonuclear symbiont *Holospora obtusa* from mass cultures of *Paramecium caudatum*. VIIth Int Congr Protozool (Abst), p 301

Fujishima M, Nagahara K, Kojima Y (1990a) Changes in morphology, buoyant density and protein composition in differentiation from the reproductive short form to the infectious long form of *Holospora obtusa*, a macronucleus-specific symbiont of the ciliate *Paramecium caudatum*. Zool Sci 7:849–860

Fujishima M, Sawabe H, Iwatsuki K (1990b) Scanning electron microscopic observations of differentiation from the reproductive short form to the infectious long form of *Holospora obtusa*. J Protozool 37:123–128

Fujishima M, Nagahara K, Kojima Y, Sayama Y (1991) Sensitivity of the infectious long form of the macronuclear endosymbiont *Holospora obtusa* of the ciliate *Paramecium caudatum* against chemical and physical factors. Eur J Protistol 27:119–126

Fujishima M, Dohra H, Kawai M (1997) Quantitative changes in periplasmic proteins of the macronuclear-specific bacterium *Holospora obtusa* in the infection process of the ciliate *Paramecium caudatum*. J Euk Microbiol 44:636–642

Fujishima M, Kawai M, Yamamoto R (2005) *Paramecium caudatum* acquires heat-shock resistance in ciliary movement by infection with the endonuclear symbiotic bacterium *Holospora obtusa*. FEMS Microbiol Lett 243:101–105

Fujishima M, Nakata K, Kodama Y (2006) *Paramecium* acquires resistance for high concentrations of various metal chlorides by infection of endonuclear symbiotic bacterium *Holospora*. Zool Sci 23(Abst):1161

Fujishima M, Iwatani K, Nakamura Y, Kodama Y (2007) Infection of *Holospora* is controlled by 89-kDa proplasmic proteins and the host actin. Protistology 5(Abstr):31

Gibson I, Bedingfield G, Horne RW, Bird B (1986) Electron microscope study of the structure of *Holospora caryophila*. Micron Microsc Acta 17:247–257

Gilbert LA, Ravindran S, Turetzky JM, Boothroyd JC, Bradley PJ (2007) *Toxoplasma gondii* targets a protein phosphatase 2C to the nucleus of infected host cells. Eukaryot Cell 6:73–83

Goldberg MB (2001) Actin-based motility of intracellular microbial pathogens. Microbiol Mol Biol Rev 65:595–625

Görtz H-D (1980) Nucleus-specific symbionts in *Paramecium caudatum*. In: Schwemmler W, Schenk HEA (eds) Endocytobiology, endocytobiosis and cell biology I. de Gruyter, Berlin, pp 381–392

Görtz H-D (1983) Endonuclear symbionts in ciliates. Int Rev Cytol 14:145–176

Görtz H-D, Dieckmann J (1980) Life cycle and infectivity of *Holospora elegans* (Hafkine), a micronucleus-specific symbiont of *Paramecium caudatum* (Ehrenberg). Protistologica 16:591–603

Görtz H-D, Fujishima M (1983) Conjugation and meiosis of *Paramecium caudatum* infected with the micronucleus-specific bacterium *Holospora elegans*. Eur J Cell Biol 32:86–91

Görtz H-D, Wiemann M (1989) Route of infection of the bacteria *Holospora elegans* and *Holospora obtusa* into the nuclei of *Paramecium caudatum*. Eur J Protistol 24:101–109

Görtz H-D, Ahlers N, Robenek H (1989) Ultrastructure of the infectious and reproductive forms of *Holospora obtusa*, a bacterium infecting the macronucleus of *Paramecium caudatum*. J Gen Microbiol 135:3079–3085

Görtz H-D, Benting J, Ansorge I, Freiburg M (1992) Cell surface proteins of the infectious form of the symbiotic bacterium *Holospora obtusa*. Symbiosis 14:391–397

Gromov BV, Ossipov DV (1981) *Holospora* (ex Hafkine 1980) nom. rev., a genus of bacteria inhabiting the nuclei of paramecia. Intern. J Syst Bacteriol 31:348–352

Gouin G, Gantelet H, Egile C, Lasa I, Ohayon H, Villiers V, Gounon P, Sansonetti PJ, Cossart P (1999) A comparative study of the actin-based motilities of the pathogenic bacteria *Listeria monocytogenes*, *Shigella flexneri* and *Rickettsia conorii*. J Cell Sci 112:1697–1708

Heckmann K (1983) Endosymbionts of *Euplotes*. Int Rev Cytol (Suppl 14):111–144

Heinzen RA (2003) Rickettsial actin-based motility: behavior and involvement of cytoskeletal regulators. Ann N Y Acad Sci 990:535–547

Heinzen RA, Grieshaber SS, Van Kirk LS, Devin CJ (1999) Dynamics of actin-based movement by *Rickettsia rickettsii* in vero cells. Infect Immun 67:4201–4207

Hori M, Fujishima M (2003) The endosymbiotic bacterium *Holospora obtusa* enhances heat-shock gene expression of the host *Paramecium caudatum*. J Eukaryot Microbiol 50:293–298

Hori M, Tomikawa I, Przybos E, Fujishima M (2006) Comparison of the evolutionary distances among syngens and sibling species of *Paramecium*. Mol Phyl Evol 38:697–704

Hori M, Fujii K, Fujishima M (2008) Micronucleus-specific bacterium *Holospora elegans* irreversibly enhances stress gene expression of the host *Paramecium caudatum*. Eukaryot Microbiol 55:515–521

Iwatani K, Dohra H, Lang BF, Burger G, Hori M, Fujishima M (2005) Translocation of an 89-kDa periplasmic protein is associated with *Holospora* infection. Biochem Biophys Res Commun 337:1198–1205

Iwatani K, Dohra H, Lang BF, Burger G, Hori M, Fujishima M (2006) Translocation of an 89-kDa periplasmic protein is associated with *Holospora* infection. VII Asian Conf Ciliate Biol (Abst) p 28

Jeon KW, Jeon MS (1976) Endosymbiosis in amoebae: recently established endosymbionts have become required cytoplasmic components. J Cell Physiol 89:337–344

Kawai M, Fujishima M (1996) Partial characterization and cryopreservation of the isolated symbiotic bacterium *Holospora recta* of the ciliate *Paramecium caudatum*. Zool Sci 13:679–684

Kawai M, Fujishima M (1997) Monoclonal antibody specific for activated form of *Holospora obtusa*, a macronucleus-specific bacterium of *Paramecium caudatum*. In: Achenk HEA, Herrmann RG, Jeon KW, Müller NE, Schwemmler W (eds) Eukaryotism and symbiosis. Springer, Berlin, pp 371–374

Kawai M, Fujishima M (2000) Invasion of the macronucleus of *Paramecium caudatum* by the bacterium *Holospora obtusa*: Fates of the bacteria and timings of invasion steps. Eur J Protostol 36:46–52

Lang BF, Brinkmann H, Koski LB, Fujishima M, Görtz H-D, Burger G (2005) On the origin of mitochondria and Rickettsia-related eukaryotic endosymbionts. Jpn J Protozool 38:171–183

Millot L, Kaltz O (2006) Cryopreservation of *Holospora undulata*, a bacterial parasite of the ciliate *Paramecium caudatum*. Cryobiology 52:161–165

Nakamura Y, Aki M, Aikawa T, Hori M, Fujishima M (2004) Differences in gene expression of the ciliate *Paramecium caudatum* caused by endonuclear symbiosis with *Holospora obtusa*, revealed using differential display reverse transcribed PCR. FEMS Microbiol Lett 240:209–213

Ossipov DV (1973) Specific infectious specificity of the omega-particles, micronuclear symbiotic bacteria of *Paramecium caudatum*. Cytology (Saint-Petersburg) 15:211–217

Ossipov DV, Podlipaev SF (1977) Electron microscope examination of early stage of infection of *Paramecium caudatum* by bacterial symbiont of the macronucleus (Jota-bacteria). Acta Protozool 31:289–308

Ossipov DV, Skoblo II, Rautian MS (1975) Iota-particles, macronuclear symbiotic bacteria of ciliate *Paramecium caudatum* clone M-115. Acta Protozool 14:263–280

Ossipov DV, Skoblo II, Borschsenius ON, Rautian MS (1980) *Holospora acuminata* – a new species of symbiotic bacterium from the micronucleus of the ciliate *Paramecium bursaria* Focke. Cytology 22:922–929

Podlipaev SF, Ossipov DV (1979) Early stages of infection of *Paramecium caudatum* micronuclei by symbiotic bacteria – omega-particles (electron microscope examination). Acta Protozool 18:465–480

Preer LB (1969) Alpha, an infectious macronuclear symbiont of *Paramecium aurelia*. J Protozool 16:570–578

Preer JR, Preer LB, Jurand A (1974) Kappa and other endosymbionts in *Paramecium aurelia*. Bact Rev 38:113–163

Rautian MS, Skoblo II, Lebedeva NA, Ossipov DV (1993) Genetics of symbiotic interactions between *Paramecium bursaria* and the intranuclear bacterium *Holospora acuminata*, natural genetic variability by infectivity and susceptibility. Acta Protozool 32:165–173

Sabaneyeva EV, Benken KA, Derlacheva ME, Skovorodkin IN, Fokin SI (2007) *Holospora* trafficking in *Paramecium*: The role of the host cell cytoskeleton. Protistology 5(Abstr):68

Saeij JPJ, Coller S, Boyle JP, Jerome ME, White MW, Boothroyd JC (2007) *Toxoplasma* co-opts host gene expression by injection of a polymorphic kinase homologue. Nature 445:324–327

Schmidt H, Freiburg M, Görtz H-D (1987) Comparison of the infectious forms of two bacterial endonucleobionts, *Holospora elegans* and *H. obtusa*, from the ciliate *Paramecium caudatum*. Microbios 49:189–197

Skoblo II, Lebedeva NA (1986) Infection of the nuclear apparatus of *Paramecium bursaria* (Ciliata) by the symbiotic bacterium *Holospora curviuscula*. Cytology 28:367–372

Skoblo II, Borchsenius ON, Lebedeva NA, Ossipov DV (1990) Symbiogenic lysis of bacteria *Holospora acuminata* in the generative nucleus of the ciliate *Paramecium bursaria*. Cytologia 32:515–519

Skovorodkin IN, Fokin SI (1991) The resistance of some stocks of ciliate *Paramecium caudatum* to the infection by bacterial endonucleobiont *Holospora undulata*. Cytology 33:98–103

Skovorodkin IN, Fokin SI, Fujishima M (2001) Fates of the endonuclear symbiotic bacteria *Holospora obtusa* and *Holospora undulata* injected into the macronucleus of *Paramecium caudatum*. Eur J Protistol 237:137–145

Smurov AO, Fokin SI (1998) Resistance of *Paramecium caudatum* infected with endonuclear bacteria *Holospora* against salinity impact. Proc Zool Inst RAS 276:175–178

Timofeyeva AS, Rautian MS (1997) *Holospora undulata*: The genome size determined by pulsed-field gel electrophoresis. Cytology 39:634–639

Tsukii Y, Harumoto T, Yazaki K (1995) Evidence for a viral macronuclear endosymbiont in *Paramecium caudatum*. J Eukaryot Microbiol 42:109–115

Tunbridge A, Stevanin T, Lee M, Marriott H, Moir J, Read R, Dockrell D (2006) Inhibition of macrophage apoptosis by *Neisseria meningitidis* requires nitric oxide detoxification mechanisms. Infect Immun 74:729–733

Wiemann M (1989) The release of *Holospora obtusa* from *Paramecium caudatum* observed with a new device for extended in vivo microscopy. J Protozool 36:176–179

Wiemann M, Görtz H-D (1991) Identification and localization of major stage-specific polypeptides of infectious *Holospora obtusa* with monoclonal antibodies. J Bacteriol 173:4842–4850

# The Killer Effect of *Paramecium* and Its Causative Agents

**Martina Schrallhammer and Michael Schweikert**

## Contents



**Abstract** Frequently, nonrelated organisms form endocytobioses resulting in organisms with different characters compared with those of the individual partners alone. Endocytic bacteria of the genus *Caedibacter* in host ciliates of the genus *Paramecium* enable their host to kill sensitive paramecia. These paramecia therefore are called “killers” and the phenomenon was named “killer trait” (Sonneborn in *Proc. Am. Philos. Soc.* 79:411–434, 1938). In their natural environment these endocytic bacteria enable their hosts to outcompete uninfected forms (Kusch et al. in *Protist* 153:47–58, 2002). Endocytobioses become more complex when the intracellular bacteria are infected with phages, such as in the three partner association *Paramecium-Caedibacter*-bacteriophage. Interaction of the different partners results in the formation of a proteinaceous intracellular structure called a “refractile

M. Schrallhammer and M. Schweikert (✉)
Abteilung Zoologie, Biologisches Institut, Universität Stuttgart, Pfaffenwaldring 57, 70569, Stuttgart, Germany
e-mail: michael.schweikert@bio.uni-stuttgart.de

M. Fujishima (ed.), *Endosymbionts in Paramecium*, Microbiology Monographs 12, DOI: 10.1007/978-3-540-92677-1_9, © Springer-Verlag Berlin Heidelberg 2009

body" (R-body) within the bacteria. Physiological tests demonstrated that R-bodies are a prerequisite for *Caedibacter* to confer the killer trait to their host, but they are not the toxin that kills sensitive paramecia (Preer et al. in *Proc. Natl. Acad. Sci. USA* 39:1228–1233, 1953). In this chapter we focus on the *Caedibacter* species, with additional brief remarks on other endocytic killer bacteria and on R-bodies produced by free-living bacteria.

## 1 Introduction

The killer trait was discovered in 1938 by Sonneborn (1938) by the discrimination of strains of paramecia he named "killers" and "sensitives." When these were put together, the sensitives died, while the killers survived for hours to days. He could show that the toxic agents are not controlled by nuclear genes, but are cytoplasmically inherited. He termed these cytoplasmic particles "kappa particles." Later, kappa particles were identified as bacteria (Preer 1950; Hamilton and Gettner 1958). After the discovery of the refractile body (R-body) (Preer and Stark 1953) all endocytobionts bearing R-bodies were named "*Caedibacter*." In this chapter we focus on the *Caedibacter* bacteria, with additional brief remarks on other endocytic killer bacteria and on R-bodies produced by free-living bacteria (Table 1).

## 2 Ecology of the Killer Trait

The killer trait is caused by bacterial endosymbionts of the genus *Caedibacter*. Paramecia bearing *Caedibacter* are called "killers." They release toxic particles (bright *Caedibacter* cells) that ultimately kill paramecia not maintaining the same bacteria ("sensitives"). Direct cell contact between killer and sensitive paramecia is not obligate for killing to occur; culture medium that has contained killers for only a few hours is toxic to sensitive paramecia (Austin 1948a, b). Ciliates infected with *Caedibacter* are resistant to the toxin of their symbiont.

The distribution of *Caedibacter* endosymbionts is worldwide, with infected *Paramecium* stocks originating from North America, Central America, Europe, Japan, and Australia (Preer et al. 1974).

### 2.1 Ecological Impact of the Killer Trait

Low numbers of *Caedibacter* cells are permanently released into the environment via the cytopyge of the killer paramecia (Nobili 1961; Preer et al. 1974). Killing occurs

**Table 1** Endocytic bacteria and associated symbiotic activity such as killer capacity and refractile body production

| Species | Old name/ synonym | Host name | Intracellular localization | Refractile bodies: Present | Refractile bodies: Maximum length (μm) | Refractile bodies: Maximum width (μm) | Refractile body type | Mode of unrolling | Killer | Extra chromo–somal elements | References |
|---|---|---|---|---|---|---|---|---|---|---|---|
| *Caedibacter caryophilus* | *Caedibacter caryophila* | *Paramecium caudatum* | Macronucleus | Yes | >20 | 0.8 | Cc | Telescopic | Killer | Phage | Schmidt et al. (1987) |
| *Caedibacter caryophilus* | *Caedibacter caryophila* | *Paramecium novaurelia* | Cytoplasm | Yes | >20 | 0.8 | Cc | Telescopic | Killer | Phage | Kusch et al. (2000) |
| *Caedibacter caryophilus* | *Caedibacter caryophila* | *Peridinium cinctum* | Cytoplasm | No | – | – | – | – | No | – | Schweikert et al. (unpublished results) |
| *Caedibacter paraconjugatus* | – | *Paramecium biaurelia* | Cytoplasm | Yes | 20 | 0.4 | 7 | From outside | Mate killer | Phage | Quackenbush (1982) |
| *Caedibacter pseudomutans* | – | *Paramecium tetraurelia* | Cytoplasm | Yes | 20 | 0.4 | 7 | From outside | Spin killer | Phage | Quackenbush (1982) |
| *Caedibacter taenios-piralis* | – | *Paramecium tetraurelia* | Cytoplasm | Yes | 20 | 0.4 | 51 | Telescopic | Killer | Plasmid | Preer and Preer (1982) |
| *caedibacter varicaedens* | – | *Paramecium biaurelia* | Cytoplasm | Yes | 20 | 0.4 | 7 | From outside | Spin killer | Phage | Quackenbush (1982) |
| *Caedibacter macronucleorum* | – | *Paramecium duboscqui* | Macronucleus | Yes | ND | ND | Unclassified | ND | Killer | – | Fokin and Görtz (1993) |
| "*Candidatus* Caedibacter acanthamoebae" | – | *Acanthamoeba polyphaga* | Cytoplasm | ND |  | – | – | – | ND | – | Horn et al. (1999) |
| *Lyticum flagellatum*[a] | Lambda | *Paramecium octaurelia* | Cytoplasm | No | – | – | – | – | Lyse killer | – | Preer and Preer (1982) |
| *Lyticum flagellatum*[a] | Lambda | *Paramecium tetraurelia* | Cytoplasm | No | – | – | – | – | Lyse killer | – | Preer and Preer (1982) |
| *Lyticum sinuosum*[a] | Sigma | *Paramecium biaurelia* | Cytoplasm | No | – | – | – | – | Lyse killer | – | Preer and Preer (1982) |
| *Marinomonas mediterranea* | – | Free living |  | Yes | ND | ND | Unclassified | Unknown | ND | – | Hernández–Romero et al. (2003) |

(continued)

**Table 1** (continued)

| Species | Old name/ synonym | Host name | Intracellular localization | Refractile bodies Present | Maximum length (μm) | Maximum width (μm) | Refractile body type | Mode of unrolling | Killer | Extra chromo– somal elements | References |
|---|---|---|---|---|---|---|---|---|---|---|---|
| *Paracaedibacter acanthamoebae*[a] | – | Acanthamobae sp. UWC9 | Cytoplasm | No | – | – | – | – | No | – | Horn et al. (1999) |
| *Paracaedibacter symbiosus*[a] | – | Acanthamobae sp. UWE39 | Cytoplasm | No | – | – | – | – | No | – | Horn et al. (1999) |
| *Pseudocaedibacter glomeratus*[a] | – | Paramecium pentaurelia | Cytoplasm | No | – | – | – | – | No | – | Fokin and Ossipov (1986) |
| *Pseudocaedibacter chlorellopellens*[a] | – | Paramecium bursaria | Cytoplasm | No |  | – | – | – | No | – | Skoblo et al. (1985) |
| *Pseudocaedibacter falsus*[a] | Nu and pi | Paramecium biaurelia | Cytoplasm | No |  | – | – | – | No | – | Quackenbush (1982) |
| –*Pseudocaedibacter falsus*[a] | Nu and pi | Paramecium pentaurelia | Cytoplasm | No | – | – | – | – | No | – | Quackenbush (1982) |
| *Pseudocaedibacter falsus*[a] | Nu and pi | Paramecium tetraurelia | Cytoplasm | No | – | – | – | – | No | – | Quackenbush (1983) |
| *Pseudocaedibacter minutus*[a] | Gamma | Paramecium octaurelia | Cytoplasm | No | – | – | – | – | Killer | – | Quackenbush (1982) |
| *Pseudocedibacter conjugatus*[a] | Mu particle | Paramecium octaurelia | Cytoplasm | No | – | – | – | – | Mate killer | – | Quackenbush (1982) |
| *Pseudocedibactr conjugats*[a] | – | Paramecium primaurelia | Cytoplasm | No | – | – | – | – | Mate killer | – | Quackenbush (1982) |
| *Pseudomonas avenae* | Acidovorax avenae | Free living | – | Yes | – | – | Pa | From outside | ND | ND | Wells and Horne (1983), Willems et al. (1992) |
| *Pseudomonas* sp. EPS 5020 | – | Free living | – | Yes | ND | ND | Unclassified | Unknown | ND | Phage | Fusté et al. (1986) |

| | | | | | | | | | | | |
|---|---|---|---|---|---|---|---|---|---|---|---|
| *Pseudomonas taeniospiralis* | *Hydrogeno-phaga taeniospiralis* | Free living | – | Yes | – | – | Pt | Telescopic | ND | Phage | Lalucat and Mayer (1978), Willems et al. (1989) |
| *Rhodospirllum centenum* | – | Free living | – | Yes | ND | ND | Unclassified | Unknown | ND | ND | Favinger et al. (1989) |

*ND* not determined

[a]For details, see Görtz and Fokin in this volume.

if a single R-body-carrying bacterium is ingested (Austin 1948b) by a *Caedibacter*-free and therefore sensitive *Paramecium*. Once ingested and incorporated into a food vacuole, the bright R-body is apparently important for toxin release (Preer et al. 1974). The milieu in the vacuole differs from the outside in respect to a lowered pH and the presence of digestive enzymes. The R-body is triggered to unroll and therewith produce a long hollow tube intended to act as a vehicle for transporting the toxin. Investigation of sensitive paramecia exposed to isolated kappa particles showed that after phagocytosis R-bodies elongate. This results in penetration of food vacuole membranes, extensive membrane breakage and disintegration, and vacuole contents are mixed with cytoplasm (Jurand et al. 1978). It was also noted that food vacuole formation ceased shortly after exposure to bright particles. *Caedibacter* cell contents are released to the cytoplasm of the sensitive *Paramecium* when the R-body unrolls. The toxin is delivered to its target site by the R-body, and its lethal action causes specific symptoms in the affected cell and finally leads to the death of the sensitive *Paramecium*.

Symbiosis between *Paramecium* and *Caedibacter* is mutualistic under natural conditions, as far as the toxic symbiont provides its host with selective advantages against competitors (Landis 1981, 1987; Kusch et al. 2002). It is not a defensive strategy of the ciliate against predators (Kusch et al. 2002; Görtz et al. 2008). Under laboratory conditions, however, *Caedibacter* is described as an energy parasite (Kusch et al. 2002) taking up ATP directly from its host via specific nucleotide transporters (Linka et al. 2003).

Within *Caedibacter*, an unusual, highly complex structure has been observed: the R-body (refractile body), visible by light microscopy. Only a certain percentage of the *Caedibacter* population within a host cell carries an R-body. These symbionts are called "brights" owing to the refractile inclusion whereas the other cells are referred to as "nonbrights" (see 3.1.1). Additionally, brights and nonbrights can be distinguished according to their size. R-body-containing bacteria are much larger than vegetative forms.

Besides these symbionts of *Paramecium,* few free-living bacteria have been reported to produce R-bodies (Table 1). Physiological tests demonstrated that R-bodies are a prerequisite for *Caedibacter* bacteria to confer the killer trait to their host, but they are not the toxin that kills sensitive paramecia (Preer et al.1953; Smith 1961; Müller 1963). The population size of endocytobionts is regulated by the growth of the host. Under maximum growth conditions host cells can divide more often and intracellular bacteria can become diluted after each cell division of the host organism. High numbers of endocytobionts can be obtained by lower division rates of the host, but overgrowth of *Caedibacter* finally causes death of the host organism (Schmidt et al. 1987; Kusch et al. 2002). New infections of hosts are rare. Distribution of endocytobionts occurs with mitotic cell division or during conjugation of the ciliates (Preer et al. 1974; Landis 1981, 1987).

The killer effect influences both the population of hosts and the population of symbionts. The competitive advantage for symbiont-carrying paramecia gives a selective advantage to paramecia with the "K-allele" phenotype. *Caedibacter* is driven in a very specialized niche, as an obligate intracellular bacterium depending completely on its host for survival. As far as these bacteria are barely infective (Preer et al. 1974; Landis 1981, 1987; Kusch et al. 2002) they strictly depend on cytoplasmic inheritance for continuity.

## 2.2 Diversity of *Caedibacter* Hosts

Previously, bacteria of the genus *Caedibacter* were found exclusively in different species of the genus *Paramecium*. Especially strains of the *Paramecium aurelia* complex, *Paramecium duboscqui*, and *Paramecium caudatum* have been found to be infected by different *Caedibacter* species so far (Pond et al. 1989). The *Caedibacter* hosts *Paramecium primaurelia, Paramecium biaurelia, Paramecium tetraurelia, Paramecium pentaurelia, Paramecium octaurelia, and Paramecium novaurelia* are regarded as sibling species in the *P. aurelia* species complex (Sonneborn 1975). Ciliates, like other protists, can be regarded as an ecosystem for endocytobionts able to invade their specific hosts (Fokin et al. 2003). As a consequence, cell compartments are potential ecological niches invaders may be able to colonize (Görtz and Fokin, this volume). Different bacteria may be adapted specifically to their host species and exclusively install an endocytobiosis with the "partner," while others have a broader host spectrum. Recently, *Caedibacter* hosts other than *Paramecium* were identified and some are not even a ciliate. *Caedibacter* species can be found in the ciliate *Didinium nasutum* (Williams 1971; Kusch et al. 2002) the dinoflagellate *Peridinum cinctum* (Schweikert and Meyer 2001; Geiger, Pfannkuchen, Eschbach & Schweikert, unpublished results), *Dileptus* species (Sonneborn 1965; Sonneborn et al. 1964), and the amoebazoa *Amoeba proteus* (Kusch et al. 2002) and *Acanthamoeba polyphaga* (Horn et al. 1999). While *Dileptus nasutum* and *A. proteus* were infected during feeding experiments or by microinjection and lost their endocytobionts after some time (Kusch et al. 2002) *Caedibacter caryophilus* was found within the dinoflagellate *P. cinctum* in natural samples and has been cultivated as an endocytobiont for more than 10 years (Schweikert and Meyer 2001).

*C. caryophilus* is able to infect the macronucleus of *P. caudatum* (Schmidt et al. 1987), the cytoplasm of *P. novaurelia* (Kusch et al. 2000), and the cytoplasm of the dinoflagellate *P. cinctum* (Schweikert, unpublished results).

Hosts are selected by the successful invasion. The criteria for the successful infection by bacteria may be cytological and genetic traits of the host rather than those of the invading bacterium.

## 2.3 Taxonomy and Phylogeny of *Caedibacter* Species

Sonneborn's studies of the killer trait gave rise to the description of one of the first examples for cytoplasmic inheritance. He discovered that the genetic determinants of that effect are located in the cytoplasm and referred to them as "plasmagenes"

(Sonneborn 1943). According to the taxonomic rules of that time, he termed these cytoplasmic factors "kappa particles," using Greek letter names.

Subsequent studies clarified that *Paramecium* killer particles are gram-negative bacteria (Hamilton and Gettner 1958), so the kappa particles had to be renamed using the binominal nomenclature according to the rules of the Bacteriological Code. As all kappas had several features in common, it was proposed to establish a new genus, namely, *Caedibacter*, and group all former kappa particles as species in this genus. For a detailed description of the genus *Caedibacter*, see Preer and Preer (1982), Görtz (2002), and Görtz and Schmidt (2005). The common features leading to the establishment of the genus *Caedibacter* are mainly morphological characters, such as (1) an obligate endosymbiotic lifestyle, (2) expression of a killer trait in their host organism, and (3) the production of R-bodies. Especially the ability to produce R-bodies has been considered as a unique characteristic to distinguish kappa particles and *Caedibacter* species from other endosymbiotic bacteria of *Paramecium*.

The first indications that species constituting the genus *Caedibacter* are not closely phylogenetically related were given by significant differences in G-C content and low DNA-DNA hybridization values among *Caedibacter* bacteria considered sister species (Dilts 1977; Quackenbush 1977, 1978). Furthermore, it turned out that the main common characteristic, the production of R-bodies, might be determined by genetically related extrachromosomal elements (Quackenbush 1983; Quackenbush et al. 1986a, b). At least in the case of *Caedibacter taeniospiralis* it was shown that R-bodies are plasmid-coded when the genes have been sequenced (Heruth et al. 1994; Jeblick and Kusch 2005).

The so-called classes of kappa (Preer et al. 1972) were arranged into four different species; a fifth was included later with the description of *C. caryophilus* in 1987. The species were defined according to their host specificity, localization within the host cell, and R-body morphology. Especially the R-body provided various criteria such as its dimensions, the shape of the ribbon tip (tapered or blunt), association with phagelike particles, and unrolling mechanism (from the inside or the outside). See also Sect. 4.1.

The five validly described (Euzéby 2008) *Caedibacter* species are *C. taeniospiralis* (Preer and Preer 1982), *C. varicaedens*, *C. pseudomutans*, *C. paraconjugatus* (Quackenbush 1982), and *C. caryophilus* (Schmidt et al. 1987; Euzéby 1997). Four species, *C. taeniospiralis, C. varicaedens, C. pseudomutans*, and *C. paraconjugatus*, are cytoplasmic endosymbionts in *P. biaurelia* or *P. tetraurelia* (sibling species in the *P. aurelia* species complex). In contrast, the fifth species, *C. caryophilus,* was first described as a macronuclear symbiont of *P. caudatum* (Table 1).

Another R-body-producing endosymbiont observed in 1993, the macronuclear symbiont of *P. duboscqui*, was named "*Caedibacter macronuleorum*" (Fokin and Görtz 1993), but, as far as the rules for prokaryotic nomenclature have changed since the description of the other species of *Caedibacter,* it is not validly described (Schrallhammer et al. 2006). Different publications raise the question of whether the characteristics used to discriminate the species of *Caedibacter* have been well chosen. The ability to produce R-bodies can be spontaneously lost (Fokin and

Görtz 1993). And also host specificity and cell localization of *Caedibacter* are not as well preserved as assumed. Kusch et al. (2000) demonstrated the presence of the same species, *C. caryophilus*, in different hosts (*P. caudatum* and *P. novaurelia*) and within different cell compartments (macronucleus and cytoplasm) (Table 1). Sequence characterization of the 16S ribosomal RNA (rRNA) gene and phylogenetic analyses of two species, *C. taeniospiralis* (Beier et al. 2002) and *C. caryophilus* (Springer et al. 1993; Schrallhammer et al. 2006), revealed that *Caedibacter* is a polyphyletic assemblage with *C. taeniospiralis* belonging to *Gammaproteobacteria*, closely related to the facultative intracellular pathogen *Francisella tularensis* and *C. caryophilus* belonging to *Alphaproteobacteria*, clustering with the obligate endosymbiont *Holospora*.

One peculiar molecular characteristic is the presence of an internal excised element found in all *C. caryophilus* 16S rRNA gene sequences, which is excised and not present in mature 16S rRNA (Springer et al. 1993; Schrallhammer et al. 2006). A similar insert, possibly also excised, is present in the 23S rRNA gene sequence of *C. caryophilus* (Genbank DQ414689; Geiger, Eschbach, Pfannkuch & Schweikert unpublished results).

# 3 Cellular and Genetical Background of the Killer Trait

The installation of the endocytobiosis of *Caedibacter* symbionts with *Paramecium* depends on the genetic constitution of the host. Only those with the codominantly expressed "K allele" can be infected; when only the "k allele" is present, infection by *Caedibacter* cannot take place (Sonneborn 1943).

## 3.1 Cellular Effects of the Killer Trait

The killer trait affects not only the sensitive *Caedibacter*-free paramecia but also influences the physiological processes of the bacteria. Its expression is disadvantageous for both parties, at least for the individuals directly involved. The sensitive *Paramecium* dies upon ingestion of a bright *Caedibacter* cell and *Caedibacter* cells lose the ability to divide after expression of the R-body.

### 3.1.1 Effects of Expressing the Killer Trait: Brights and Nonbrights

*Caedibacter* populations are subdivided into two different types of cell morphology, by phase-contrast microscopy recognizable as bright and nonbright cells. The cells possessing an R-body are called "brights" because of the refractile properties of that structure which appears as a bright spot inside the bacterium. *Caedibacter* without an R-body are referred to as "nonbrights" (Preer and Stark 1953). Bright cells arise from the nonbright forms (Preer et al. 1953). The change from nonbright to bright appears to be lethal: *Caedibacter* harboring an R-body do not have the capacity to reproduce. Only nonbrights are reproductive and undergo cell division by normal binary fission.

In general, R-bodies are produced just by a fraction of the *Caedibacter* population. The proportion of bright particles typically varies from 1 to 35%, but occasionally it is as high as 50% depending upon the particular *Caedibacter* strain, the nutritional status of the host, and presumably different exogenous factors. It has been shown that the percentage of bright particles in *C. taeniospiralis* populations increases after exposure to UV light (Preer et al. 1974). Brights can be easily separated from non-brights by Percoll gradient centrifugation (Pilhofer et al. 2007) owing to the higher density of the R-body-harboring bacteria. Killing activity correlates directly with exposure of sensitive paramecia to bright particles but is unrelated to exposure to nonbright particles (Preer et al. 1953). This relationship was confirmed in later studies with purified bright and nonbright particles (Müller 1963).

### 3.1.2 Symptoms Caused by the Killer Trait in Sensitive Paramecia

Toxicity tests with isolated R-bodies (Preer et al. 1953) showed that the R-body is not the toxin that kills the sensitive paramecia, but the presence of R-bodies is essential for killing to occur. The special shape of R-bodies, their functional behavior, and the pH-dependent induced unrolling suggest their role as a toxin transmitter, releasing it from the bacterial cytoplasm into the cytoplasm of the *Paramecium* (Fig. 1). Different strains of killer *Caedibacter* cause sensitive paramecia to die in different ways. Lethality is specific to the species of *Caedibacter*, not the host (Austin 1948a, b; Pond et al. 1989) (Table 1). However, although the lethal toxins may vary in some ways, there are common features in most types of killing mediated by *Caedibacter* sp., e.g., the specific prelethal symptoms caused in the sensitive *Paramecium*. Four different categories of killing have been described (Sonneborn 1959; Pond et al. 1989), namely, hump killing, spin killing, vacuolization, and

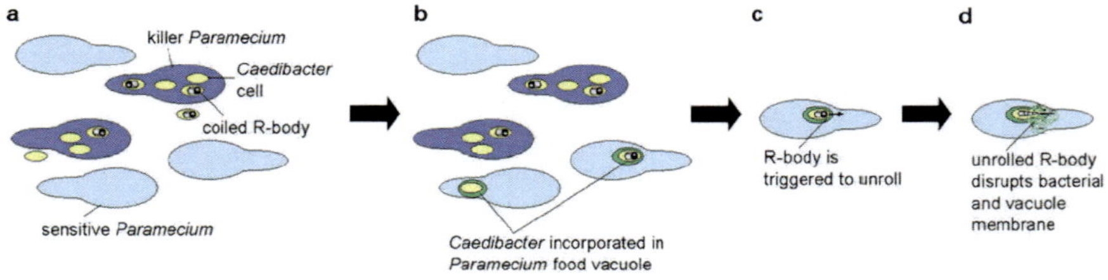


**Fig. 1** Killer trait of *Paramecium.* **a** Mixed population of sensitive paramecia and killer paramecia. Killers harbor and release *Caedibacter* cells; some of the bacteria contain a refractile body (R-body). **b** Released *Caedibacter* are ingested and incorporated in food vacuoles by sensitive paramecia. **c** Low pH and the presence of digestive enzymes in the vacuole trigger the R-body to unroll, producing a long hollow tube. **d** The uncoiled R-body penetrates bacterial and vacuole membranes; this leads to extensive membrane breakage and release of bacterial cell contents

paralysis. For each of the basic groups, prelethal symptoms in detail are (1) hump killing (*C. taeniospiralis*), a large aboral blister develops in affected cells; (2) spin killing (*C. pseudomutans, C. varicaedens* 7); affected cells swim with a direction of rotation opposite to the normal direction of rotation; (3) vacuolization (*C. varicaedens* 562), large vacuoles develop in affected cells; and (4) paralysis (*C. caryophilus*), the affected cells cease swimming and only occasionally show weak avoidance reactions.

The nature and the mechanism of action of the toxin have not been identified yet (Preer et al. 1974; Kusch and Görtz 2006). Though the toxins have not been identified in any of the many killer symbionts known, there are good indications that these toxins interact with biological membranes and disturb the osmoregulatory mechanisms of sensitive paramecia (Görtz et al. 2008). Preer and Preer (1967) have reported evidence that the toxin involved in the killer trait is proteinaceous. Gibson (unpublished results reported in Pond et al. 1989) observed with protein chromatography that the toxin in cultures of *P. tetraurelia* 51 is a protein of 20–25 kDa.

In a study by Quackenbush and Burbach (1983) plasmid DNA isolated from *C. taeniospiralis* was cloned into *Escherichia coli*. Clones carrying recombinant plasmids were screened for toxicity toward sensitive strains of paramecia and for the ability to produce R-bodies (see Sect. 3.2). None of the clones appeared to be toxic. It was assumed by comparison of prelethal effects of hump killing, spin killing, and vacuolization (Jurand et al. 1971) that the toxins involved in the killer trait are related in that they affect osmoregulatory properties of biological membranes. Sequence analysis of the plasmid pKAP298 of *C. taeniospiralis* (Jeblick and Kusch 2005) revealed as a putative toxin a protein with homology to ATPases of the Soj/ParA family and membrane-associated ATPases involved in eukaryotic ATPase-dependent ion carriers. This protein could affect the *Paramecium* cell membrane and might disturb the osmoregulation.

## 3.2 Genetic Determinants of R-Bodies and Toxins

In all *Caedibacter* species mobile genetic elements, e.g., plasmids or bacteriophages, have been detected (Pond et al. 1989). Those plasmids and bacteriophages are considered to be responsible for the ability of *Caedibacter* sp. to produce R-bodies and toxicity. Therefore, this symbiosis is called a "three-partner association" comprising *Paramecium–Caedibacter*–bacteriophage.

The plasmids of *C. taeniospiralis* are termed "pKAP" followed by the strain number in which the respective plasmid occurs. Plasmid sizes vary among *C. taeniospiralis* strains, ranging from 41.5 to 49 kbp. These size differences are introduced by the presence of 1.5- and/or 7.5-kbp transposons inserted at several sites in the plasmids (Dilts and Quackenbush 1986; Quackenbush et al. 1986a, b).

These entangled relations enticed Quackenbush et al. (1986a) to formulate the following title: "Extrachromosomal elements of extrachromosomal elements of *Paramecium* and their extrachromosomal elements". Here the genome of

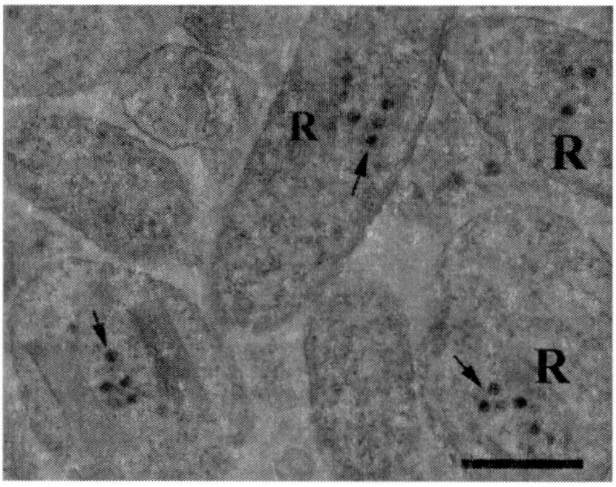


**Fig. 2** Ultrastructure of *Caedibacter caryophilus* inside the macronucleus of *Paramecium caudatum*. Bright forms with an R-body (*R*) display larger dimensions in comparison to vegetative nonbright forms. *Arrow* phage capsids. *Scale bar* 1 μm

*Caedibacter* is identified as an extrachromosomal element of *Paramecium*, the plasmids (or phage genomes) as extrachromosomal elements of *Caedibacter*, and the transposons are understood as extrachromosomal elements of the plasmids.

Quackenbush and Burbach (1983) did physical mapping of the 49-kbp plasmid pKAP47 (*C. taeniospiralis* 47). They determined the region involved in R-body synthesis by cloning the responsible fragment of 2.7 kbp in *E. coli* which produced R-bodies. Toxicity of these clones was not observed (see Sect. 3.1.1). The plasmid region coding for R-bodies was sequenced and three independently transcribed genes (*rebA*, *rebB*, and *rebC*) were characterized (Heruth et al. 1994). A fourth open reading frame with appropriate regulatory sequences was found within the *reb* locus (termed "*rebD*"), but no evidence suggests that this putative gene is expressed in *E. coli*. The three predicted *reb* genes code for relatively small proteins, namely, RebA (18 kDa), RebB (13 kDa), and RebC (10 kDa). *C. taeniospiralis* plasmid pKAP298 was completely sequenced (Jeblick and Kusch 2005). Sixty-three open reading frames were detected, 23 with homologies to genes with known function. Most notable are beside the four *reb* genes six putative phage genes and ten identified transposases or partial transposases. The high density of mobile elements and the presence of several genes coding for phage components indicate that pKAP298 evolved from a bacteriophage (Jeblick and Kusch 2005). This original phage most probably belonged to the *Caudovirales*.

As a putative toxin of the hump killer *C. taeniospiralis* a protein with homology to ATPases of the Soj/ParA family and membrane-associated ATPases involved in eukaryotic ATPase-dependent ion carriers was identified (see Sect. 3.1.2).

In *Caedibacter* species other than *C. taeniospiralis* no plasmids have been observed, but all except *C. taeniospiralis* contain bacteriophages or phagelike particles, in general in close spatial association with the R-bodies (Fig. 4; see Sect. 4.2). Early observations suggested that genetic determinants of R-bodies are located on plasmids or bacteriophages that have lost the ability to lyse their hosting bacteria upon maturation of the virions (Preer and Preer 1967; Preer et al. 1974). Sonneborn (1959) proposed that R-body synthesis is linked to prophage induction.

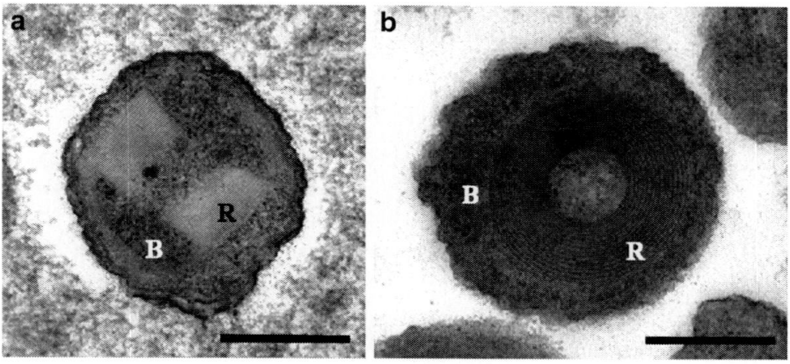


**Fig. 3** Electron micrograph of a coiled R-body inside *C. caryophilus*. **a** Longitudinal section. **b** Transverse section. *B* bacterial cell, *R* R-body. *Scale bar* 1 μm

Synthesis of R-bodies can be induced by treatments used for prophage induction such as starvation of the host, UV irradiation (Preer et al. 1974), and treatment with mitomycin C (Pond et al. 1989).

The hypothesis of an involvement of bacteriophages in R-body production is supported by following observations: (1) the presence of phagelike particles closely associated with the central core of the R-body; (2) the transformation of nonbright particles into bright particles involving the simultaneous expression of R-bodies and toxin; and (3) *Caedibacter* cells that contain R-bodies are not themselves reproductively active, but rather arise from the reproductively active nonbright kappa particles (Sonneborn 1959).

It was proposed that nonbright *Caedibacter* cells might be lysogenic, with prophage induction resulting in loss of reproductive ability and expression of bright characteristics (i.e., R-body production and killing activity).

## 4 Structure and Function of R-Bodies

R-bodies are proteinaceous ribbons which are typically coiled within the bacterial cell, visible as bright inclusion bodies using phase contrast microscopy (Preer and Stark 1953). The convoluted R-body represents a hollow cylinder. Although R-bodies have been investigated intensively in the past, the detailed structure and function of the R-body are not yet completely understood.

### 4.1 Composition and Structure of R-Bodies

R-bodies can be regarded as bacterial inclusion bodies. Major ultrastructural components are a sheath, numerous capsids of icosaedral phages, capsomeres (electron microscopically visible substructures of capsids), or phage tail-like structures

(Preer et al. 1972). The nomenclature of R-body types follows the bacterial name or the *Paramecium* host strain number in which they were discovered first (Pond et al. 1989; Table 1).

R-body sheaths are very stable ribbons with different dimensions according to the host bacteria (Pond et al. 1989; Lalucat 1989) (Table 1). They are extremely resistant to denaturation and solubilization (Preer and Preer 1967; Pond et al. 1989). Harsh treatment, generally known to denature proteins, does not affect the R-body sheath. Protocols such as 1–10% sodium dodecyl sulfate, 8 M urea, 5% β-mercaptoethanol, 5% dithiothreitol, temperature up to 100°C for up to 1 h, 8.6 M guanidine hydrochloride at 37°C for 2 h, or 6 M guanidine thiocyanate, did not lead to a noticeable R-body dissociation (Pond et al. 1989). The unusual stability suggests covalent linkage of proteins within the sheath. Disulfide bridges have been discussed, but this possibility was not supported after translation of *reb* sequences into protein sequences (Heruth et al. 1994). Types of covalent linkage other than disulfide bridges are known for the bacteriophage lambda (Hendrix and Casjens 1974; Harding 1985). The R-body sheath may be synthesized from a single structural protein (Pond et al. 1989), but involvement of other proteins (Reb proteins or others of the infected bacterium) cannot be ruled out.

Up to know four different types of R-bodies according to size and way of unrolling have been identified (Pond et al. 1989) (Table 1). Their dimensions range from less than 10 to 20 μm (ribbon length) and from 0.25 to 0.8 μm (ribbon width).The average thickness of the sheath in the convoluted state is 11–16 nm (Lalucat 1989). R-body length may differ, but ribbon morphology and ribbon width are conserved according to the bacterial species. R-bodies are composed of proteins and the R-body sheath comprises at least two layers of proteins (Laculat 1989). The coiled R-body is surrounded by a fibrous core material (Laculat 1989) possibly involved in sheath assembly (Lalucat and Mayer 1978; Wells and Horne 1983).

A membranous structure embraces the convoluted R-body of *C. taeniospiralis* and *C. caryophilus* (Fig. 3). This structure might be connected to the plasma membrane of the bacterium and resembles the "outer envelope" reported in *Marinomonas mediterranea* (Hernández- Romero et al. 2003). It can be speculated that the sheath is a product of aberrant assembly of bacteriophages (Preer et al. 1974; Pond et al. 1989). Regardless of the validity of such speculations, understanding the bacteriophage assembly process may shed light on R-body assembly and vice versa (Pond et al. 1989). No regular spacing arrangement was detected with optical diffraction methods (Lalucat 1989), but both surfaces of the sheath are different in texture. The inner surface appears to be rough, whereas the outer surface seems to be smooth (Wells and Horne 1993).

R-bodies have a characteristic mode of unrolling, triggered by different physical and chemical conditions in vitro, e.g., by a low pH. In their natural environment, R-bodies unroll after the containing bacterium has been incorporated into the phagosome of a sensitive *Paramecium* and the phagosome turns functionally into a lysosome. Bacterial cell lysis caused by acidification of the vacuole lumen leads to a rapid change of pH in the medium surrounding the R-body. This can be simulated in vitro by changing the pH in the solution in which isolated R-bodies are floating.

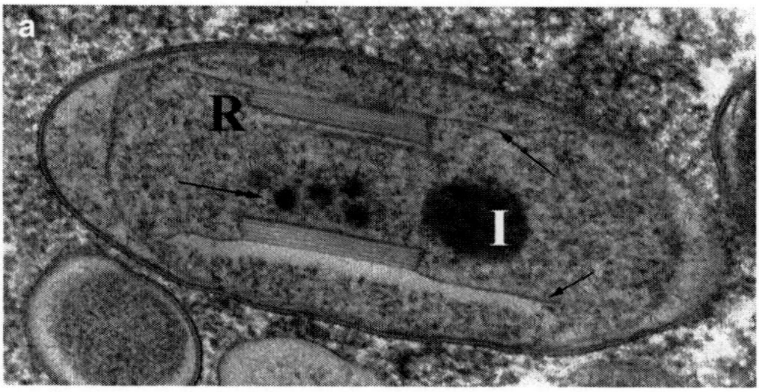


**Fig. 4** Transmission electron micrograph of *C. caryophilus* in the macronucleus of *P. caudatum*. High-pressure frozen and freeze-substituted sample. *R* R-body, *double arrow* phage capsids, *arrows* thin membranous structure, *I* inclusion body. *Scale bar* 1 μm

Thereby unrolling is induced and can be visualized light-microscopically. Both unrolling and recoiling are reversible and a change of pH seems to be sufficient. These processes are not dependent on the availability of ATP. The pH-sensitivity of the R-bodies was confirmed by different approaches in the past, while only one reference reports the reenrolling of previously uncoiled R-bodies. Our results show that changes in pH are crucial and sufficient for reversible uncoiling and reenrolling of the sheath of isolated R-bodies from *C. caryophilus*. The reversibility of the unrolling process may indicate that a conformal change of at least one component of the R-body sheath occurs.

## 4.2 Spatial Association with Bacteriophages

The presence of bacteriophages spatially associated with the R-body inside different *Caedibacter* species (Table 1) is well known from electron micrographs (Figs. 3–5). Phage-induction experiments such as using UV irradiation and certain physiological conditions such as starvation of the host (Sonneborn 1959; Preer et al. 1974) can increase the ratio of R-body-carrying *Caedibacter*. In most *Caedibacter* species the R-body structure is closely associated with bacteriophages or phagelike particles (Pond et al. 1989). As a result of this, speculations about the nature of R-body components led to phage capsid proteins. R-body genes may be translated into phage capsid-like proteins able to self-assemble into functional R-bodies. Although defective phages can be found scattered throughout the cytoplasm of the bacterium, they are much more frequently found in the central cavity of the hollow cylinder built by the R-body ribbon (Preer and Jurand 1968). Sometimes, helical or platelet-like particles associated with the sheath were found in *C. taeniospiralis* and *C. varicaedens* (Pond et al. 1989). They were described as monomers or multimers of capsomeres. Another interpretation is that the helical structures may resemble phage tails in different stages of assembly.

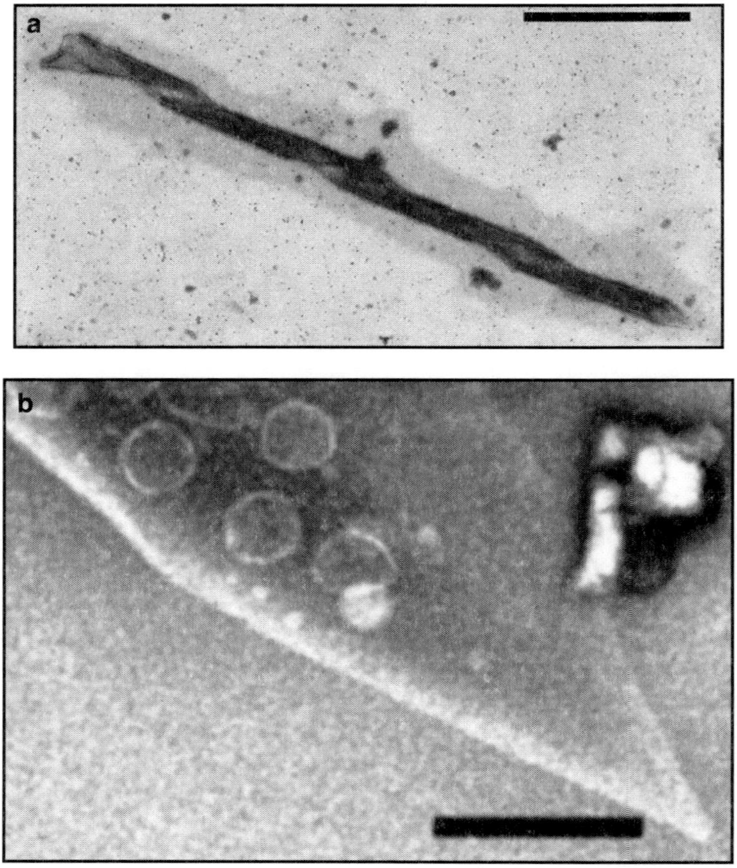


**Fig. 5** **a** Completely uncoiled R-body ribbon. Blunt end (*upper left*) and tapered end (*lower right*). Negative contrast. **b** Details of the tapered end. *P* phage capsids, *arrows* capsomeres. *Scale bars* 10 μm (**a**) and 1 μm (**b**)

## 5 Concluding Remarks and Further Perspectives

The *Paramecium–Caedibacter*–bacteriophage system is a tripartite association which results in the formation of a highly complex structure, the R-body. This structure is a precondition for the expression of a sophisticated and specialized action, the killer trait. The ecological function of the R-body, together with the toxin, is to raise the fitness of the host *Paramecium* while eliminating competitors. An increased host fitness positively effects the endosymbiotic *Caedibacter* and subsequently also the bacteriophages. Future research may reveal a broader distribution of *Caedibacter* species and a greater diversity of *Caedibacter*–hosts, improving our knowledge of *Caedibacter*-host interactions. The current taxonomy of the genus *Caedibacter* is hampered by limits of the morphological traits used for classification which apparently do not reflect phylogentic relationships. A revision of the genus *Caedibacter* is needed. With the discovery of new hosts for *Caedibacter* we may get a more detailed picture of the ecological impact of these bacteria in the environment. A closer look at R-body and killer trait evolution especially regarding their transmission to far-related bacteria could give further hints to the origin of the extrachromosomal elements coding for the R-body and the still unidentified toxin. Involvement of different extrachromosomal elements in R-body synthesis is particularly interesting since it provides a possible explanation for how R-bodies can be produced by phylogenetically distant bacteria (*Gammaproteobacteria* and *Alphaproteobacteria*) assuming that R-body coding elements probably have a common evolutionary origin. R-bodies can be regarded as nanostructures with distinct properties according to their function. Generally, morphological diversity of proteinaceous structures in protists, optimized during evolution, is a rich reservoir of potential tools for nanotechnology if we can succeed in acquiring information about their development, composition, assembly, spatial structure, and function.

## References


Austin ML (1948a) The killing substance paramecin: activity of single particles. Am Nat 82:51–59

Austin ML (1948b) The killing action and rate of production of single particles of paramecin 51. Physiol Zool 21:69–86

Beier CL, Horn M, Michel R, Schweikert M, Görtz H-D, Wagner M (2002) The genus *Caedibacter* comprises endosymbionts of *Paramecium* spp. related to the *Rickettsiales (Alphaproteobacteria)* and to *Francisella tularensis (Gammaproteobacteria)*. Appl Environ Microbiol 68:6043–6050

Dilts JA (1977) Chromosomal and extrachromosomal deoxyribonucleic acid from four bacterial endosymbionts derived from stock 51 of *Paramecium tetraurelia*. J Bacteriol 129:888–894

Dilts JA, Quackenbush RL (1986) A mutation in the R body-coding sequences destroys expression of the killer trait in *P. tetraurelia*. Science 232:641–643

Euzéby JP (1997) List of Bacterial Names with Standing in Nomenclature: a folder available on the Internet. Int J Syst Bacteriol 47:590–592

Euzéby JP (2008) List of prokaryotic names with standing in nomenclature. http://www.bacterio.net. Accessed 15 Sep 2008

Favinger J, Stadtwald R, Gest H (1989) *Rhodospirillum centenum*, sp. nov., a thermotolerant cyst-forming anoxygenic photosynthetic bacterium. Antonie Van Leuwenhoek 55:291–296

Fokin SI, Görtz H-D (1993) *Caedibacter macronucleorum* sp. nov., a bacterium inabiting the macronucleus if *Paramecium duboscqui*. Arch Protistkd 143:319–324

Fokin SI, Ossipov DV (1986) *Pseudocaedibacter glomeratus* n. sp.–a cytoplasmic symbiont of the ciliate *Paramecium pentaurelia*. Cytologia 28:1000–1004

Fokin SI, Schweikert M, Görtz H-D, Fujishima M (2003) Bacterial endocytobionts of Ciliophora. Diversity and some interactions with the host. Eur J Protistol 39:475–480

Fusté MC, Simon-Pujol MD, Marques AM, Guinea J, Congregado F (1986) Isolation of free living bacterium containing R-bodies. J Gen Microbiol 132:2801–2805

Görtz H-D (2002) Symbiotic associations between ciliates and Prokaryotes. In: The prokaryotes: an evolving electronic resource for the microbial community, 3rd edn. Springer, New York, http://141.150.157.117.8080/prokPUB/chaprender/jsp/showchap/jsp?chapnum=355

Görtz H-D, Schmidt HJ (2005) Family III. *Holosporaceae* fam. nov. In: Brenner DJ, Krieg NR, Staley JT, Garrity GM (eds) Bergey's manual of systematic bacteriology, vol 2, 2nd edn. Springer, New York, pp 146–149

Görtz HD, Rosati G, Schweikert M, Schrallhammer M, Omura G, Suzaki T (2008) Implications of microbial symbionts on defense and competition of hosts-ciliates. In: White JF, Torres MS (ed) Defensive mutualism in microbial symbiosis. CRC, Boca Raton

Hamilton LD, Gettner ME (1958) Fine structure of kappa in *Paramecium aurelia*. J Biophys Biochem Cytol 4:122–123

Harding JJ (1985) Nonenzymatic covalent posttranslational modification of proteins in vivo. Adv Protein Chem 37:247–334

Heruth DC, Pond FR, Dilts JA, Quackenbush RL (1994) Characterization of genetic determinants for R body synthesis and assembly in *Caedibacter taeniospiralis* 47 and 166. J Bacteriol 176:3559–3567

Hernández-Romero D, Lucas-Elío P, López-Serrano D, Solano F, Sanchez-Amat A (2003) *Marinomonas mediterranea* is a lysogenic bacterium that synthesizes R-bodies. Microbiology 149:2679–2686

Hendrix R, Casjens S (1974) Protein fusion: a novel reaction in bacteriophage self assembly. Proc Natl Acad Sci U S A 71:1451–1455

Horn M, Fritsche TR, Gautom RK, Schleifer K-H, Wagner M (1999) Novel bacterial endosymbionts of *Acanthamoeba* spp. related to the *Paramecium caudatum* symbiont *Caedibacter caryophilus*. Environ Microbiol 1:357–367

Jeblick J, Kusch J (2005) Sequence transcription activity and evolutionary origin of the R-body coding plasmid pKAP298 from the intracellular parasitic bacterium *Caedibacter taeniospiralis*. J Mol Evol 60:164–173

Jurand A, Rudman BM, Preer JR Jr (1971) Prelethal effects of killing action by stock 7 of *Paramecium aurelia*. J Exp Zool 177:365–388

Jurand A, Preer JR Jr, Rudman BM (1978) Further investigations on the prelethal effects of the killing action of kappa-containing killer stocks of *Paramecium aurelia*. J Exp Zool 206:25–48

Kusch J, Görtz H-D (2006) Towards an understanding of the killer trait: *Caedibacter* endocytobionts in *Paramecium*. Progr Cell Subcell Biol 41:61–57

Kusch J, Stremmel M, Breiner H-W, Adams V, Schweikert M, Schmidt HJ (2000) The toxic symbiont *Caedibacter caryophila* in the cytoplasm of *Paramecium novaurelia*. Microb Ecol 40:330–335

Kusch J, Czubatinsky L, Wegmann S, Hübner M, Alter M, Albrecht P (2002) Competetive advantages of *Caedibacter*-infected *Paramecia*. Protist 153:47–58

Landis WG (1981) The ecology, role of the killer trait, and interactions of five species of *Paramecium aurelia* komplex inhabiting the littoral zone. Can J Zool 59:1734–1743

Landis WG (1987) Factors determining the frequency of the killer trait within populations of the *Paramecium aurelia* complex. Genetics 115:197–205

Lalucat J (1989) Analysis of refractile (R) bodies. In: Mayer F (ed) Methods in microbiology: electron microscopy in microbiology, vol 20. Academic, San Diego, pp 79–90

Lalucat J, Mayer F (1978) Spiral bodies–intracytoplasmic membraneous structures in a hydrogen oxydizing bacterium. Zeitschr Allg Mikrobiol 18:517–521

Linka N, Hurka H, Lang BF, Burger G, Winkler HH, Stamme C, Urbany C, Seil I, Kusch J, Neuhaus HE (2003) Phylogenetic relationships of non-mitochondrial nucleotide transport proteins in bacteria and eukaryotes. Gene 306:27–35

Müller JA (1962) Induced physiological and morphological changes in the B particle and R body from killer paramecia. J Protozool 9:26

Müller JA (1963) Separation of kappa particles with infective activity from those with killing activity and identification of the infective particles in Paramecium aurelia. Exp Cell Res 30:492–508

Nobili R (1961) Alcune considerazioni sulla liberazione e sull'azione delle particelle kappa in individui amacronucleati di *Paramecium aurelia*, stock 51, syngen 4. Atti Soc Toscana Sci Nat Mem P V Ser B 1961:158–172

Pilhofer M, Bauer AP, Schrallhammer M, Richter L, Ludwig W, Schleifer KH, Petroni G (2007) Characterization of bacterial operons consisting of two tubulins and a kinesin-like gene by the novel Two-Step Gene Walking method. Nucleic Acids Res 35(20):e135. doi:10.1093/nar/gkm836

Pond FR, Gibson I, Lalucat J, Quackenbush RL (1989) R-body-producing bacteria. Microbiol Rev 53:25–67

Preer JR Jr (1950) Microscopically visible bodies in the cytoplasm of the "killer" strains of *Paramecium aurelia*. Genetics 35:344–362

Preer JR Jr, Jurand A (1968) The relationship between virus-like particles and R-bodies of *Paramecium aurelia*. Genet Res 12:331–340

Preer JR Jr, Preer LB (1967) Virus-like bodies in killer paramecia. Proc Natl Acad Sci U S A 58:1774–1781

Preer JR Jr, Preer LB (1982) Revival of names of protozoan endosymbionts and proposal of *Holospora caryophila* nom. nov. Int J Syst Bacteriol 32:140–141

Preer JR Jr, Stark PS (1953) Cytolgical observations on the cytoplasmic factor "kappa" in *Paramecium aurelia*. Exp Cell Res 5:478–499

Preer JR Jr, Siegel RW, Stark PS (1953) The relationship between kappa and paramecin in *Paramecium aurelia*. Proc Natl Acad Sci U S A 39:1228–1233

Preer LB, Jurand A, Preer JR Jr, Rudman BM (1972) The classes of kappa in *Paramecium aurelia*. J Cell Sci 11:581–600

Preer LB, Rudman BM, Preer JR Jr, Jurand J (1974) Induction of R bodies by ultraviolet light in killer paramecia. J Gen Microbiol 80:209–215

Quackenbush RL (1977) Phylogenetic relationships of bacterial endosymbionts of *Paramecium aurelia*: polynucleotide sequence relationships of 51 kappa and its mutants. J Bacteriol 129:895–900

Quackenbush RL (1978) Genetic relationships among bacterial endosymbionts of *Paramecium aurelia*. Deoxyribonucleotide sequence relationships among members of *Caedibacter*. J Gen Microbiol 108:181–187

Quackenbush RL (1982) IN: Validation of the publication of new names and new combinations previously published outside the IJSB. List no. 8. Int J Syst Bacteriol 32:266–268

Quackenbush RL (1983) Plasmids from bacterial endosymbionts of humpkiller paramecia. Plasmid 9:298–306

Quackenbush RL, Burbach JA (1983) Cloning and expression of DNA sequences associated with the killer trait of *Paramecium tetraurelia* stock 47. Proc Natl Acad Sci U S A 80:250–254

Quackenbush RL, Cox BJ, Kanabrocki JA (1986a) Extrachromosomal elements of extrachromosomal elements of paramecia, and their extrachromosomal elements. In: Hollaender A et al. (ed), Extrachromosomal elements of lower eukaryotes. Plenum, New York, pp 265–278

Quackenbush RL, Cox BJ, Kanabrocki JA (1986b) Extrachromosomal elements of extrachromosomal elements of *Paramecium* and their extrachromosomal elements. Basic Life Sci 40:265–278

Quackenbush RL, Dilts JA, Cox BJ (1986) Transposonlike elements in *Caedibacter taeniospiralis*. J Bacteriol 166:349–352

Schmidt HJ, Görtz H-D, Qackenbusch RL (1987) *Caedibacter caryophila* sp. nov., a killer symbiont inhabiting the macronucleus of *Paramecium caudatum*. Int J Syst Bacteriol 37:459–462

Schrallhammer M, Fokin SI, Schleifer KH, Petroni G (2006) Molecular characterization of the obligate endosymbiont "*Caedibacter macronucleorum*" Fokin and Görtz 1993 and of its host *Paramecium duboscqui* strain Ku4-8. J Eukaryot Microbiol 53:499–506

Schweikert M and Meyer B (2001) Characterization of intracellular bacteria in the freshwater dinoflagellate *Peridinium cintum*. Protoplasma 217:177–184

Skoblo II, Borchsenius ON, Lebedeva NA, Ossipov DV (1985) A new species of symbiotic bacteria of *Paramecium bursaria*, Ciliophora, Protozoa. Cytologia 27:1292–1297

Smith JE (1961) Purification of kappa particles of *Paramecium aurelia*, stock 51. Am Zool 1:390

Sonneborn TM (1938) Mating types in *Paramecium aurelia*: diverse conditions for mating in different stocks; occurrence, number and interrelations of the types. Proc Am Philos Soc 79:411–434

Sonneborn TM (1943) Gene and cytoplasm. I. The determination and inheritance of the killer character in variety 4 of *Paramecium aurelia*. Proc Natl Acad Sci U S A 29:329–338

Sonneborn TM (1959) Kappa and related particles in *Paramecium*. Adv Virus Res 6:229–356

Sonneborn TM (1965) The metagon: RNA and cytoplasmic inheritance. Am Nat 99:279–307

Sonneborn TM, Gibson I, Schneller MV (1964) Killer particles and metagons of *Paramecium* grown in *Didinium*. Science 144:567–568

Springer N, Ludwig W, Amann R, Schmidt HJ, Görtz H-D Schleifer KH (1993) Occurrence of fragmented 16S rRNA in an obligate bacterial endosymbiont of *Paramecium caudatum*. Proc Natl Acad Sci U S A 90:9892–9895

Wells B, Horne RW (1983) The ultrastructure of *Pseudomonas avenae* II. intracellular refractile (R-body) structure. Micron Microsc Acta 14:329–344

Willems A, Busse J, Goor M, Pot B, Falsen E, Jantzen E, Hoste B, Gillis M, Kersters K, Auling G, De Ley J (1989) *Hydrogenophaga*, a new genus of hydrogen-oxidising bacteria that includes Hydrogenophaga flava comb. nov. (formerly *Pseudomonas flava), Hydrogenophaga palleronii* (formerly *Pseudomonas palleronii*), *Hydrogenophaga pseudoflava* (formerly Pseudomonas pseudoflava and "*Pseudomonas carboxydoflava*"), and *Hydrogenophaga taeniospiralis* (formerly Pseudomonas taeniospiralis). Int J Syst Bacteriol 39:319–333

Willems A, Goor M, Thielemans S, Gillis M, Kersters K, De Ley J (1992) Transfer of several phytopathogenic *Pseudomonas* species to *Acidovorax avenae* subsp. avenae subsp. nov., comb. nov., *Acidovorax avenae* subsp. *citrulli, Acidovorax avenae* subsp. *cattleyae*, and *Acidovorax konjaci*. Int J Syst Bacteriol 42:107–119

Williams J (1971) The growth in vitro of killer particles from *Paramecium aurelia* and the axenic culture of this protozoan. J Gen Microbiol 68:253–262

# Index